FUNDAMENTAL FORMULAS OF PHYSICS

EDITED BY

DONALD H. MENZEL
DIRECTOR, HARVARD COLLEGE OBSERVATORY

In Two Volumes

VOLUME ONE

DOVER PUBLICATIONS, INC.
NEW YORK

Published in Canada by General Publishing Company, Ltd., 30 Lesmill Road, Don Mills, Toronto, Ontario.
Published in the United Kingdom by Constable and Company, Ltd., 10 Orange Street, London WC2H 7EG.

This Dover edition, first published in 1960, is an unabridged and revised version of the work originally published in 1955 by Prentice-Hall, Inc. The first edition appeared in one volume, but this Dover edition is divided into two volumes.

International Standard Book Number: 0-486-60595-7
Library of Congress Catalog Card Number: 60-51149

Manufactured in the United States of America
Dover Publications, Inc.
31 East 2nd Street
Mineola, N.Y. 11501

PREFACE

A survey of physical scientists, made several years ago, indicated the need for a comprehensive reference book on the fundamental formulas of mathematical physics. Such a book, the survey showed, should be broad, covering, in addition to basic physics, certain cross-field disciplines where physics touches upon chemistry, astronomy, meteorology, biology, and electronics.

The present volume represents an attempt to fill the indicated need. I am deeply indebted to the individual authors, who have contributed time and effort to select and assemble formulas within their special fields. Each author has had full freedom to organize his material in a form most suitable for the subject matter covered. In consequence, the styles and modes of presentation exhibit wide variety. Some authors considered a mere listing of the basic formulas as giving ample coverage. Others felt the necessity of adding appreciable explanatory text.

The independence of the authors has, inevitably, resulted in a certain amount of overlap. However, since conventional notation may vary for the different fields, the duplication of formulas should be helpful rather than confusing.

In the main, authors have emphasized the significant formulas, without attempting to develop them from basic principles. Apart from this omission, each chapter stands as a brief summary or short textbook of the field represented. In certain instances, the authors have included material not heretofore available.

The book, therefore, should fill needs other than its intended primary function of reference and guide for research. A student may find it a handy aid for review of familiar field or for gaining rapid insight into the techniques of new ones. The teacher will find it a useful guide in the broad field of physics. The chemist, the astronomer, the meteorologist, the biologist, and the engineer should derive valuable aid from the general sections as well as from the cross-field chapters in their specialties. For example, the chapter on Electromagnetic Theory has been designed to meet the needs of both engineers and physicists. The handy conversion factors facilitate rapid conversion from Gaussian to MKS units or vice versa.

In a work of this magnitude, some errors will have inevitably crept in. I should appreciate it, if readers would call them to my attention.

DONALD H. MENZEL
Harvard College Observatory
Cambridge, Mass.

CONTENTS

FUNDAMENTAL FORMULAS *of* PHYSICS

Chapter 1

BASIC MATHEMATICAL FORMULAS

By Philip Franklin

Professor of Mathematics
Massachusetts Institute of Technology

Certain parts of pure mathematics are easily recognized as necessary tools of theoretical physics, and the results most frequently applied may be summarized as a compilation of formulas and theorems. But any physicist who hopes or expects that any moderate sized compilation of mathematical results will satisfy all his needs is doomed to disappointment. In fact no collection, short of a library, could begin to fulfill all the demands that the physicist will make upon the pure mathematician. And these demands grow constantly with the increasing size of the field of mathematical physics.

The following compilation of formulas is, therefore, intended to be representative rather than comprehensive. To conserve space, certain elementary formulas and extensive tables of indefinite integrals have been omitted, since these are available in many well-known mathematical handbooks.

A select list of reference books, arranged by subject matter, is given at the end of Chapter 1. This bibliography may assist the reader in checking formulas, or in pursuing details or extensions beyond the material given here.

As a preliminary check on the formulas to be included, a tentative list was submitted to a number of physicists. These included all the authors of the various chapters of this book. They contributed numerous suggestions, which have been followed as far as space requirements have permitted.

1. Algebra

1.1. Quadratic equations. If $a \neq 0$, the roots of

$$ax^2 + bx + c = 0 \tag{1}$$

are

$$x_1 = \frac{-b + (b^2 - 4ac)^{1/2}}{2a} = \frac{-2c}{b + (b^2 - 4ac)^{1/2}} \tag{2}$$

$$x_2 = \frac{-b - (b^2 - 4ac)^{1/2}}{2a} = \frac{2c}{-b + (b^2 - 4ac)^{1/2}} \tag{3}$$

When $b^2 \gg |4ac|$, the second form for x_1 when $b > 0$, and for x_2 when $b < 0$ is easier to compute with precision, since it involves the sum, instead of the difference, of two nearly equal terms.

1.2. Logarithms. Let ln represent natural logarithm, or $\log_e$ where $e = 2.71828$. Then for $M > 0$, $N > 0$:

$$\left.\begin{aligned} &\ln MN = \ln M + \ln N, \quad \ln M^p = p \ln M, \quad \ln 1 = 0 \\ &\ln \frac{M}{N} = \ln M - \ln N, \quad \ln \frac{1}{N} = -\ln N, \quad \ln e = 1 \end{aligned}\right\} \quad (1)$$

In this book $\log_{10}$ will be written simply log. For conversion between base e and base 10

$$\ln M = 2.3026 \log M, \qquad \log M = 0.43429 \ln M \qquad (2)$$

1.3. Binomial theorem. For n any positive integer,

$$(a + b)^n = a^n + na^{n-1}b + \frac{n(n-1)}{2!} a^{n-2}b^2$$

$$+ \frac{n(n-1)(n-2)}{3!} a^{n-3}b^3 + \dots + nab^{n-1} + b^n \qquad (1)$$

or

$$(a + b)^n = \sum_{r=0}^{n} {}^nC_r a^{n-r}b^r = \sum_{r=0}^{n} \frac{n!}{r!(n-r)!} a^{n-r}b^r \qquad (2)$$

where $n!$ (factorial n) is defined by

$$n! = 1\cdot 2\cdot 3\cdot \dots \cdot (n-1)n \quad \text{for} \quad n > 1 \quad \text{and} \quad 1! = 1, \quad 0! = 1 \qquad (3)$$

1.4. Multinomial theorem. For n any positive integer, the general term in the expansion of $(a_1 + a_2 + \dots + a_k)^n$ is

$$\frac{n!}{r_1! r_2! r_3! \dots r_k!} a_1^{r_1} a_2^{r_2} a_3^{r_3} \dots a_k^{r_k} \qquad (1)$$

where $r_1, r_2, \dots r_k$ are positive integers such that $r_1 + r_2 + \dots + r_k = n$.

1.5. Proportion

$$\frac{a}{A} = \frac{b}{B} \quad \text{or} \quad \frac{a}{b} = \frac{A}{B} \quad \text{if and only if} \quad aB = bA. \qquad (1)$$

If $a/A = b/B = c/C = k$, then for any weighting factors p, q, r each fraction equals

$$k = \frac{pa + qb + rc}{pA + qB + rC} \qquad (2)$$

1.6. Progressions. Let l be the last term. Then the sum of the arithmetic progression to n terms

$$s = a + (a + d) + (a + 2d) + \dots + [a + (n - 1)d]$$

is

$$s = \frac{n}{2}(a + l), \tag{1}$$

where $l = a + (n - 1)d$, and the sum of the geometric progression to n terms

$$s = a + ar + ar^2 + \dots + ar^{n-1} = a\left(\frac{1 - r^n}{1 - r}\right) \tag{2}$$

1.7. Algebraic equations. The general equation of the nth degree

$$P(x) = a_0x^n + a_1x^{n-1} + a_2x^{n-2} + \dots + a_{n-1}x + a_n = 0 \tag{1}$$

has n roots. If the roots of $P(x) = 0$, or zeros of $P(x)$, are $r_1, r_2, \dots, r_n$, then

$$P(x) = a_0(x - r_1)(x - r_2) \dots (x - r_n)$$

and the symmetric functions of the roots

$$\sum r_i = -\frac{a_1}{a_0}, \quad \sum r_ir_j = \frac{a_2}{a_0}, \quad \sum r_ir_jr_k = -\frac{a_3}{a_0}, \quad \dots,$$

$$r_1r_2 \dots r_n = (-1)^n \frac{a_n}{a_0} \tag{2}$$

If m roots are equal to r, then r is a multiple root of order m. In this case $P(x) = (x - r)^mQ_1(x)$ and $P'(x) = (x - r)^{m-1}Q_2(x)$, where $P'(x) = dP/dx$, § 3.1. Thus r will be a multiple root of $P'(x) = 0$ of order $m - 1$. All multiple or repeated roots will be zeros of the greatest common divisor of $P(x)$ and $P'(x)$.

If r is known to be a zero of $P(x)$, so that $P(r) = 0$, then $(x - r)$ is a factor of $P(r)$, and dividing $P(x)$ by $(x - r)$ will lead to a depressed equation with degree lowered by one.

When $P(a)$ and $P(b)$ have opposite signs, a real root lies between a and b. The interpolated value $c = a - P(a)(b - a)/[P(b) - P(a)]$, and calculation of $P(c)$ will lead to new closer limits.

If c_1 is an approximate root, $c_2 = c_1 - P(c)/P'(c)$ is Newton's improved approximation. We may repeat this process. Newton's procedure can be applied to transcendental equations. If an approximate complex root can be found, it can be improved by Newton's method.

1.8. Determinants. The determinant of the nth order

$$D = \begin{vmatrix} a_{11} & a_{12} & \dots & a_{1n} \\ a_{21} & a_{22} & \dots & a_{2n} \\ \dots & \dots & \dots & \dots \\ a_{n1} & a_{n2} & \dots & a_{nn} \end{vmatrix} \tag{1}$$

is defined to be the sum of $n!$ terms

$$D = \Sigma\,(\pm)\, a_{1i}a_{2j}a_{3k} \dots a_{nl} \tag{2}$$

In each term the second subscripts $ijk...l$ are one order or permutation of the numbers $123...n$. The even permutations, which contain an even number of inversions, are given the plus sign. The odd permutations, which contain an odd number of inversions, are given the minus sign.

The cofactor A_{ij} of the element a_{ij} is $(-1)^{i+j}$ times the determinant of the $(n-1)$st order obtained from D by deleting the ith row and jth column.

The value of a determinant D is unchanged if the corresponding rows and columns are interchanged, or if, to each element of any row (or column), is added m times the corresponding element in another row (or column).

If any two rows (or columns) are interchanged, D is multiplied by (-1). If each element of any one row (or column) is multiplied by m, then D is multiplied by m.

If any two rows (or columns) are equal, or proportional, $D = 0$.

$$D = a_{1j}A_{1j} + a_{2j}A_{2j} + \dots + a_{nj}A_{nj} \tag{3}$$

$$0 = a_{1j}A_{1k} + a_{2j}A_{2k} + \dots + a_{nj}A_{nk}, \quad k \neq j \tag{4}$$

where j and k are any two of the integers $1, 2, \dots, n$.

1.9. Linear equations. The solution of the system

$$\begin{aligned} a_{11}x_1 + a_{12}x_2 + \dots + a_{1n}x_n &= c_1 \\ a_{21}x_1 + a_{22}x_2 + \dots + a_{2n}x_n &= c_2 \\ \dots \quad \dots \quad \dots \quad \dots \quad \dots & \dots \\ a_{n1}x_1 + a_{n2}x_2 + \dots + a_{nn}x_n &= c_n \end{aligned} \tag{1}$$

is unique if the determinant of §1.8, $D \neq 0$. The solution is

$$x_1 = \frac{C_1}{D}, \qquad x_2 = \frac{C_2}{D}, \qquad \dots, \qquad x_n = \frac{C_n}{D} \tag{2}$$

where C_k is the nth order determinant obtained from D by replacing the elements of its kth column $a_{1k}, a_{2k}, \dots, a_{nk}$ by $c_1, c_2, \dots, c_n$.

When all the $c_j = 0$, the system is homogeneous. In this case, if $D = 0$, but not all the A_{ij} are zero, the ratios of the x_i satisfy

$$\frac{x_1}{A_{11}} = \frac{x_2}{A_{12}} = \dots = \frac{x_n}{A_{1n}} \tag{3}$$

A homogeneous system has no nonzero solutions if $D \neq 0$.

2. Trigonometry

2.1. Angles. Angles are measured either in degrees or in radians, units such that

$$2 \text{ right angles} = 180 \text{ degrees} = \pi \text{ radians}$$

where $\pi = 3.14159$. For conversion between degrees and radians,

$$1 \text{ degree} = 0.017453 \text{ radian}; \quad 1 \text{ radian} = 57.296 \text{ degrees}$$

2.2. Trigonometric functions. For a single angle A, the equations

$$\left.\begin{aligned} \tan A &= \frac{\sin A}{\cos A}, & \cot A &= \frac{\cos A}{\sin A}, \\ \sec A &= \frac{1}{\cos A}, & \csc A &= \frac{1}{\sin A} \end{aligned}\right\} \tag{1}$$

define the tangent, cotangent, secant, and cosecant in terms of the sine and cosine. Between the six functions we have the relations

$$\left.\begin{aligned} \sin^2 A + \cos^2 A &= 1, & \cot A &= \frac{1}{\tan A} \\ \sec^2 A &= 1 + \tan^2 A, & \csc^2 A &= 1 + \cot^2 A \end{aligned}\right\} \tag{2}$$

2.3. Functions of sums and differences

$$\left.\begin{aligned} \sin (A + B) &= \sin A \cos B + \cos A \sin B \\ \sin (A - B) &= \sin A \cos B - \cos A \sin B \\ \cos (A + B) &= \cos A \cos B - \sin A \sin B \\ \cos (A - B) &= \cos A \cos B + \sin A \sin B \end{aligned}\right\} \tag{1}$$

$$\tan (A + B) = \frac{\tan A + \tan B}{1 - \tan A \tan B}, \qquad \tan (A - B) = \frac{\tan A - \tan B}{1 + \tan A \tan B} \tag{2}$$

2.4. Addition theorems

$$\left.\begin{aligned} \sin A + \sin B &= 2 \sin \tfrac{1}{2}(A + B) \cos \tfrac{1}{2}(A - B) \\ \sin A - \sin B &= 2 \cos \tfrac{1}{2}(A + B) \sin \tfrac{1}{2}(A - B) \\ \cos A + \cos B &= 2 \cos \tfrac{1}{2}(A + B) \cos \tfrac{1}{2}(A - B) \\ \cos A - \cos B &= -2 \sin \tfrac{1}{2}(A + B) \sin \tfrac{1}{2}(A - B) \end{aligned}\right\} \tag{1}$$

2.5. Multiple angles

$$\sin 2A = 2 \sin A \cos A, \qquad \cos 2A = \cos^2 A - \sin^2 A \tag{1}$$

$$\sin^2 A = \tfrac{1}{2} - \tfrac{1}{2} \cos 2A, \qquad \cos^2 A = \tfrac{1}{2} + \tfrac{1}{2} \cos 2A \tag{2}$$

$$\sin \frac{A}{2} = \pm \sqrt{\frac{1 - \cos A}{2}}, \qquad \cos \frac{A}{2} = \pm \sqrt{\frac{1 + \cos A}{2}} \tag{3}$$

$$\left.\begin{aligned} \tan 2A &= \frac{2 \tan A}{1 - \tan^2 A}, \\ \tan \frac{A}{2} &= \pm \sqrt{\frac{1 - \cos A}{1 + \cos A}} = \frac{1 - \cos A}{\sin A} = \frac{\sin A}{1 + \cos A} \end{aligned}\right\} \tag{4}$$

$$\left.\begin{aligned} \sin 3A &= 3 \sin A - 4 \sin^3 A, \\ \cos 3A &= 4 \cos^3 A - 3 \cos A \end{aligned}\right\} \tag{5}$$

$$\left.\begin{aligned} \sin 4A &= \sin A(8 \cos^3 A - 4 \cos A), \\ \cos 4A &= 8 \cos^4 A - 8 \cos^2 A + 1 \end{aligned}\right\} \tag{6}$$

$$\left.\begin{aligned} \sin^4 A &= \tfrac{1}{8}(3 - 4 \cos 2A + \cos 4A), \\ \cos^4 A &= \tfrac{1}{8}(3 + 4 \cos 2A + \cos 4A) \end{aligned}\right\} \tag{7}$$

$$\cos nA \cos mA = \tfrac{1}{2} \cos (n - m)A + \tfrac{1}{2} \cos (n + m)A \tag{8}$$

$$\sin nA \sin mA = \tfrac{1}{2} \cos (n - m)A - \tfrac{1}{2} \cos (n + m)A \tag{9}$$

$$\cos nA \sin mA = \tfrac{1}{2} \sin (n + m)A - \tfrac{1}{2} \sin (n - m)A \tag{10}$$

2.6. Direction cosines. The direction cosines of a line in space are the cosines of the angles α, β, γ which a parallel line through the origin O makes with the coordinate axes OX, OY, OZ. For the line segment P_1P_2, joining the points $P_1 = (x_1,y_1,z_1)$ and $P_2 = (x_2,y_2,z_2)$ the direction cosines are

$$\cos \alpha = \frac{x_2 - x_1}{d}, \qquad \cos \beta = \frac{y_2 - y_1}{d}, \qquad \cos \gamma = \frac{z_2 - z_1}{d} \tag{1}$$

where d is the distance between P_1 and P_2, so that

$$d = \sqrt{(x_2 - x_1)^2 + (y_2 - y_1)^2 + (z_2 - z_1)^2} \tag{2}$$

For any set of direction cosines, $\cos^2 \alpha + \cos^2 \beta + \cos^2 \gamma = 1$.

If a, b, c are direction ratios for a line, or numbers proportional to the direction cosines, then

$$\left.\begin{aligned} \cos\alpha = \frac{a}{\sqrt{a^2+b^2+c^2}}, \qquad \cos\beta = \frac{b}{\sqrt{a^2+b^2+c^2}}, \\ \cos\gamma = \frac{c}{\sqrt{a^2+b^2+c^2}} \end{aligned}\right\} \tag{3}$$

The angle θ between two lines with direction angles $\alpha_1, \beta_1, \gamma_1$ and $\alpha_2, \beta_2, \gamma_2$ may be found from

$$\cos\theta = \cos\alpha_1 \cos\alpha_2 + \cos\beta_1 \cos\beta_2 + \cos\gamma_1 \cos\gamma_2 \tag{4}$$

The equation of a plane is $Ax + By + Cz = D$. A, B, C are direction ratios for any normal line perpendicular to the plane. The distance from the plane to $P_1 = (x_1, y_1, z_1)$ is

$$\frac{Ax_1 + By_1 + Cz_1 - D}{\sqrt{A^2+B^2+C^2}} \tag{5}$$

The equations of a straight line with direction ratios A, B, C through the point $P_1 = (x_1, y_1, z_1)$ are

$$\frac{x - x_1}{A} = \frac{y - y_1}{B} = \frac{z - z_1}{C} \tag{6}$$

The angle θ between two planes whose normals have direction ratios A_1, B_1, C_1 and A_2, B_2, C_2, or between two lines with these direction ratios may be found from

$$\cos\theta = \frac{A_1A_2 + B_1B_2 + C_1C_2}{\sqrt{A_1^2+B_1^2+C_1^2}\ \sqrt{A_2^2+B_2^2+C_2^2}} \tag{7}$$

2.7. Plane right triangle. $C = 90^\circ$, opposite c.

$$\left.\begin{aligned} c = \sqrt{a^2+b^2}, \qquad & A = \tan^{-1}\frac{b}{a} \\ b = c\cos A, \qquad & a = c\sin A \end{aligned}\right\} \tag{1}$$

2.8. Amplitude and phase

$$a\cos\omega t + b\sin\omega t = c\sin(\omega t + A) \tag{1}$$

where the *amplitude* c and *phase* A are related to the constants a and b by the equations of §2.7.

2.9. Plane oblique triangle. Sides a, b, c opposite A, B, C.

$$A + B + C = 180 \text{ degrees} \quad \text{or} \quad \pi \text{ radians}$$

Law of sines:
$$\frac{a}{\sin A} = \frac{b}{\sin B} = \frac{c}{\sin C} \tag{1}$$

Law of cosines:
$$c^2 = a^2 + b^2 - 2ab \cos C \tag{2}$$

$$\text{Area } K = \tfrac{1}{2}ab \sin C = \sqrt{s(s-a)(s-b)(s-c)} \tag{3}$$

where $s = \frac{1}{2}(a + b + c)$.

2.10. Spherical right triangle. $C = 90°$, opposite c.

$$\sin a = \sin A \sin c, \qquad \sin b = \sin B \sin c \tag{1}$$

$$\sin a = \tan b \cot B, \qquad \sin b = \tan a \cot A \tag{2}$$

$$\cos A = \cos a \sin B, \qquad \cos B = \cos b \sin A \tag{3}$$

$$\cos A = \tan b \cot c, \qquad \cos B = \tan a \cot c \tag{4}$$

$$\cos c = \cot A \cot B, \qquad \cos c = \cos a \cos b \tag{5}$$

2.11. Spherical oblique triangle. Sides a, b, c opposites A, B, C.

$$0° < a + b + c < 360°, \qquad 180° < A + B + C < 540°$$

Law of sines:
$$\frac{\sin a}{\sin A} = \frac{\sin b}{\sin B} = \frac{\sin c}{\sin C} \tag{1}$$

Law of cosines:
$$\cos c = \cos a \cos b + \sin a \sin b \cos C \tag{2}$$

$$\cos C = -\cos A \cos B + \sin A \sin B \cos c \tag{3}$$

Spherical excess:
$$E = A + B + C - 180°$$

If R is the radius of the sphere upon which the triangle lies, its area $K = \pi R^2 E/180$. If

$$s = \tfrac{1}{2}(a + b + c),$$

$$\tan \frac{E}{4} = \sqrt{\tan \tfrac{1}{2}s \tan \tfrac{1}{2}(s-a) \tan \tfrac{1}{2}(s-b) \tan \tfrac{1}{2}(s-c)} \tag{4}$$

2.12. Hyperbolic functions. For a single number x, the equations

$$\sinh x = \frac{e^x - e^{-x}}{2}, \qquad \cosh x = \frac{e^x + e^{-x}}{2} \tag{1}$$

define the hyperbolic sine and hyperbolic cosine in terms of the exponential function. And the equations

$$\left.\begin{aligned} \tanh x &= \frac{\sinh x}{\cosh x}, & \coth x &= \frac{\cosh x}{\sinh x}, \\ \operatorname{sech} x &= \frac{1}{\cosh x}, & \operatorname{csch} x &= \frac{1}{\sinh x} \end{aligned}\right\} \quad (2)$$

define the hyperbolic tangent, hyperbolic cotangent, hyperbolic secant and hyperbolic cosecant, respectively, in terms of the hyperbolic sine and hyperbolic cosine. The four functions $\sinh x$, $\tanh x$, $\coth x$, and $\operatorname{csch} x$ are odd functions, so that $\sinh(-x) = -\sinh x$ and $\tanh(-x) = -\tanh x$. The functions $\cosh x$ and $\operatorname{sech} x$ are even, so that $\cosh(-x) = \cosh x$. Between the six hyperbolic functions we have the relations

$$\left.\begin{aligned} \cosh^2 x - \sinh^2 x &= 1, & \coth x &= \frac{1}{\tanh x} \\ \operatorname{sech}^2 x + \tanh^2 x &= 1, & \coth^2 x - \operatorname{csch}^2 x &= 1 \end{aligned}\right\} \quad (3)$$

2.13. Functions of sums and differences

$$\left.\begin{aligned} \sinh(x+y) &= \sinh x \cosh y + \cosh x \sinh y \\ \sinh(x-y) &= \sinh x \cosh y - \cosh x \sinh y \\ \cosh(x+y) &= \cosh x \cosh y + \sinh x \sinh y \\ \cosh(x-y) &= \cosh x \cosh y - \sinh x \sinh y \end{aligned}\right\} \quad (1)$$

2.14. Multiple arguments

$$\sinh 2x = 2 \sinh x \cosh x, \quad \cosh 2x = \cosh^2 x + \sinh^2 x \quad (1)$$

$$\sinh^2 x = \tfrac{1}{2} \cosh 2x - \tfrac{1}{2}, \quad \cosh^2 x = \tfrac{1}{2} \cosh 2x + \tfrac{1}{2} \quad (2)$$

2.15. Sine, cosine, and complex exponential function. $i^2 = -1$.

$$e^{ix} = \cos x + i \sin x, \qquad e^{-ix} = \cos x - i \sin x \quad (1)$$

$$\cos x = \frac{e^{ix} + e^{-ix}}{2}, \qquad \sin x = \frac{e^{ix} - e^{-ix}}{2i} \quad (2)$$

2.16. Trigonometric and hyperbolic functions

$$\left.\begin{aligned} \cos ix &= \cosh x, & \cosh ix &= \cos x \\ \sin ix &= i \sinh x, & \sinh ix &= i \sin x \\ \tan ix &= i \tanh x, & \tanh ix &= i \tan x \end{aligned}\right\} \quad (1)$$

2.17. Sine and cosine of complex arguments

$$\sin (x + iy) = \sin x \cosh y + i \cos x \sinh y \tag{1}$$

$$\cos (x + iy) = \cos x \cosh y - i \sin x \sinh y \tag{2}$$

$$e^{x+iy} = e^x \cos y + ie^x \sin y \tag{3}$$

2.18. Inverse functions and logarithms

$$\cos^{-1} x = -i \ln(x + i\sqrt{1 - x^2}), \qquad \cosh^{-1} x = \ln(x + \sqrt{x^2 - 1}) \tag{1}$$

$$\sin^{-1} x = -i \ln(ix + \sqrt{1 - x^2}), \qquad \sinh^{-1} x = \ln(x + \sqrt{x^2 + 1}) \tag{2}$$

$$\tan^{-1} x = -\frac{i}{2} \ln \frac{1 + ix}{1 - ix}, \qquad \tanh^{-1} x = \frac{1}{2} \ln \frac{1 + x}{1 - x} \tag{3}$$

3. Differential Calculus

3.1. The derivative. Let $y = f(x)$ be a function of x. Then the derivative dy/dx is defined by the equation

$$\frac{dy}{dx} = \lim_{\Delta x \to 0} \frac{\Delta y}{\Delta x} = \lim_{\Delta x \to 0} = \frac{f(x + \Delta x) - f(x)}{\Delta x} \tag{1}$$

Alternative notations for the derivative are $f'(x)$, $D_x y$, y', and $\dot{y}$ is used for dy/dt, when $y = F(t)$.

3.2. Higher derivatives. The second derivative is

$$\frac{d^2y}{dx^2} = \frac{d}{dx}\left(\frac{dy}{dx}\right) = \frac{d}{dx} f'(x) = f''(x) \tag{1}$$

The nth derivative is

$$\frac{d^ny}{dx^n} = \frac{d}{dx}\left(\frac{d^{n-1}y}{dx^{n-1}}\right) = \frac{d}{dx} f^{(n-1)}(x) = f^{(n)}(x) \tag{2}$$

3.3. Partial derivatives. Let $z = f(x,y)$ be a function of two variables. Then the partial derivative $\partial z/\partial x$ is defined by the equation

$$\frac{\partial z}{\partial x} = \lim_{\Delta x \to 0} \frac{f(x + \Delta x, y) - f(x,y)}{\Delta x} = f_x(x,y) = z_x = \left.\frac{\partial z}{\partial x}\right|_y \tag{1}$$

Similarly the derivative $\partial z/\partial y$ is formed with y varying and x fixed.

The second partial derivatives are defined by

$$\frac{\partial^2 z}{\partial x^2} = \frac{\partial}{\partial x}\left(\frac{\partial z}{\partial x}\right), \qquad \frac{\partial^2 z}{\partial y^2} = \frac{\partial}{\partial y}\left(\frac{\partial z}{\partial y}\right),$$
$$\frac{\partial^2 z}{\partial x\, \partial y} = \frac{\partial}{\partial x}\left(\frac{\partial z}{\partial y}\right) = \frac{\partial}{\partial y}\left(\frac{\partial z}{\partial x}\right) = \frac{\partial^2 z}{\partial y\ \partial x} \tag{2}$$

The same process defines partial derivatives of any order, and when the highest derivatives involved are continuous, the result is independent of the order in which the differentiations are performed.

For a function of more than two variables, we may form partial derivatives by varying the independent quantities one at a time. Thus for $u = f(x,y,z)$ the first derivatives are $\partial u/\partial x$, $\partial u/\partial y$, $\partial u/\partial z$.

3.4. Derivatives of functions

Inverse functions:

$$y = y(x), \qquad x = x(y), \qquad \frac{dy}{dx} = \frac{1}{dx/dy}, \qquad \frac{d^2y}{dx^2} = \frac{-\,d^2x/dy^2}{(dx/dy)^3} \tag{1}$$

Chain rule: $\quad y = f(u), \qquad u = g(x), \qquad \dfrac{dy}{dx} = \dfrac{dy}{du}\dfrac{du}{dx}$ (2)

Implicit function: $\quad f(x,y) = 0, \qquad \dfrac{dy}{dx} = -\dfrac{\partial f/\partial x}{\partial f/\partial y}$ (3)

Parameter: $\quad y = y(t), \qquad x = x(t)$

Let

$$\dot{y} = \frac{dy}{dt}, \qquad \dot{x} = \frac{dx}{dt} \tag{4}$$

Then

$$\frac{dy}{dx} = \frac{\dot{y}}{\dot{x}}, \qquad \frac{d^2y}{dx^2} = \frac{\dot{x}\ddot{y} - \dot{y}\ddot{x}}{\dot{x}^3}$$

Linearity:

$$\frac{d}{dx}(au + bv) = a\frac{du}{dx} + b\frac{dv}{dx} \tag{5}$$

3.5. Products

$$(uv)' = uv' + vu', \qquad (uv)'' = uv'' + 2u'v' + vu'' \tag{1}$$

Leibniz rule:

$$(uv)^{(n)} = uv^{(n)} + {}^nC_1u'v^{(n-1)} + \dots + {}^nC_ru^{(r)}v^{(n-r)} + \dots + u^{(n)}v \tag{2}$$

The ${}^nC_r$ are defined in §1.3.

3.6. Powers and quotients

$$(u^n)' = nu^{n-1}u', \qquad \left(\frac{1}{v}\right)' = (v^{-1})' = \frac{-v'}{v^2}, \qquad \left(\frac{u}{v}\right)' = \frac{vu' - uv'}{v^2} \tag{1}$$

3.7. Logarithmic differentiation $\dfrac{d(\ln y)}{dx} = \dfrac{y'}{y}.$

$$y = \frac{uv}{w}, \qquad y' = y\frac{d(\ln y)}{dx} = y\left(\frac{u'}{u} + \frac{v'}{v} - \frac{w'}{w}\right) \tag{1}$$

$$y = \frac{u^a v^b}{w^c}, \qquad y' = y\frac{d(\ln y)}{dx} = y\left(a\frac{u'}{u} + b\frac{v'}{v} - c\frac{w'}{w}\right) \tag{2}$$

$$y = u^v, \qquad \frac{d}{dx}(u^v) = u^v(v \ln u)' = u^v\left(v\frac{u'}{u} + v' \ln u\right) \tag{3}$$

3.8. Polynomials

$$\frac{da}{dx} = 0, \qquad \frac{d(bx)}{dx} = b, \qquad \frac{d(cx^n)}{dx} = ncx^{n-1} \tag{1}$$

$$\frac{d}{dx}(a_n x^n + \dots + a_r x^r + \dots + a_1 x + a_0)$$
$$= na_n x^{n-1} + \dots + ra_r x^{r-1} + \dots + a_1 \tag{2}$$

3.9. Exponentials and logarithms. We write ln for $\log_e$, where $e = 2.71828$, as in §1.2.

$$\frac{de^x}{dx} = e^x, \qquad \frac{da^x}{dx} = (\ln a)a^x \tag{1}$$

$$\frac{d \ln x}{dx} = \frac{1}{x}, \qquad \frac{d \log_a x}{dx} = \frac{\log_a e}{x} \tag{2}$$

3.10. Trigonometric functions

$$\frac{d}{dx}\sin x = \cos x, \qquad \frac{d}{dx}\tan x = \sec^2 x, \qquad \frac{d}{dx}\sec x = \tan x \sec x \tag{1}$$

$$\frac{d}{dx}\cos x = -\sin x, \qquad \frac{d}{dx}\cot x = -\csc^2 x, \qquad \frac{d}{dx}\csc x = -\cot x \csc x \tag{2}$$

3.11. Inverse trigonometric functions

$$\frac{d}{dx}\sin^{-1} x = \frac{1}{\sqrt{1 - x^2}}, \qquad \frac{d}{dx}\tan^{-1} x = \frac{1}{1 + x^2} \tag{1}$$

$$\frac{d}{dx}\cos^{-1} x = \frac{-1}{\sqrt{1 - x^2}}, \qquad \frac{d}{dx}\cot^{-1} x = \frac{-1}{1 + x^2} \tag{2}$$

3.12. Hyperbolic functions

$$\left.\begin{aligned}&\frac{d}{dx}\sinh x = \cosh x, \qquad \frac{d}{dx}\tanh x = \operatorname{sech}^2 x,\\ &\frac{d}{dx}\operatorname{sech} x = -\tanh x \operatorname{sech} x\end{aligned}\right\} \tag{1}$$

$$\left.\begin{aligned}&\frac{d}{dx}\cosh x = \sinh x, \qquad \frac{d}{dx}\coth x = -\operatorname{csch}^2 x,\\ &\frac{d}{dx}\operatorname{csch} x = -\coth x \operatorname{csch} x\end{aligned}\right\} \tag{2}$$

3.13. Inverse hyperbolic functions

$$\frac{d}{dx}\sinh^{-1} x = \frac{1}{\sqrt{x^2+1}}, \qquad \frac{d}{dx}\tanh^{-1} x = \frac{1}{1-x^2} \tag{1}$$

$$\frac{d}{dx}\cosh^{-1} x = \frac{1}{\sqrt{x^2-1}}, \qquad \frac{d}{dx}\coth^{-1} x = \frac{-1}{x^2-1} \tag{2}$$

3.14. Differential. Let $y = f(x)$, and Δx be the increment in x. Then

$$dx = \Delta x \qquad \text{and} \qquad dy = f'(x)dx = \left(\frac{dy}{dx}\right)dx \tag{1}$$

Parameter:

$$y = y(t), \qquad x = x(t), \qquad dx = \dot{x}\,dt, \qquad dy = \dot{y}\,dt \tag{2}$$

First differentials are independent of the choice of independent variable.

3.15. Total differential. For two independent variables $z = f(x,y)$,

$$dz = \frac{\partial z}{\partial x}dx + \frac{\partial z}{\partial y}dy \qquad \text{and} \qquad \frac{dz}{dt} = \frac{\partial z}{\partial x}\frac{dx}{dt} + \frac{\partial z}{\partial y}\frac{dy}{dt} \tag{1}$$

For three variables, $u = f(x,y,z)$,

$$du = \frac{\partial u}{\partial x}dx + \frac{\partial u}{\partial y}dy + \frac{\partial u}{\partial z}dz \tag{2}$$

$$\frac{du}{dt} = \frac{\partial u}{\partial x}\frac{dx}{dt} + \frac{\partial u}{\partial y}\frac{dy}{dt} + \frac{\partial u}{\partial z}\frac{dz}{dt}, \qquad \frac{\partial u}{\partial v} = \frac{\partial u}{\partial x}\frac{\partial x}{\partial v} + \frac{\partial u}{\partial y}\frac{\partial y}{\partial v} + \frac{\partial u}{\partial z}\frac{\partial z}{\partial v} \tag{3}$$

and similarly for functions of more variables.

3.16. Exact differential. The condition that as in § 3.15 for some $f(x,y)$,

$$A(x,y)dx + B(x,y)dy = dz,$$

with $z = f(x,y)$ is

$$\frac{\partial A}{\partial y} = \frac{\partial B}{\partial x} \tag{1}$$

for some $f(x,y,z)$ to make

$$A(x,y,z)dx + B(x,y,z)dy + C(x,y,z)dz = du,$$

with $u = f(x,y,z)$, the condition is

$$\frac{\partial C}{\partial y} = \frac{\partial B}{\partial z}, \qquad \frac{\partial A}{\partial z} = \frac{\partial C}{\partial x}, \qquad \frac{\partial B}{\partial x} = \frac{\partial A}{\partial y} \tag{2}$$

3.17. Maximum and minimum values. Let $y = f(x)$ be regular in the interval a, b. Then at a relative maximum, $f(x_1)$ with $a < x_1 < b$, $f'(x)$ decreases from plus to minus as x increases through x_1. At a relative minimum, $f(x_2)$ with $a < x_2 < b$, $f'(x)$ increases from minus to plus as x increases through x_2. Thus the largest and smallest values of $f(x)$ will be included in the set $f(a)$, $f(b)$, and $f(x_k)$, where $f'(x_k) = 0$.

3.18. Points of inflection. The graph of $y = f(x)$ is concave downward in any interval throughout which $f''(x)$ is negative. The graph is concave upward in any interval throughout which $f''(x)$ is positive. At any point $y_3 = f(x_3)$ such that $f''(x)$ changes sign as x increases through x_3, the graph of $y = f(x)$ is said to have a point of inflection.

3.19. Increasing absolute value. Let $OP = s(t)$, the distance along OX. Then the velocity $v = ds/dt = \dot{s}$, and the acceleration $a = dv/dt = \dot{v} = d^2s/dt^2 = \ddot{s}$. Then s increases when v is positive, and v increases when a is positive. The distance from O, or $|s|$ increases when $|s|^2 = s^2$ increases, so that $|s|$ increases when $s\dot{s}$ is positive, or when s and v have the same sign. The speed or $|v|$ increases when $|v|^2 = v^2$ increases, so that $|v|$ increases when $v\dot{v}$ is positive, or when v and a have the same sign.

3.20. Arc length. In the triangle formed by dx, dy, and ds, the differential of arc length, the angle opposite dy is the slope angle, and the angle opposite ds is 90°. Thus

$$dx = ds \cos \tau, \qquad dy = ds \sin \tau, \qquad \frac{dy}{dx} = \tan \tau, \text{ and } ds^2 = dx^2 + dy^2, \tag{1}$$

$$ds = \sqrt{1 + y'^2}\, dx = \sqrt{\dot{x}^2 + \dot{y}^2}\; dt \tag{2}$$

In polar coordinates, $r = r(\theta)$,

$$ds^2 = dr^2 + r^2 d\theta^2$$

and

$$ds = \sqrt{r'^2 + r^2}\, d\theta = \sqrt{\dot{r}^2 + r^2\dot{\theta}^2}\; dt \tag{3}$$

3.21. Curvature. $R = ds/d\tau$ is the radius of curvature, and the curvature $K = 1/R$ is given by

$$K = \frac{1}{R} = \frac{y''}{(1 + y'^2)^{3/2}} = \frac{\dot{x}\ddot{y} - \dot{y}\ddot{x}}{(\dot{x}^2 + \dot{y}^2)^{3/2}} \tag{1}$$

In polar coordinates $r = r(\theta)$,

$$K = \frac{r^2 + 2r'^2 - rr''}{(r^2 + r'^2)^{3/2}} \tag{2}$$

3.22. Acceleration in plane motion. The velocity vector has x and y components $\dot{x} = dx/dt$ and $\dot{y} = dy/dt$, and magnitude $v = \sqrt{\dot{x}^2 + \dot{y}^2}$. The acceleration vector has x and y components $\ddot{x} = d^2x/dt^2$ and $\ddot{y} = d^2y/dt^2$. It has a tangential component $\dot{v} = dv/dt$ and a normal component v^2/R. As in §3.21, $1/R = K$. The magnitude of the acceleration is

$$a = \sqrt{\ddot{x}^2 + \ddot{y}^2} = \sqrt{\dot{v}^2 + \frac{v^4}{R^2}} \tag{1}$$

3.23. Theorem of the mean. Let $f(x)$ and $F(x)$ be regular in the interval a, b. Then for at least one value x_k with $a < x_k < b$,

Rolle's theorem:

If $$f(a) = 0, \quad f(b) = 0, \quad f'(x_1) = 0 \tag{1}$$

Law of the mean:

$$f(b) = f(a) + (b - a)f'(x_2) \tag{2}$$

Cauchy's mean value theorem:

If $$F'(x) \neq 0 \text{ inside } a, b, \tag{3}$$

$$\frac{f(b) - f(a)}{F(b) - F(a)} = \frac{f'(x_3)}{F'(x_3)}$$

3.24. Indeterminate forms. Let $f(x)$ and $F(x)$ each approach zero as x approaches a. We briefly describe the evaluation of $\lim f(x)/F(x)$ as the indeterminate form 0/0. When $f(x)$ and $F(x)$ are each analytic at a, and series in terms of powers of $(x - a)$ can be easily found, the limit may be found by using these series.

l'Hospital's rule: If the limit on the right exists, then when $f(x) \to 0$, $F(x) \to 0$,

$$\lim_{x \to a} \frac{f(x)}{F(x)} = \lim_{x \to a} \frac{f'(x)}{F'(x)} \tag{1}$$

In this rule x may approach a from one side, $x \to a+$ or $x \to a-$, and it applies when in place of $x \to a$, $x \to +\infty$, or $x \to -\infty$. The rule in any form also applies when as $x \to a$, $f(x)$ and $F(x)$ each tend to infinity. This is the indeterminate from ∞/∞.

By writing $f = 1/(1/f)$ for one of the factors, we may reduce the form $0 \cdot \infty$ to $0/0$ or ∞/∞ and then use l'Hospital's rule. A similar procedure with each term sometimes reduces the form $\infty - \infty$ to $0/0$.

If the evaluation of $\lim f^F$ leads to an indeterminate form 0^0, 1^∞, ∞^0, the evaluation of $L = \lim F \ln f$ leads to a form $0 \cdot \infty$ or $\infty \cdot 0$. This may be found as indicated above, and then $\lim f^F = e^L$.

3.25. Taylor's theorem. Let $f(x)$ be analytic at a. Then

$$f(x) = f(a) + f'(a)\frac{(x-a)}{1!} + f''(a)\frac{(x-a)^2}{2!} + f'''(a)\frac{(x-a)^3}{3!}$$

$$+ \dots + f^{(n-1)}(a)\frac{(x-a)^{n-1}}{(n-1)!} + \dots \tag{1}$$

For real values, the remainder after the term with $f^{(n-1)}(a)$ is

$$R_n = f^{(n)}(x_1)\frac{(x-a)^n}{n!} \tag{2}$$

where x_1 is a suitably chosen value in the interval a, x. An alternative form is

$$f(a+h) = f(a) + f'(a)\frac{h}{1!} + f''(a)\frac{h^2}{2!} + f'''(a)\frac{h^3}{3!} + \dots + f^{(n)}(a)\frac{h^n}{n!} + \dots \tag{3}$$

The special case when $a = 0$ is called Maclaurin's series.

$$f(x) = f(0) + f'(0)\frac{x}{1!} + f''(0)\frac{x^2}{2!} + f'''(0)\frac{x^3}{3!} + \dots + f^{(n)}(0)\frac{x^n}{n!} + \dots \tag{4}$$

For computation these series are usually used with $(x - a)$ or h small, and the remainder has the order of magnitude of the first term neglected.

Let $f(x,y)$ be analytic at (a,b). Then in the notation of § 3.3,

$$f(x,y) = f(a,b) + (x-a)f_x(a,b) + (y-b)f_y(a,b)$$

$$+ \frac{1}{2!}[(x-a)^2 f_{xx}(a,b) + 2(x-a)(y-b)f_{xy}(a,b) + (y-b)^2 f_{yy}(a,b)]$$

$$+ \dots + \frac{1}{n!}\left[(x-a)\frac{\partial}{\partial X} + (y-b)\frac{\partial}{\partial Y}\right]^n f(X,Y)\Big|_{\substack{X=a \\ Y=b}} + \dots \tag{5}$$

An alternative form is

$$f(a+h,\, y+k) = f(a,b) + hf_x(a,b) + kf_y(a,b)$$

$$+ \dots + \frac{1}{n!}\left(h\frac{\partial}{\partial x} + k\frac{\partial}{\partial y}\right)^n f(x,y)\Big|_{\substack{x=a\\y=b}} + \dots \tag{6}$$

The special case when $a = 0$, $b = 0$, Maclaurin's series is

$$f(x,y) = f(0,0) + xf_x(0,0) + yf_y(0,0) + \dots$$

$$+ \frac{1}{n!}\left(h\frac{\partial}{\partial x} + k\frac{\partial}{\partial y}\right)^n f(x,y)\Big|_{\substack{x=0\\y=0}} + \dots \tag{7}$$

And similar expansions hold for any number of variables.

3.26. Differentiation of integrals

$$\frac{d}{dx}\int_a^x f(x)dx = \frac{d}{dx}\int_a^x f(t)dt = f(x), \qquad \frac{d}{dx}\int_x^b f(t)dt = -f(x) \tag{1}$$

$$\frac{d}{dx}\int_a^b f(x,t)dt = \int_a^b \frac{\partial}{\partial x} f(x,t)dt \tag{2}$$

$$\frac{d}{dx}\int_a^x f(x,t)dt = f(x,x) + \int_a^x \frac{\partial}{\partial x} f(x,t)dt \tag{3}$$

If $f(x,x)$ is infinite or otherwise singular one may use

$$\frac{d}{dx}\int_a^x f(x,t)dt = \frac{1}{x-a}\int_a^x \left[(x-a)\frac{\partial f}{\partial x} + (t-a)\frac{\partial f}{\partial t} + f\right]dt \tag{4}$$

$$\frac{d}{dx}\int_{u(x)}^{v(x)} f(x,t)dt = v'(x)f[x,v(x)] - u'(x)f[x,u(x)] + \int_{u(x)}^{v(x)} \frac{\partial}{\partial x} f(x,t)dt \tag{5}$$

4. Integral Calculus

4.1. Indefinite integral. With respect to x, the indefinite integral of $f(x)$ is $F(x)$ provided that $dF/dx = F'(x) = f(x)$. The indefinite integral of the differential $f(x)dx$ is $F(x)$ if $dF(x) = f(x)dx$.

$$\int f(x)dx = F(x) \quad \text{if} \quad F'(x) = f(x) \quad \text{or} \quad dF(x) = f(x)dx \tag{1}$$

For any constant C, $F(x) + C$ is also a possible indefinite integral. By using all values of C, and any one $F(x)$, we obtain all possible indefinite integrals :

$$\int f(x)dx = F(x) + C.$$

4.2. Indefinite integrals of functions

$$\int dF(x) = F(x) + C, \quad d\int f(x)dx = f(x)dx \tag{1}$$

Linearity:

$$\int (au + bv)dx = a\int u\,dx + b\int v\,dx \quad (a \text{ and } b \text{ not both zero}) \tag{2}$$

Integration by parts:

$$\int u\,dv = uv - \int v\,du \tag{3}$$

$$\int u\frac{dv}{dx}dx = uv - \int v\frac{du}{dx}dx \tag{4}$$

Substitution:

$$\int f(y)\,dx = \int f(y)\frac{dx}{dy}dy = \int \frac{f(y)}{dy/dx}dy \tag{5}$$

4.3. Polynomials

$$\int 0\cdot dx = C, \quad \int a\,dx = ax + C, \quad \int bx^n\,dx = \frac{1}{n+1}bx^{n+1} \tag{1}$$

$$\left.\begin{aligned}&\int (a_n x^n + \dots + a_r x^r + \dots + a_1 x + a_0)dx \\ &\quad = \frac{1}{n+1}a_n x^{n+1} + \dots + \frac{1}{r+1}a_r x^{r+1} + \dots \frac{1}{2}a_1 x^2 + a_0 x + C\end{aligned}\right\} \tag{2}$$

4.4. Simple rational fractions

$$\int \frac{A}{x-r}dx = A\ln|x-r| + C, \quad \int \frac{A}{(x-r)^k}dx = A\frac{(x-r)^{1-k}}{1-k} + C \tag{1}$$

$$\left.\begin{aligned}\int \frac{2A(x-a)-2Bb}{(x-a)^2+b^2}dx &= 2\,\mathrm{Re}\int \frac{A+Bi}{x-a-bi}dx \\ &= A\ln[(x-a)^2+b^2] - 2B\tan^{-1}\frac{x-b}{b} + C\end{aligned}\right\} \tag{2}$$

$$\int \frac{dx}{x^2+a^2} = \frac{1}{a}\tan^{-1}\frac{x}{a} + C, \quad \int \frac{dx}{x^2-a^2} = \frac{1}{2a}\ln\left|\frac{x-a}{x+a}\right| + C \tag{3}$$

$$\int \frac{dx}{x} = \ln|x| + C \tag{4}$$

4.5. Rational functions. To integrate $R(x)dx$ where $R(x)$ is a rational function of x, first write $R(x)$ as the quotient of two polynomials $P_1(x)/D(x)$.

If the degree of $P_1(x)$ is the same or greater than the degree of $D(x)$, by division find polynomials $Q(x)$ and $P(x)$ such that

$$\int R(x)dx = \int \frac{P_1(x)}{D(x)}\,dx = \int Q(x)dx + \int \frac{P(x)}{D(x)}\,dx \qquad (1)$$

where $P(x)/D(x)$ is a proper fraction, that is, has the degree of $P(x)$ less than the degree of $D(x)$. Integrate $Q(x)dx$ as in § 4.3.

To integrate the proper fraction $P(x)/D(x)$, decompose it into partial fractions as follows. Suppose first that the roots of $D(x) = 0$ are all distinct. Let $D(x)$ be of the nth degree and call the roots $r_1, r_2, \ldots, r_n$. Then

$$\frac{P(x)}{D(x)} = \frac{A_1}{x-r_1} + \frac{A_2}{x-r_2} + \ldots + \frac{A_n}{x-r_n} \quad \text{with} \quad A_k = \frac{P(r_k)}{D'(r_k)} \qquad (2)$$

Assume that $P(x)$ and $D(x)$ have real coefficients. Then either r and A are real and we use the first equation of § 4.4, or complex roots occur in conjugate pairs, $a + bi$, $a - bi$, with conjugate numerators $A + Bi$, $A - Bi$, and we use the third equation (2) of § 4.4.

If the equation $D(x) = 0$ has multiple roots, each factor $(x - r)^m$ in $D(x)$ will lead to a series of fractions

$$\frac{P(x)}{D(x)} = \frac{P(x)}{(x-r)^m D_1(x)} = \frac{A'}{(x-r)} + \frac{A''}{(x-r)^2} + \ldots + \frac{A^{(m)}}{(x-r)^m} + \ldots \qquad (3)$$

To determine the m constants, solve the system of equations

$$P(r) = A^{(m)}D_1(r), \quad P'(r) = A^{(m)}D_1'(r) + A^{(m-1)} \qquad (4)$$

$$P''(r) = A^{(m)}D_1''(r) + 2A^{(m-1)}D_1'(r) + 2\cdot 1A^{(m-2)}D_1(r), \quad \ldots \qquad (5)$$

$$\left.\begin{aligned} P^{(m-1)}(r) = A^{(m)}D_1{}^{(m-1)}(r) &+ (m-1)A^{(m-1)}D_1'(r) \\ &+ (m-1)(m-2)A^{(m-2)}D_1''(r) \\ &+ \ldots + (m-1)!\,A'D_1{}^{(m-1)}(r) \end{aligned}\right\} \qquad (6)$$

To integrate the simple fractions, use the first and second equations (1) of § 4.4. Conjugate complex roots will lead to conjugate fractions which may be combined in pairs to give a real result after integration.

4.6. Trigonometric functions

$$\int \sin ax\,dx = -\frac{1}{a}\cos ax + C, \quad \int \sin^2 ax\,dx = \frac{x}{2} - \frac{\sin 2ax}{4a} + C \qquad (1)$$

$$\int \cos ax\,dx = \frac{1}{a}\sin ax + C, \quad \int \cos^2 ax\,dx = \frac{x}{2} + \frac{\sin 2ax}{4a} + C \qquad (2)$$

$$\left.\begin{aligned}\int \frac{dx}{\sin ax} = \int \csc ax\, dx &= \frac{1}{a} \ln \left| \tan \frac{ax}{2} \right| + C \\ &= \frac{1}{a} \ln | \csc ax - \cot ax | + C\end{aligned}\right\} \quad (3)$$

$$\left.\begin{aligned}\int \frac{dx}{\cos ax} = \int \sec ax\, dx &= \frac{1}{a} \ln \left| \tan \left(\frac{ax}{2} + \frac{\pi}{4} \right) \right| + C \\ &= \frac{1}{a} \ln | \sec ax + \tan ax | + C\end{aligned}\right\} \quad (4)$$

$$\left.\begin{aligned}&\int \frac{dx}{\sin^2 ax} = \int \csc^2 ax\, dx = -\frac{1}{a} \cot ax + C \\ &\int \tan ax\, dx = -\frac{1}{a} \ln | \cos ax | + C\end{aligned}\right\} \quad (5)$$

$$\left.\begin{aligned}&\int \frac{dx}{\cos^2 ax} = \int \sec^2 ax\, dx = \frac{1}{a} \tan ax + C \\ &\int \cot ax\, dx = \frac{1}{a} \ln | \sin ax | + C\end{aligned}\right\} \quad (6)$$

4.7. Exponential and hyperbolic functions

$$\int e^{ax}\, dx = \frac{1}{a} e^{ax} + C, \quad \int b^{ax}\, dx = \frac{1}{a \ln b} b^{ax} + C \quad (1)$$

$$\int \sinh ax\, dx = \frac{1}{a} \cosh ax + C, \quad \int \cosh ax\, dx = \frac{1}{a} \sinh ax + C \quad (2)$$

4.8. Radicals

$$\int \frac{dx}{\sqrt{a^2 - x^2}} = \sin^{-1} \left(\frac{x}{a} \right) + C = -\cos^{-1} \left(\frac{x}{a} \right) + C_1 \quad (1)$$

$$\int \frac{dx}{\sqrt{x^2 + A}} = \ln | x + \sqrt{x^2 + A} |^{*} + C \quad (2)$$

$$\int \sqrt{a^2 - x^2}\, dx = \frac{x}{2} \sqrt{a^2 - x^2} + \frac{a^2}{2} \sin^{-1} \frac{x}{a} + C \quad (3)$$

$$\int \sqrt{x^2 + A}\, dx = \frac{x}{2} \sqrt{x^2 + A} + \frac{A}{2} \ln | x + \sqrt{x^2 + A} |^{*} + C \quad (4)$$

* For A positive, $A = a^2$, $\ln (x + \sqrt{x^2 + a^2}) = \sinh^{-1} x/a + \ln a$ (5)
For A negative, $A = -a^2$, $\ln (x + \sqrt{x^2 - a^2}) = \cosh^{-1} x/a + \ln a$ (6)
In each case $\ln a$ or $(A/2) \ln a$ may be combined with the constant C.

4.9. Products

$$\int e^{ax} \sin bx\, dx = \frac{1}{a^2 + b^2} e^{ax}(a \sin bx - b \cos bx) + C \tag{1}$$

$$\int e^{ax} \cos bx\, dx = \frac{1}{a^2 + b^2} e^{ax}(a \cos bx + b \sin bx) + C \tag{2}$$

$$\int x e^{ax}\, dx = \frac{1}{a^2} e^{ax}(ax - 1) + C, \quad \int \ln ax\, dx = x \ln ax - x + C \tag{3}$$

We may use § 2.15 to express factors like $\sin ax$, $\cos bx$ in terms of complex exponentials. And we may use § 2.12 to express factors like $\sinh ax$, $\cosh bx$ in terms of exponentials. Thus any product of such factors, or product times e^{kx}, may be reduced to a sum of exponential terms like that in the first equation of § 4.7. If powers of x are also present we are led to

$$\int x^n e^{ax}\, dx = \frac{1}{a} x^n e^{ax} - \frac{n}{a} \int x^{n-1} e^{ax}\, dx \tag{4}$$

This reduction formula may be used repeatedly until the power of x disappears when n is a positive integer.

Let $f(x)$ be the natural logarithm, an inverse trigonometric function, or an inverse hyperbolic function of some simple function of x, for example, $\ln (ax + b)$, $\sin^{-1} x$, $\tanh^{-1} x^2$. Then for $n = 0, 1, 2, \ldots$ the equation

$$\int x^n f(x) dx = \frac{1}{n+1} x^{n+1} f(x) - \frac{1}{n+1} \int x^{n+1} f'(x) dx \tag{5}$$

often leads to a simpler integral.

4.10. Trigonometric or exponential integrands. Any rational function of trigonometric functions of x may be written in the form $R(\sin x, \cos x)$, and the substitution $t = \tan x/2$ makes

$$R(\sin x, \cos x) dx = R\left(\frac{2t}{1 + t^2}, \frac{1 - t^2}{1 + t^2}\right) \frac{2dt}{1 + t^2} \tag{1}$$

The new integrand is a rational function of t, discussed in § 4.5. If the integrand is a rational function of e^{ax}, let $t = e^{ax}$ and

$$R(e^{ax}) dx = R(t) \frac{1}{at}\, dt \tag{2}$$

This may be applied to $R(\sin x, \cos x)$ with $a = i$, if § 2.15 is used.

4.11. Algebraic integrands. Let p and q be integers and $a \neq 0$; then the substitution $t = (ax + b)^{1/q}$ makes

$$R[x, (ax + b)^{p/q}]dx = R\left(\frac{t^q - b}{a}, t^p\right) \frac{q}{a} t^{q-1}\, dt \tag{1}$$

The new integrand is a rational function of t, discussed in § 4.5.

For certain integrands of the form $R(x, \sqrt{px^2 + qx + r})$ the following transformations are useful.

$y = \dfrac{1}{x}$ makes

$$\frac{dx}{x\sqrt{px^2 + qx + r}} = -\frac{dy}{\sqrt{ry^2 + qy + p}} \tag{2}$$

$y = \dfrac{1}{x - s}$ makes

$$\frac{dx}{(x - s)\sqrt{px^2 + qx + r}} = -\frac{dy}{\sqrt{(ps^2 + qs + r)y^2 + (2ps + q)y + p}} \tag{3}$$

$y = x + \dfrac{q}{2p}$ makes

$$R(x, \sqrt{px^2 + qx + r})dx = R\left(y - \frac{q}{2p}, \sqrt{py^2 + \frac{4pr - q^2}{4p}}\right)dy \tag{4}$$

Simple integrals of this type are given in § 4.8. Also note that

$$R(x, \sqrt{a^2 - x^2})\, dx = R(a \sin t, a \cos t) a \cos t\, dt \quad \text{if} \quad x = a \sin t \tag{5}$$

$$R(x, \sqrt{x^2 - a^2})\, dx = R(a \sec t, a \tan t) a \tan t \sec t\, dt \quad \text{if} \quad x = a \sec t \tag{6}$$

$$R(x, \sqrt{x^2 + a^2})\, dx = R(a \tan t, a \sec t) a \sec^2 t\, dt \quad \text{if} \quad x = a \tan t \tag{7}$$

The new integrands in t are essentially those treated in § 4.10.

4.12. Definite integral. Let $x_1, x_2, \ldots x_{n-1}$ be points of subdivision of the interval a,b such that $x_k < x_{k+1}$, with $a = x_0$, $b = x_n$. Let $\delta_k = x_k - x_{k-1}$, and d_M be the maximum value of δ_k. Then

$$\int_a^b f(x)dx = \lim_{d_M \to 0} [f(\xi_1)\delta_1 + f(\xi_2)\delta_2 + \ldots + f(\xi_n)\delta_n] \tag{1}$$

For each choice of n, and the x_k, the ξ_k are any values such that $x_{k-1} \leqq \xi_k \leqq x_k$. The limit exists for $f(x)$ regular on a, b :

$$\int_a^b f(x)dx = \left[\int f(x)dx\right]_a^b = F(x)\Big|_a^b = F(b) - F(a) \tag{2}$$

where $F(x)$ is any function whose derivative with respect to x is $f(x)$, as in § 4.1.

4.13. Approximation rules. Let the interval a,b be divided into n equal parts each of length h so that $b - a = nh$. And let $f(a + kh) = y_k$; $k = 0, 1, 2, \ldots, n$. Then the *trapezoidal rule* is

$$\int_a^b f(x)dx = \frac{h}{2}(y_1 + 2y_2 + 2y_3 + \ldots + 2y_{n-1} + y_n) - R_n \tag{1}$$

where

$$R_n = \tfrac{1}{12}(b - a)h^2 f''(x_1) \tag{2}$$

for some suitable x_1 in a,b.

The more accurate *Simpson's rule* requires that n be even, and is

$$\left.\begin{aligned}\int_a^b f(x)dx = \frac{h}{3}(y_1 + 4y_2 + 2y_3 + 4y_4 \\ + \ldots + 2y_{n-2} + 4y_{n-1} + y_n) - R_n\end{aligned}\right\} \tag{3}$$

where

$$R_n = \tfrac{1}{90}(b - a)h^4 f^{\mathrm{IV}}(x_2) \tag{4}$$

for some suitable x_2 in a,b.

In Gauss' Method we let $x = \frac{1}{2}(a + b) + \frac{1}{2}(b - a)w$. Then

$$\int_a^b f(x)dx = \frac{b-a}{2}\int_{-1}^1 g(w)dw = \sum_{m=1}^n R_m g(w_m) \tag{5}$$

The w_m are the n roots of

$$P_n(x) = 0 \tag{6}$$

(see § 8.2), and for $r = 0, 1, 2, \ldots n - 1$.

$$\sum_{m=1}^n R_m \left(\frac{1 + w_m}{2}\right)^r = \frac{1}{r+1} \tag{7}$$

In particular,

$n = 2$: $w_1 = - w_2 = 0.57735$, $R_1 = R_2 = \frac{1}{2}$, $E(x^4) = \frac{4}{45}$

$n = 3$: $w_1 = - w_3 = 0.77460$, $w_2 = 0$,
$R_1 = R_3 = \frac{5}{18}$, $R_2 = \frac{4}{9}$, $E(x^6) = \frac{4}{175}$

$n = 4$: $w_1 = - w_4 = 0.86114$, $w_2 = - w_3 = 0.33998$,
$R_1 = R_4 = 0.17393$, $R_2 = R_3 = 0.32607$, $E(x^8) = \frac{64}{11025}$

The formula is exact if $f(x)$ is a polynomial of degree not exceeding $2n - 1$. The error for a higher degree polynomial may be estimated from the given value of

$$E(x^{2n}) = \frac{1}{2}\int_{-1}^1 w^{2n}\,dw - \sum_{m=1}^n R_m {w_m}^{2n}$$

4.14. Linearity properties. *Linearity* in the *integrand:*

$$\int_a^b [Af(x) + Bg(x)]dx = A\int_a^b f(x)dx + B\int_a^b g(x)dx \qquad (1)$$

Linearity with respect to the *interval:*

$$\int_a^b f(x)dx = -\int_b^a f(x)dx$$

$$\int_a^b f(x)dx = \int_a^c f(x)dx + \int_c^b f(x)dx \qquad (2)$$

4.15. Mean values. The mean value of $f(x)$ on the interval a,b is

$$\bar{f} = \frac{1}{b-a}\int_a^b f(x)dx \qquad (1)$$

If $f(x)$ is *continuous* on a,b then for some x_1 on a,b the mean value

$$\bar{f} = f(x_1) \quad \text{or} \quad \int_a^b f(x)dx = (b-a)f(x_1) \quad \textit{mean value theorem} \qquad (2)$$

The root-mean-square value of $f(x)$ on the interval a,b is

$$\bar{\bar{f}} = \sqrt{\frac{1}{b-a}\int_a^b [f(x)]^2 dx} \qquad (3)$$

4.16. Inequalities. Assume that $a < b$. If $f(x) < g(x)$ in a,b, then

$$\int_a^b f(x)dx < \int_a^b g(x)dx \qquad (1)$$

If $m < f(x) < M$ in a,b then

$$m(b-a) < \int_a^b f(x)dx < M(b-a) \qquad (2)$$

If $|f(x)| \leqq M$ in a,b (or on the complex path of integration C of length L), then

$$\left|\int_a^b f(x)dx\right| \leqq M(b-a) \quad \text{or} \quad \left|\int_C f(z)dz\right| \leqq ML$$

Schwarz's inequality:

$$\left[\int_a^b f(x)g(x)dx\right]^2 \leqq \int_a^b [f(x)]^2 dx \int_a^b [g(x)]^2 dx$$

With the notation of § 4.15, and

$$p = fg, \quad \bar{p} \leqq \bar{\bar{f}}\bar{\bar{g}}, \quad \text{and} \quad \bar{f} \leqq \bar{\bar{f}}$$

4.17. Improper integrals

$$\left.\begin{aligned}\int_a^\infty f(x)dx &= \lim_{t\to\infty}\int_a^t f(x)dx\\ \int_{-\infty}^b f(x)dx &= \lim_{t\to\infty}\int_{-t}^b f(x)dx\end{aligned}\right\} \qquad (1)$$

$$\int_{-\infty}^\infty f(x)dx = \int_{-\infty}^c f(x)dx + \int_c^\infty f(x)dx \qquad (2)$$

If $f(x)$ becomes infinite or has a singularity at $x = b$ with $b \neq a$,

$$\int_a^b f(x)dx = \lim_{h\to 0}\int_a^{b-h} f(x)dx \qquad (3)$$

4.18. Definite integrals of functions. Let $u(x)$ and $v(x)$ be continuous in a,b. Then

$$\int_a^b du(x) = u(b) - u(a) \qquad (1)$$

$$\int_a^b u(x)v'(x)dx = u(x)v(x)\Big|_a^b - \int_a^b v(x)u'(x)dx \qquad (2)$$

Let $t = g(u)$, $a = g(c)$, $b = g(d)$, and $g'(u)$ maintain its sign in c,d. Then

$$\int_a^b f(t)dt = \int_c^d f[g(u)]g'(u)du \qquad (3)$$

4.19. Plane area

$$A = \iint dx\,dy = \int_{x_1}^{x_2} dx \int_{y_1(x)}^{y_2(x)} dy = \int_{x_1}^{x_2}(y_2 - y_1)dx \qquad (1)$$

$$A = \iint r\,dr\,d\theta = \int_{\theta_1}^{\theta_2} d\theta \int_{r_1(\theta)}^{r_2(\theta)} r\,dr = \frac{1}{2}\int_{\theta_1}^{\theta_2}(r_2^2 - r_1^2)d\theta \qquad (2)$$

4.20. Length of arc. See § 3.20. For a plane curve,

$$s = \int_{x_1}^{x_2}\sqrt{1 + y'^2}\,dx = \int_{t_1}^{t_2}\sqrt{\dot{x}^2 + \dot{y}^2}\,dt \qquad (1)$$

$$s = \int_{\theta_1}^{\theta_2}\sqrt{r'^2 + r^2}\,d\theta = \int_{t_1}^{t_2}\sqrt{\dot{r}^2 + r^2\dot{\theta}^2}\,dt \qquad (2)$$

4.21. Volumes. With suitable limits on the iterated and triple integrals,

$$V = \int_{x_1}^{x_2} A(x)dx = \int dx\int[z_2(x,y) - z_1(x,y)]dy = \int d\theta\int(z_2 - z_1)\,r\,dr \qquad (1)$$

$$V = \iiint dx\,dy\,dz = \iiint r\,dr\,d\theta\,dz = \iiint r^2\sin\phi\,d\theta\,d\phi\,dr \qquad (2)$$

4.22. Curves and surfaces in space

$$s = \int_{t_1}^{t_2} \sqrt{\dot{x}^2 + \dot{y}^2 + \dot{z}^2}\, dt \tag{1}$$

where $x = x(t)$, $y = y(t)$, $z = z(t)$.

$$A = \iint \sqrt{f_x^{\,2} + f_y^{\,2} + 1}\, dx\, dy \tag{2}$$

where $z = f(x,y)$.

4.23. Change of variables in multiple integrals.

If

$$F[x(u,v), y(u,v)] = G(u,v) \tag{1}$$

and the limits are suitably related, then

$$\iint F(x,y)dx\, dy = \iint G(u,v) \frac{\partial(x,y)}{\partial(u,v)} du\, dv \tag{2}$$

Similarly if

$$F[x(u,v,w), y(u,v,w), z(u,v,w)] = G(u,v,w) \tag{3}$$

then

$$\iiint F(x,y,z)dx\, dy\, dz = \iiint G(u,v,w) \frac{\partial(x,y,z)}{\partial(u,v,w)} du\, dv\, dw \tag{4}$$

The Jacobians inserted here are

$$\frac{\partial(x,y)}{\partial(u,v)} = \begin{vmatrix} x_u & y_u \\ x_v & y_v \end{vmatrix} = x_u y_v - x_v y_u, \quad \frac{\partial(x,y,z)}{\partial(u,v,w)} = \begin{vmatrix} x_u & y_u & z_u \\ x_v & y_v & z_v \\ x_w & y_w & z_w \end{vmatrix} \tag{5}$$

If the Jacobian determinant is not zero, the transformation $x = x(u,v)$, $y = y(u,v)$ has an inverse transformation. When there is a functional relation

$$F[x(u,v), y(u,v)] = 0 \tag{6}$$

the Jacobian $\partial(x,y)/\partial(u,v)$ is identically zero. Also

$$\frac{\partial(x,y)}{\partial(X,Y)} \frac{\partial(X,Y)}{\partial(u,v)} = \frac{\partial(x,y)}{\partial(u,v)}, \quad \frac{\partial(x,y)}{\partial(u,v)} = \frac{1}{\partial(u,v)/\partial(x,y)} \tag{7}$$

Similar results hold for any number of variables.

4.24. Mass and density.

Let ρ be the variable (or constant) density of a curve, area, or volume. Then the total mass is given by

$$M = \int dm$$

where $dm = \rho\, ds$, $dm = \rho\, dA$, or $dm = \rho\, dV$. For the differentials ds, dA, dV, use the integrands from §§ 4.19-4.22. Mean density is

$$\bar{\rho} = \frac{\int \rho\, ds}{\int ds}, \quad \bar{\rho} = \frac{\int \rho\, dA}{\int dA}, \quad \bar{\rho} = \frac{\int \rho\, dV}{\int dV}$$

4.25. Moment and center of gravity. Let x, y, z be the coordinates of the center of gravity of one of the elements of mass dm of § 4.24. Then the moments about the coordinate planes are

$$M_{yz} = \int x\, dm, \quad M_{zx} = \int y\, dm, \quad M_{xy} = \int z\, dm \tag{1}$$

And the center of gravity of M is $\bar{x}$, $\bar{y}$, $\bar{z}$ where

$$\bar{x} = \frac{\int x\, dm}{\int dm}, \quad \bar{y} = \frac{\int y\, dm}{\int dm}, \quad \bar{z} = \frac{\int z\, dm}{\int dm} \tag{2}$$

For a mass M composed of several, e.g., three, parts M_1, M_2, M_3, the masses and moments are additive so that

$$\bar{x} = \frac{M_{yz}}{M} = \frac{\bar{x}_1 M_1 + \bar{x}_2 M_2 + \bar{x}_3 M_3}{M_1 + M_2 + M_3} \tag{3}$$

and similarly for $\bar{y}$ and $\bar{z}$.

4.26. Moment of inertia and radius of gyration. Let x, y, z represent root-mean-square values of coordinates for any one of the elements of mass dm of § 4.24. Then the second moments about the coordinate planes are

$$I_{yz} = \int x^2\, dm, \quad I_{zx} = \int y^2\, dm, \quad I_{xy} = \int z^2\, dm \tag{1}$$

The moments of inertia about the coordinate axes are

$$I_x = I_{zx} + I_{xy} = \int (y^2 + z^2) dm \tag{2}$$

and similarly for I_y and I_z. For a figure in the xy plane, $z = 0$, $I_{xy} = 0$, and if $I_z = I_0$,

$$I_0 = I_x + I_y = \int (x^2 + y^2) dm = \int r^2\, dm \tag{3}$$

For any moment of inertia, as I_x, the corresponding radius of gyration k is defined by

$$Mk^2 = I_x \quad \text{or} \quad k = \sqrt{I_x/M} \tag{4}$$

For a mass M composed of several, e.g., three, parts M_1, M_2, M_3, the masses and moments of inertia are additive, so that

$$k = \sqrt{\frac{k_1{}^2 M_1 + k_2{}^2 M_2 + k_3{}^2 M_3}{k_1{}^2 + k_2{}^2 + k_3{}^2}} \tag{5}$$

If I is the moment of inertia of M about any axis, and I_g is the moment of inertia of M about a parallel axis through the center of gravity, and the distance between the parallel axes is d, then

$$I = I_g + Md^2 \tag{6}$$

5. Differential Equations

5.1. Classification. Any equation that involves differentials or derivatives is a *differential equation*. If the equation contains any partial derivatives, it is a *partial differential equation*. If it does not, it is an *ordinary differential equation*.

The *order* of the differential equation is the same as the order of the derivative of highest order in the equation.

Suppose that a differential equation is reducible to a form in which each member is a polynomial in all the derivatives that occur. Then the *degree* of the equation is the largest exponent of the highest derivative in the reduced form.

A differential equation is *linear* if it is a first-degree algebraic equation in the set of variables made up of the dependent variables together with all of their derivatives.

5.2. Solutions. Consider a single ordinary differential equation with dependent variable y and independent variable x. Then $y = f(x)$ is an explicit solution of the differential equation if this equation is identically satisfied in x when we substitute $f(x)$, $f'(x)$, etc. for y, dy/dx, etc. therein. Any implicit relation $F(x,y) = 0$ is a solution if when solved for y it leads to explicit solutions. Every ordinary differential equation of the nth order admits of a *general solution* containing n independent constants.

5.3. First-order and first degree. Consider an ordinary differential equation of the first order and first degree. Any such equation may be written in one of the forms

$$M\,dx + N\,dy = 0, \quad M + N\frac{dy}{dx} = 0, \quad \text{or} \quad \frac{dy}{dx} = -\frac{M}{N} \tag{1}$$

where $M(x,y)$ and $N(x,y)$ are functions of x and y. We indicate how to recognize and solve certain special types in §§ 5.4-5.8.

5.4. Variables separable. Here the M and N in § 5.3 are products of factors, where each factor is either a function of x alone, or a function of y

alone. Divide the equation by the factor of M containing y and by the factor of N containing x. Then the differential equation is

$$f(x)dx = g(y)dy \tag{1}$$

in which the variables are *separated*. The solution is

$$\int f(x)dx = \int g(y)dy + c \quad \text{or} \quad \int_{x_0}^{x} f(x)dx = \int_{y_0}^{y} g(y)dy \tag{2}$$

if $y = y_0$ when $x = x_0$.

5.5. Linear in y. Here N in § 5.3 is a function of x alone, and M is a first-degree polynomial in y. Thus the equation is

$$A(x)\frac{dy}{dx} + B(x)y = C(x) \quad \text{or} \quad \frac{dy}{dx} + P(x)y = Q(x) \tag{1}$$

after division by $A(x)$ and with a new notation. Calculate

$$\int P(x)dx \quad \text{and} \quad I(x) = e^{\int P dx} \tag{2}$$

with any constant in the integral. The solution of the equation is

$$y = \frac{1}{I(x)}\left[\int I(x)Q(x)dx + c\right] \tag{3}$$

If $y = y_0$ when $x = x_0$, the solution is found from

$$I(x)y - I(x_0)y_0 = \int_{x_0}^{x} I(x)Q(x)dx \tag{4}$$

5.6. Reducible to linear. To solve the Bernoulli equation

$$A(x)\frac{dy}{dx} + B(x)y = C(x)y^n, \quad n \neq 1 \quad \text{or} \quad 0 \tag{1}$$

multiply by $(1 - n)y^{-n}/A(x)$ and make the substitution

$$u = y^{1-n}, \quad \frac{du}{dx} = (1 - n)y^{-n}\frac{dy}{dx} \tag{2}$$

The resulting equation is like that of § 5.5, with u in place of y.

If M in § 5.3 is a function of y alone, and M is a first-degree polynomial in x, the differential equation is

$$A(y)\frac{dx}{dy} + B(y)x = C(y) \tag{3}$$

and may be solved as in § 5.5 with the roles of x and y interchanged.

Whenever we observe new variables $u(x,y)$ and $t(x,y)$ which reduce a given differential equation to the form

$$A(t)\frac{du}{dt} + B(t)u = C(t) \tag{4}$$

we may solve as in § 5.5 with t and u in the roles of x and y.

5.7. Homogeneous. Here the M and N in § 5.3 are each homogeneous of the same degree, that is of the same dimension when x and y are each assigned the dimension one. In this case use

$$y = vx, \quad dy = v\,dx + x\,dv \tag{1}$$

to eliminate y and dy. The new equation in x and v will be separable as in § 5.4. If more convenient, let $x = uy$.

5.8. Exact equations. The equation of § 5.3 will be exact if

$$\partial M/\partial y = \partial N/\partial x$$

Integrate $M\,dx$, regarding y as constant and adding an unknown function of y, say $f(y)$. Differentiate the result with respect to y and equate the new result to N. From the resulting equation determine the unknown function of y. The solution is then

$$\int M\,dx + f(y) + c = 0 \tag{1}$$

If more convenient, interchange M and N and also x and y in the above process.

5.9. First order and higher degree. The general differential equation of the first order is

$$F\left(x,y,\frac{dy}{dx}\right) = 0, \quad \text{or} \quad F(x,y,p) = 0 \tag{1}$$

if we write a single letter p in place of dy/dx. For special solvable types see §§ 5.10-5.11.

5.10. Equations solvable for *p*. Some equations § 5.9 may be easily solved for p. The resulting first-degree equations may often be integrated by the methods of §§ 5.3-5.8.

5.11. Clairaut's form. Clairaut's equation is

$$y = x\frac{dy}{dx} + f\left(\frac{dy}{dx}\right) \quad \text{or} \quad y = px + f(p) \tag{1}$$

where f is any function of one variable. The general solution is

$$y = cx + f(c) \tag{2}$$

This represents a family of straight lines. They usually have an envelope given in terms of the parameter c by

$$x = -f'(c), \quad y = f(c) - cf'(c) \tag{3}$$

These are the parametric equations of the *singular solution.*

5.12. Second order. The general differential equation of the second order is

$$F\left(x,y,\frac{dy}{dx},\frac{d^2y}{dx^2}\right) = 0 \tag{4}$$

If either of the letters x or y is absent from the function F, the substitution $dy/dx = p$ reduces the equation to one of the first order. If the letter y is missing, we put

$$\frac{dy}{dx} = p, \quad \frac{d^2y}{dx^2} = \frac{dp}{dx} \tag{5}$$

and

$$F\left(x,\frac{dy}{dx},\frac{d^2y}{dx^2}\right) = 0 \quad \text{becomes} \quad F\left(x,p,\frac{dp}{dx}\right) = 0 \tag{6}$$

The solution of this first-order equation in the variables x and p may be found as in § 5.3 or § 5.9. Suppose it is written in the form $p = G(x,c_1)$; then

$$y = \int G(x,c_1)dx + c_2 \tag{7}$$

is the general solution of the second-order equation with y missing.

If the letter x is missing, we put

$$\frac{dy}{dx} = p, \quad \frac{d^2y}{dx^2} = p\frac{dp}{dy} \tag{8}$$

and

$$f\left(y,\frac{dy}{dx},\frac{d^2y}{dx^2}\right) = 0 \quad \text{becomes} \quad F\left(y,p,p\frac{dp}{dy}\right) = 0 \tag{9}$$

The solution of this first-order equation in the variables y and p may be found as in § 5.3 and § 5.9. Suppose it is written in the form $p = G(y,c_1)$; then

$$x = \int \frac{dy}{G(y,c_1)} + c_2 \tag{10}$$

is the general solution of the second-order equation with x missing.

5.13. Linear equations. The general linear differential equation of the second order is

$$A(x)\frac{d^2y}{dx^2} + B(x)\frac{dy}{dx} + C(x)y = E(x), \quad \text{with} \quad A(x) \neq 0 \tag{1}$$

We get the corresponding homogeneous equation by replacing $E(x)$ by zero. Two solutions u_1, u_2 of the homogeneous equation are linearly independent and constitute a fundamental system if their Wronskian determinant $W(x)$ is not zero, where

$$W(x) = u_1u_2' - u_2u_1' = \begin{vmatrix} u_1 & u_2 \\ u_1' & u_2' \end{vmatrix} \tag{2}$$

When $W(x) \neq 0$, u_1 and u_2 constitute a *fundamental system* of solutions of the homogeneous equation. And if u is any particular solution of the non-homogeneous equation with $E(x)$ present, the general solution of this equation is given by

$$y = u + c_1u_1 + c_2u_2 \tag{3}$$

In this form c_1 and c_2 are arbitrary constants, u is called the *particular integral*, and $c_1u_1 + c_2u_2$ is the *complementary function.*

Similarly for the linear equation of the nth order

$$A_n(x)\frac{d^ny}{dx^n} + A_{n-1}(x)\frac{d^{n-1}y}{dx^{n-1}} + \ldots + A_1(x)\frac{dy}{dx} + A_0(x)y = E(x) \tag{4}$$

the general solution is

$$y = u + c_1u_1 + c_2u_2 + \ldots + c_nu_n \tag{5}$$

Here u, the particular integral, is any solution of the given equation, and $u_1, u_2, \ldots, u_n$ form a fundamental system of solutions of the homogeneous equation obtained by replacing $E(x)$ by zero, or a set of solutions of the homogeneous equation whose Wronskian determinant $W(x)$ is not zero, where

$$W(x) = \begin{vmatrix} u_1 & u_2 & \ldots & u_n \\ u_1' & u_2' & \ldots & u_n' \\ \ldots & \ldots & \ldots & \ldots \\ u_1^{(n-1)} & u_2^{(n-1)} & \ldots & u_n^{(n-1)} \end{vmatrix} \tag{6}$$

For any n functions u_j, $W(x) = 0$ if some one u is linearly dependent on the others, as $u_n = k_1u_1 + k_2u_2 + \ldots + k_{n-1}u_{n-1}$ with the coefficients k_i constant.

And for n solutions of a linear differential equation of the nth order, if $W(x)$ is not zero, the solutions are linearly independent.

5.14. Constant coefficients. To solve the homogeneous equation

$$A\frac{d^2y}{dx^2} + B\frac{dy}{dx} + Cy = 0 \tag{1}$$

where A, B, and C are constants, find the roots of the auxiliary equation $Ap^2 + Bp + C = 0$. If the roots are unequal quantities r and s, the solution is $y = c_1e^{rx} + c_2e^{sx}$.

When the coefficients A, B, C are real, and the roots are complex, they will occur as a conjugate pair $a + bi$ and $a - bi$. In this case the real form of the solution is $y = e^{ax}(c_1 \cos bx + c_2 \sin bx)$ or by § 2.8, $y = C_1e^{ax} \sin (bx + C_2)$.

If the roots are equal and are r,r the solution is $y = e^{rx}(c_1 + c_2x)$.

Similarly, to solve the homogeneous equation of the nth order

$$A_n\frac{d^ny}{dx^n} + A_{n-1}\frac{d^{n-1}y}{dx^{n-1}} + \ldots + A_1\frac{dy}{dx} + A_0 = 0$$

where A_n, A_{n-1}, ..., A_0 are constants, find the roots of the auxiliary equation

$$A_np^n + A_{n-1}p^{n-1} + \ldots + A_1p + A_0 = 0 \tag{2}$$

For each distinct root r there is a term ce^{rx} in the solution. The terms of the solution are to be added together.

When r occurs twice among the n roots of the auxiliary equation, the corresponding term is $e^{rx}(c_1 + c_2x)$.

When r occurs three times, the corresponding term is $e^{rx}(c_1 + c_2x + c_3x^2)$, and so forth.

When there is a pair of conjugate complex roots $a + bi$ and $a - bi$, the real form of the terms in the solution is $e^{ax}(c_1 \cos bx + c_2 \sin bx)$.

When the same pair occurs twice, the corresponding term is $e^{ax}[(c_1 + c_2x) \cos bx + (d_1 + d_2x) \sin bx]$, and so forth.

Consider next the general nonhomogeneous linear differential equation of order n, with constant coefficients, or

$$A_n\frac{d^ny}{dx^n} + A_{n-1}\frac{d^{n-1}y}{dx^{n-1}} + \ldots + A_1\frac{dy}{dx} + A_0 = E(x)$$

By § 5.13 we may solve this by adding any particular integral to the complementary function, or general solution of the homogeneous equation obtained by replacing $E(x)$ by zero. The complementary function may be found from the rules just given in this section. And the particular integral may then be found by the methods of §§ 5.15-5.16.

5.15. Undetermined coefficients. In the last equation of § 5.14 let the right member $E(x)$ be a sum of terms each of which is of the type

$$k, \quad k \cos bx, \quad k \sin bx, \quad ke^{ax}, \quad kx \tag{1}$$

or, more generally,

$$kx^m e^{ax}, \quad kx^m e^{ax} \cos bx, \quad \text{or} \quad kx^m e^{ax} \sin bx \tag{2}$$

Here m is zero or a positive integer, and a and b are any real numbers. Then the form of the particular integral I may be predicted by the following rules.

Case I. $E(x)$ is a single term T. Let D be written for d/dx, so that the given equation is $P(D)y = E(x)$, where

$$P(D) = A_n D^n + A_{n-1} D^{n-1} + \ldots + A_1 D + A_0 \tag{3}$$

With the term T associate the simplest polynomial $Q(D)$ such that $Q(D)T = 0$. For the particular types k, etc., $Q(D)$ will be

$$D, \quad D^2 + b^2, \quad D^2 + b^2, \quad D - a, \quad D^2 \tag{4}$$

and for the general types $kx^m e^{ax}$, etc., $Q(D)$ will be

$$(D - a)^{m+1}, \quad (D^2 - 2aD + a^2 + b^2)^{m+1}, \quad (D^2 - 2aD + a^2 + b^2)^{m+1} \tag{5}$$

Thus $Q(D)$ will always be some power of a first- or second-degree factor,

$$Q(D) = F^q, \quad F = D - a, \quad \text{or} \quad F = D^2 - 2aD + a^2 + b^2 \tag{6}$$

Use § 5.14 to find the terms in the solution of $P(D)y = 0$, and also the terms in the solution of $Q(D)P(D)y = 0$. Then assume that the particular integral I is a linear combination with unknown coefficients of those terms in the solution of $Q(D)P(D)y = 0$ which are not in the solution of $P(D)y = 0$. Thus if $Q(D) = F^q$, and F is *not* a factor of $P(D)$, assume

$$I = (Ax^{q-1} + Bx^{q-2} + \ldots + L)e^{ax} \tag{7}$$

when $F = D - a$, and

$$I = (Ax^{q-1} + Bx^{q-2} + \ldots + L)e^{ax} \cos bx$$

$$+ (Mx^{q-1} + Nx^{q-2} + \ldots + R)e^{ax} \sin bx$$

when

$$F = D^2 - 2aD + a^2 + b^2 \tag{8}$$

When F *is* a factor of $P(D)$ and the highest power of F which is a divisor of $P(D)$ is F^p, try the I above multiplied by x^p.

Case II. $E(x)$ is a sum of terms. With each term in $E(x)$ associate a polynomial $Q(D) = F^q$ as before. Arrange in one group all terms that have the same F. The particular integral of the given equation will be the sum of solutions of equations each of which has one group on the right. For any one such equation, the form of the particular integral is given as for Case I with q the highest power of F associated with any term of the group on the right.

After the form has been found, in Case I or Case II, the unknown coefficients follow when we substitute back in the given differential equation, equating coefficients of like terms, and solve the resulting system of simultaneous equations.

5.16. Variation of parameters. Whenever a fundamental system of solutions $u_1, u_2, \ldots, u_n$ for the homogeneous equation is known, a particular integral of

$$A_n(x)\frac{d^ny}{dx^n} + A_{n-1}(x)\frac{d^{n-1}y}{dx^{n-1}} + \ldots + A_1(x)\frac{dy}{dx} + A_0(x)y = E(x) \tag{1}$$

may be found in the form

$$y = \sum_{k=1}^{n} v_k u_k \tag{2}$$

Here the v_k are functions of x found by integrating their derivatives v_k', and these derivatives are the solutions of the following n simultaneous equations :

$$\left.\begin{array}{l} \sum_{k=1}^{n} v_k' u_k = 0, \quad \sum_{k=1}^{n} v_k' u_k' = 0, \quad \sum_{k=1}^{n} v_k' u_k'' = 0, \quad \ldots, \\ \sum_{k=1}^{n} v_k' u_k^{(n-2)} = 0, \quad A_n(x) \sum_{k=1}^{n} v_k' u_k^{(n-1)} = E(x) \end{array}\right\} \tag{3}$$

To find the v_k from $v_k = \int v_k'\, dx + c_k$, any choice of the constants will lead to a particular integral. The special choice

$$v_k = \int_0^x v_k'\, dx$$

leads to the particular integral having $y, y', y'', \ldots, y^{(n-1)}$ each equal to zero when $x = 0$.

5.17. The Cauchy-Euler "homogeneous linear equation." This has the form

$$k_n x^n \frac{d^n y}{dx^n} + k_{n-1} x^{n-1} \frac{d^{n-1} y}{dx^{n-1}} + \dots + k_1 x \frac{dy}{dx} + k_0 y = F(x) \tag{1}$$

The substitution $x = e^t$, which makes

$$x \frac{dy}{dx} = \frac{dy}{dt}, \quad x^k \frac{d^k y}{dx^k} = \left(\frac{d}{dt} - k + 1\right) \dots \left(\frac{d}{dt} - 2\right)\left(\frac{d}{dt} - 1\right)\frac{dy}{dt} \tag{2}$$

transforms this into a linear differential equation with constant coefficients. By §§ 5.14 to 5.16 its solution may be found in the form $y = g(t)$, leading to $y = g(\ln x)$ as the solution of the given Cauchy-Euler equation.

5.18. Simultaneous differential equations. A system of two equations in x,y depending on t, if linear and with constant coefficients, may be written

$$\left.\begin{aligned} f_1(D)x + g_1(D)y &= E_1(t) \\ f_2(D)x + g_2(D)y &= E_2(t) \end{aligned}\right\} \tag{1}$$

where $D = d/dt$. By §§ 5.14 to 5.16 find the solution of

$$[g_2(D)f_1(D) - g_1(D)f_2(D)]x = g_2(D)E_1(t) - g_1(D)E_2(t) \tag{2}$$

in the form

$$x = u + c_1 u_1 + c_2 u_2 + \dots + c_m u_m, \tag{3}$$

and the solution of

$$[g_2(D)f_1(D) - g_1(D)f_2(D)]y = f_1(D)E_2(t) - f_2(D)E_1(t) \tag{4}$$

in the form

$$y = v + c_1' u_1 + c_2' u_2 + \dots + c_m' u_m \tag{5}$$

Here u, v, u_1, ..., u_m are functions of t. The constants c_k and c_k' are not independent, and the relations between them must be found by substitution of x and y in the original equations. In the general case, and the usual choice of u and v, these relations may be used to determine the c_k' in terms of the c_k.

Consider the special homogeneous linear system of n equations of the first order, with the coefficients c_{ks} constant,

$$\frac{dy_k}{dt} = c_{k1} y_1 + c_{k2} y_2 + \dots + c_{kn} y_n, \quad k = 1, 2, \dots, n \tag{6}$$

Let the equation

$$\begin{vmatrix} c_{11}-r & c_{12} & \dots & c_{1n} \\ c_{21} & c_{22}-r & \dots & c_{2n} \\ \dots & \dots & \dots & \dots \\ c_{n1} & c_{n2} & \dots & c_{nn}-r \end{vmatrix} = 0 \tag{7}$$

have n distinct roots $r_1, r_2, \dots, r_n$. Then the solution is

$$y_k = a_{k1}e^{r_1 t} + a_{k2}e^{r_2 t} + \dots + a_{kn}e^{r_n t} \tag{8}$$

where for each s the ratios of the coefficients a_{ks} are found from

$$\left.\begin{aligned} (c_{11}-r_s)a_{1s} + c_{12}a_{2s} + \dots + c_{1n}a_{ns} &= 0 \\ c_{21}a_{1s} + (c_{22}-r_s)a_{2s} + \dots + c_{2n}a_{ns} &= 0 \\ \dots \quad \dots \quad \dots \quad \dots \quad \dots \quad \dots \quad \dots \quad & \dots \\ c_{n1}a_{1s} + c_{n2}a_{2s} + \dots + (c_{nn}-r_s)a_{ns} &= 0 \end{aligned}\right\} \tag{9}$$

5.19. First-order partial differential equations. When linear in the derivatives of the two dependent variables, the equation is

$$A(x,y,u)p + B(x,y,u)q = C(x,y,u) \tag{1}$$

where

$$p = \frac{\partial u}{\partial x} = u_x \quad \text{and} \quad q = \frac{\partial u}{\partial y} = u_y \tag{2}$$

The system of differential equations for the characteristic curves is

$$\frac{dx}{A} = \frac{dy}{B} = \frac{du}{C} \tag{3}$$

If this is solved in the form

$$f(x,y,u) = c_1 \quad \text{and} \quad g(x,y,u) = c_2 \tag{4}$$

the solution of the partial differential equation is given by

$$F(f,g) = 0, \quad f = G(g), \quad \text{or} \quad g = H(f) \tag{5}$$

where F, G, H are arbitrary functions.

5.20. Second-order partial differential equations. When linear in the second derivatives of the two dependent variables, the equation is

$$R(x,y)r + 2S(x,y)s + T(x,y)t = V(x,y,p,q,u) \tag{1}$$

where

$$p = u_x, \quad q = u_y, \quad r = u_{xx}, \quad s = u_{xy}, \quad t = u_{yy} \qquad (2$$

The ordinary differential equation for the characteristics is

$$R\left(\frac{dy}{dx}\right)^2 - 2S\frac{dy}{dx} + T = 0 \tag{3}$$

By § 5.10 its solution may be found in the form

$$f(x,y) = a, \quad g(x,y) = b \tag{4}$$

These are the equations of the two families of characteristic curves.

The type of equation depends on the sign of the determinant

$$\begin{vmatrix} R & S \\ S & T \end{vmatrix} = RT - S^2 \tag{5}$$

for any x,y region under consideration. If $RT - S^2 < 0$, or $RT < S^2$, the equation is hyperbolic. In this case the characteristics are real, and if the parameters a and b are used as new variables, we obtain the first normal form

$$u_{ab} = F(a,b,u,u_a,u_b) \tag{6}$$

A second normal form for the hyperbolic type results from the substitution

$$a = X + Y, \quad b = X - Y \tag{7}$$

This is

$$u_{XX} - u_{YY} = G(X,Y,u,u_X,u_Y) \tag{8}$$

If $RT - S^2 = 0$, or $RT = S^2$, the equation is parabolic. In this case the two families of characteristics are real and coincident. We here use $X = a = b$ and Y any second independent function of x,y as new variables. This gives the normal form

$$u_{XX} = F(X,Y,u,u_X,u_Y) \tag{9}$$

If $RT - S^2 > 0$, or $RT > S^2$, the equation is elliptic. Here the characteristics are not real, but use of the substitution

$$a = X + iY, \quad b = X - iY \tag{10}$$

leads to the real normal form

$$u_{XX} + u_{YY} = G(X,Y,u,u_X,u_Y) \tag{11}$$

The wave equation, $u_{tt} - c^2 u_{xx} = 0$, is hyperbolic and admits the general solution

$$u = f(x - ct) + g(x + ct) \tag{12}$$

The heat equation

$$a^2 u_{xx} = u_t \tag{13}$$

is parabolic and admits no general solution in terms of arbitrary functions.

Laplace's equation, $u_{xx} + u_{yy} = 0$, is elliptic and admits the general solution

$$u = f(x + iy) + g(x - iy). \tag{14}$$

5.21. Runge-Kutta method of finding numerical solutions. For the differential equation $dy/dx = f(x,y)$ of § 5.3 the solution may be found step by step. We start with x_0,y_0, first compute x_1,y_1, then x_2,y_2, and so on up to x_N,y_N. Here $x_{n+1} - x_n = h$ is small, and at each step we find

$$k_1 = f(x_n,y_n)h, \quad k_2 = f\left(x_n + \frac{h}{2}, y_n + \frac{k_1}{2}\right)h,$$

$$k_3 = f\left(x_n + \frac{h}{2}, y + \frac{k_2}{2}\right)h, \quad k_4 = f(x_n + h, y_n + k_3)h.$$

Then

$$x_{n+1} = x_n + h, \quad y_{n+1} = y_n + \Delta y,$$

where

$$\Delta y = \tfrac{1}{6}(k_1 + 2k_2 + 2k_3 + k_4)$$

6. Vector Analysis

6.1. Scalars. A scalar S is a real number capable of representation by a signed coordinate on a scale.

6.2. Vectors. A vector V is a quantity which is determined by a length and a direction. The vector V may be graphically represented by any line segment having this length and this direction $V = OA$.

6.3. Components. Let $\boldsymbol{i}, \boldsymbol{j}, \boldsymbol{k}$ represent three vectors of unit length along the three mutually perpendicular lines OX, OY, OZ, respectively, which are taken as the positive coordinate axes. Let $\boldsymbol{V}$ be a vector in space, and a, b, c the projections of $\boldsymbol{V}$ on the three lines OX, OY, OZ. Then

$$\boldsymbol{V} = a\boldsymbol{i} + b\boldsymbol{j} + c\boldsymbol{k} \tag{1}$$

where a, b, c are the components of $\boldsymbol{V}$, while $a\boldsymbol{i}$, $b\boldsymbol{j}$, $c\boldsymbol{k}$ are the component vectors. The magnitude of $\boldsymbol{V}$ is

$$|\boldsymbol{V}| = V = \sqrt{a^2 + b^2 + c^2} \tag{2}$$

And the direction cosines of $\boldsymbol{V}$ are $a/|\boldsymbol{V}|$, $b/|\boldsymbol{V}|$, $c/|\boldsymbol{V}|$ when $|\boldsymbol{V}| \neq 0$. When $|\boldsymbol{V}| = 0$, we have the null vector

$$0 = 0\boldsymbol{i} + 0\boldsymbol{j} + 0\boldsymbol{k} \tag{3}$$

which has length zero and no determinate direction.

6.4. Sums and products by scalars. These may be formed by corresponding operations on the components. Thus if

$$V_1 = a_1 i + b_1 j + c_1 k \quad \text{and} \quad V_2 = a_2 i + b_2 j + c_2 k \tag{1}$$

$$V_1 + V_2 = (a_1 + a_2)i + (b_1 + b_2)j + (c_1 + c_2)k \tag{2}$$

$$SV = VS = (Sa)i + (Sb)j + (Sc)k \tag{3}$$

$$(S_1 + S_2)V = S_1 V + S_2 V, \quad S(V_1 + V_2) = SV_1 + SV_2 \tag{4}$$

6.5. The scalar or dot product, $V_1 \cdot V_2$. This is defined as

$$V_1 \cdot V_2 = |V_1|\,|V_2| \cos\theta \tag{1}$$

where θ is any angle from V_1 to V_2.

$$V_1 \cdot V_2 = V_2 \cdot V_1 = a_1 a_2 + b_1 b_2 + c_1 c_2, \quad V_1 \cdot V_1 = |V_1|^2 \tag{2}$$

$$V_1 \cdot (V_2 + V_3) = V_1 \cdot V_2 + V_1 \cdot V_3 \tag{3}$$

$$(V_1 + V_2) \cdot V_3 = V_1 \cdot V_2 + V_2 \cdot V_3 \tag{4}$$

$$i \cdot i = j \cdot j = k \cdot k = 1, \quad i \cdot j = j \cdot k = k \cdot i = j \cdot i = k \cdot j = i \cdot k = 0 \tag{5}$$

6.6. The vector or cross product, $V_1 \times V_2$. This is defined as

$$V_1 \times V_2 = |V_1|\,|V_2| \sin\theta\, u$$

where θ is some angle not exceeding 180° from V_1 to V_2 and u is a unit vector perpendicular to the plane of V_1 and V_2 and so directed that a right-threaded screw along u will advance when turned through the angle θ. For $\theta = 0$ or 180°, V_1 and V_2 do not determine a plane, but $\sin\theta = 0$ makes the product the null vector.

$$\left.\begin{aligned} V_1 \times V_2 = -V_2 \times V_1 &= \begin{vmatrix} i & j & k \\ a_1 & b_1 & c_1 \\ a_2 & b_2 & c_2 \end{vmatrix} \\ &= (b_1 c_2 - b_2 c_1)i + (c_1 a_2 - c_2 a_1)j + (a_1 b_2 - a_2 b_1)k \end{aligned}\right\} \tag{1}$$

$$V_1 \times (V_2 + V_3) = V_1 \times V_2 + V_1 \times V_3 \tag{2}$$

$$(V_1 + V_2) \times V_3 = V_1 \times V_3 + V_2 \times V_3 \tag{3}$$

$$V_1 \times (V_2 \times V_3) = (V_1 \cdot V_3)V_2 - (V_1 \cdot V_2)V_3 \tag{4}$$

$$(V_1 \times V_2) \times V_3 = (V_1 \cdot V_3)V_2 - (V_2 \cdot V_3)V_1 \tag{5}$$

$$\left.\begin{aligned} i \times i = 0,\ \ j \times j = 0,\ \ k \times k = 0,\ \ i \times j = k,\ \ j \times k = i, \\ k \times i = j,\ \ j \times i = -k,\ \ k \times j = -i,\ \ i \times k = -j \end{aligned}\right\} \tag{6}$$

6.7. The triple scalar product. The volume of the parallelepiped having V_1, V_2, V_3 as three of its edges is

$$V_1 \cdot (V_2 \times V_3) = (V_1 \times V_2) \cdot V_3 = V_2 \cdot (V_3 \times V_1) = \begin{vmatrix} a_1 & a_2 & a_3 \\ b_1 & b_2 & b_3 \\ c_1 & c_2 & c_3 \end{vmatrix} \tag{1}$$

if we so number the vectors that the cross product makes an acute angle with the vector outside the parenthesis.

6.8. The derivative. Let a vector r have its components variable but functions of a parameter t, so that

$$r(t) = x(t)i + y(t)j + z(t)k \quad \text{or} \quad r = xi + yj + zk \tag{1}$$

To differentiate r, we merely differentiate each component. Thus

$$\frac{dr}{dt} = r'(t) = x'(t)i + y'(t)j + z'(t)k = \frac{dx}{dt}i + \frac{dy}{dt}j + \frac{dz}{dt}k \tag{2}$$

For two vectors,

$$r_1 = x_1 i + y_1 j + z_1 k \quad \text{and} \quad r_2 = x_2 i + y_2 j + z_2 k,$$

$$\frac{d}{dt}(r_1 + r_2) = \frac{dr_1}{dt} + \frac{dr_2}{dt} \tag{3}$$

$$\frac{d}{dt}(r_1 \cdot r_2) = r_2 \cdot \frac{dr_1}{dt} + r_1 \cdot \frac{dr_2}{dt} \tag{4}$$

$$\frac{d}{dt}(r_1 \times r_2) = \frac{dr_1}{dt} \times r_2 + r_1 \times \frac{dr_2}{dt} = -r_2 \times \frac{dr_1}{dt} + r_1 \times \frac{dr_2}{dt} \tag{5}$$

$$r \cdot \frac{dr}{dt} = |r| \frac{d|r|}{dt} \tag{6}$$

Hence if $|r|$ is constant, $r \cdot dr/dt = 0$

6.9. The Frenet formulas. Let P_0 be a fixed and P a variable point on a curve in space. Take s, the arc length P_0P, as the parameter. Then the vector from the origin O to P is

$$\overrightarrow{OP} = r(s) = x(s)i + y(s)j + z(s)k \tag{1}$$

With each point P we associate three mutually perpendicular unit vectors $\boldsymbol{t}, \boldsymbol{p}$, and $\boldsymbol{b}$. These satisfy the equations:

$$\frac{d\boldsymbol{r}}{ds} = \boldsymbol{t}, \quad \frac{d\boldsymbol{t}}{ds} = \frac{1}{\rho}\boldsymbol{p}, \quad \frac{d\boldsymbol{b}}{ds} = -\frac{1}{\tau}\boldsymbol{p}, \quad \frac{d\boldsymbol{p}}{ds} = \frac{1}{\tau}\boldsymbol{b} - \frac{1}{\rho}\boldsymbol{t} \tag{2}$$

The tangent vector $\boldsymbol{t}$ has the direction of $\boldsymbol{r}'(s)$, the principal normal vector $\boldsymbol{p}$ has the direction of $\boldsymbol{r}''(s)$, and the binormal vector $\boldsymbol{b}$ has the direction of $\boldsymbol{t} \times \boldsymbol{p} = \boldsymbol{b}$. The curvature $1/\rho$ is the length of $\boldsymbol{r}''(s) = d\boldsymbol{t}/ds$. The torsion $1/\tau$ is determined from $d\boldsymbol{b}/ds = -\boldsymbol{p}/\tau$.

6.10. Curves with parameter t. When the parameter is t,

$$\overrightarrow{OP} = \boldsymbol{r}(t) = x(t)\boldsymbol{i} + y(t)\boldsymbol{j} + z(t)\boldsymbol{k} \tag{1}$$

Using dots for t derivatives we form

$$\dot{\boldsymbol{r}} = \dot{x}\boldsymbol{i} + \dot{y}\boldsymbol{j} + \dot{z}\boldsymbol{k} \quad \text{and} \quad \dot{\boldsymbol{r}} \times \ddot{\boldsymbol{r}} = \begin{vmatrix} \dot{y} & \dot{z} \\ \ddot{y} & \ddot{z} \end{vmatrix} \boldsymbol{i} + \begin{vmatrix} \dot{z} & \dot{x} \\ \ddot{z} & \ddot{x} \end{vmatrix} \boldsymbol{j} + \begin{vmatrix} \dot{x} & \dot{y} \\ \ddot{x} & \ddot{y} \end{vmatrix} \boldsymbol{k} \tag{2}$$

Then for the unit vectors of § 6.9, the tangent $\boldsymbol{t}$ has the direction of $\dot{\boldsymbol{r}}$, the binormal $\boldsymbol{b}$ has the direction of $\dot{\boldsymbol{r}} \times \ddot{\boldsymbol{r}}$, and the principal normal $\boldsymbol{p}$ has the direction of $(S_1\dot{\boldsymbol{r}} \times \ddot{\boldsymbol{r}}) \times (S_2\dot{\boldsymbol{r}})$, where S_1 and S_2 are any scalar factors. The curvature and torsion may be found from

$$\frac{1}{\rho} = \frac{\dot{\boldsymbol{r}} \times \ddot{\boldsymbol{r}}}{\dot{\boldsymbol{r}}^3}, \quad \frac{1}{\tau} = \frac{(\dot{\boldsymbol{r}} \times \ddot{\boldsymbol{r}} \cdot \dddot{\boldsymbol{r}})}{\dot{\boldsymbol{r}}^2} \tag{3}$$

When the parameter t is the time, the velocity vector $\boldsymbol{v} = v\boldsymbol{t} = \dot{\boldsymbol{r}}$, the speed

$$v = |\,\boldsymbol{v}\,| = \sqrt{\dot{x}^2 + \dot{y}^2 + \dot{z}^2} \tag{4}$$

and the acceleration vector

$$\boldsymbol{a} = \dot{\boldsymbol{v}} = \ddot{\boldsymbol{r}} = \dot{v}\boldsymbol{t} + \frac{v^2}{\rho}\boldsymbol{p}$$

6.11. Relative motion. A varying coordinate system with fixed origin O is instantaneously rotating about some line OL with angular velocity ω. Let $\boldsymbol{w}$ be a vector of length ω along OL. For a variable point P and $\boldsymbol{i}, \boldsymbol{j}, \boldsymbol{k}$ in the moving system

$$\overrightarrow{OP} = \boldsymbol{r} = x\boldsymbol{i} + y\boldsymbol{j} + z\boldsymbol{k} \quad \text{and} \quad \dot{\boldsymbol{r}} = \boldsymbol{v}_{\text{rel}} = \dot{x}\boldsymbol{i} + \dot{y}\boldsymbol{j} + \dot{z}\boldsymbol{k} \tag{1}$$

The absolute velocity of P, $\boldsymbol{v}_{\text{abs}}$ is given by

$$\boldsymbol{v}_{\text{abs}} = \boldsymbol{v}_{\text{rel}} + \boldsymbol{w} \times \boldsymbol{r} = \dot{\boldsymbol{r}} + \boldsymbol{w} \times \boldsymbol{r} \tag{2}$$

where $\boldsymbol{v}_{\text{rel}}$ is the velocity relative to the moving system.

6.12. The symbolic vector del. The equation

$$\nabla = \frac{\partial}{\partial x}\boldsymbol{i} + \frac{\partial}{\partial y}\boldsymbol{j} + \frac{\partial}{\partial z}\boldsymbol{k} \tag{1}$$

defines the vector differential operator ∇, read *del.*

The *gradient* of a scalar function $f(x,y,z)$, ∇f or grad f is

$$\operatorname{grad} f = \nabla f = \frac{\partial f}{\partial x}\boldsymbol{i} + \frac{\partial f}{\partial y}\boldsymbol{j} + \frac{\partial f}{\partial z}\boldsymbol{k} \tag{2}$$

The gradient ∇f extends in the direction in which the derivative df/ds is a maximum, and $|\nabla f|$ is equal to that maximum. In a direction making an angle θ with the gradient, $df/ds = |\nabla f| \cos \theta$.

We may apply del to a vector function of position $\boldsymbol{Q}$,

$$\boldsymbol{Q}(x,y,z) = Q_1(x,y,z)\boldsymbol{i} + Q_2(x,y,z)\boldsymbol{j} + Q_3(x,y,z)\boldsymbol{k} \tag{3}$$

The *divergence* of the vector function $\boldsymbol{Q}$, $\nabla \cdot \boldsymbol{Q}$ or div $\boldsymbol{Q}$ is

$$\operatorname{div} \boldsymbol{Q} = \nabla \cdot \boldsymbol{Q} = \frac{\partial Q_1}{\partial x} + \frac{\partial Q_2}{\partial y} + \frac{\partial Q_3}{\partial z} \tag{4}$$

For a fluid, let $\boldsymbol{Q}$ equal the velocity vector multiplied by the scalar density. Then div $\boldsymbol{Q}$ is the rate of flow outward per unit volume at a point.

The curl or rotation of the vector function $\boldsymbol{Q}$, $\nabla \times \boldsymbol{Q}$ is

$$\left.\begin{aligned} \operatorname{curl} \boldsymbol{Q} = \operatorname{rot} \boldsymbol{Q} = \nabla \times \boldsymbol{Q} &= \begin{vmatrix} \boldsymbol{i} & \boldsymbol{j} & \boldsymbol{k} \\ \dfrac{\partial}{\partial x} & \dfrac{\partial}{\partial y} & \dfrac{\partial}{\partial z} \\ Q_1 & Q_2 & Q_3 \end{vmatrix} \\ &= \left(\frac{\partial Q_3}{\partial y} - \frac{\partial Q_2}{\partial z}\right)\boldsymbol{i} + \left(\frac{\partial Q_1}{\partial z} - \frac{\partial Q_3}{\partial x}\right)\boldsymbol{j} + \left(\frac{\partial Q_2}{\partial x} - \frac{\partial Q_1}{\partial y}\right)\boldsymbol{k} \end{aligned}\right\} \tag{5}$$

For a fluid, let $\boldsymbol{Q}$ equal the velocity vector times the density. When the motion is analyzed into a dilatation and a rigid displacement, for the latter the angular velocity at a point or vorticity is $\frac{1}{2}$ curl $\boldsymbol{Q}$.

In terms of del the Laplacian operator is ∇^2, and

$$\nabla^2 f = \nabla \cdot \nabla f = \operatorname{div} \operatorname{grad} f = \frac{\partial^2 f}{\partial x^2} + \frac{\partial^2 f}{\partial y^2} + \frac{\partial^2 f}{\partial z^2} \tag{6}$$

$$\nabla \cdot (\nabla \times \boldsymbol{Q}) = \operatorname{div} \operatorname{curl} \boldsymbol{Q} = 0, \quad \nabla \times (\nabla f) = \operatorname{curl} \operatorname{grad} f = 0 \tag{7}$$

And $\nabla \times \boldsymbol{Q} = \operatorname{curl} \boldsymbol{Q} = 0$ is a necessary and sufficient condition for $\boldsymbol{Q}$ to be the gradient of some function f,

$$\boldsymbol{Q} = \nabla f \tag{8}$$

$$\text{curl curl } \boldsymbol{Q} = \boldsymbol{\nabla} \times (\boldsymbol{\nabla} \times \boldsymbol{Q}) = \text{grad div } \boldsymbol{Q} - \boldsymbol{\nabla}^2 \boldsymbol{Q} = \boldsymbol{\nabla}(\boldsymbol{\nabla} \cdot \boldsymbol{Q}) - \boldsymbol{\nabla} \cdot \boldsymbol{\nabla} \boldsymbol{Q} \quad (9)$$

where

$$\boldsymbol{\nabla}^2 \boldsymbol{Q} = \boldsymbol{\nabla} \cdot \boldsymbol{\nabla} \boldsymbol{Q} = (\boldsymbol{\nabla}^2 Q_1)\boldsymbol{i} + (\boldsymbol{\nabla}^2 Q_2)\boldsymbol{j} + (\boldsymbol{\nabla}^2 Q_3)\boldsymbol{k} \quad (10)$$

6.13. The divergence theorem. Let S consist of one or more closed surfaces that collectively bound a Volume V. Then

$$\int_V \text{div } \boldsymbol{Q} \, dV = \int_V \boldsymbol{\nabla} \cdot \boldsymbol{Q} \, dV = \int_S \boldsymbol{Q} \cdot \boldsymbol{n} \, dS \quad (1)$$

where $\boldsymbol{n}$ is the unit normal to S, drawn outward.

6.14. Green's theorem in a plane. Let B be a simple closed curve in the x,y plane which bounds an area A, with the positive direction so chosen that it and the inner normal are related like OX and OY. Then

$$\int_B (M \, dx + N \, dy) = \int_A \left(\frac{\partial N}{\partial x} - \frac{\partial M}{\partial y} \right) dA \quad (2)$$

for M and N any two functions of x and y.

6.15. Stokes's theorem. Let S be a portion of a surface in space bounded by a simple closed curve B. If $\boldsymbol{m}$ is the direction into S perpendicular to B, the positive direction for B is related to $\boldsymbol{m}$ and $\boldsymbol{n}$, the positive normal to S, like OX, OY, and OZ. Then

$$\int_S (\text{curl } \boldsymbol{Q}) \cdot \boldsymbol{n} \, dS = \int_S \boldsymbol{n} \cdot (\boldsymbol{\nabla} \times \boldsymbol{Q}) dS = \int_B \boldsymbol{Q} \cdot d\boldsymbol{s} \quad (1)$$

6.16. Curvilinear coordinates. For an orthogonal system

$$ds^2 = h_1{}^2 du^2 + h_2{}^2 dv^2 + h_3{}^2 dw^2, \quad dV = h_1 h_2 h_3 \, du \, dv \, dw \quad (1)$$

In terms of three unit vectors $\boldsymbol{i}_1$, $\boldsymbol{i}_2$, $\boldsymbol{i}_3$ in the direction of increasing u, v, w, respectively, we have

$$\text{grad} f = \boldsymbol{\nabla} f = \frac{1}{h_1} \frac{\partial f}{\partial u} \boldsymbol{i}_1 + \frac{1}{h_2} \frac{\partial f}{\partial v} \boldsymbol{i}_2 + \frac{1}{h_3} \frac{\partial f}{\partial w} \boldsymbol{i}_3 \quad (2)$$

$$\text{div } \boldsymbol{Q} = \boldsymbol{\nabla} \cdot \boldsymbol{Q} = \frac{1}{h_1 h_2 h_3} \left[\frac{\partial}{\partial u} (h_2 h_3 Q_1) + \frac{\partial}{\partial v} (h_3 h_1 Q_2) + \frac{\partial}{\partial w} (h_1 h_2 Q_3) \right] \quad (3)$$

$$\left.\begin{aligned}\text{curl}\,\boldsymbol{Q} = \nabla \times \boldsymbol{Q} &= \frac{1}{h_1h_2h_3}\begin{vmatrix} h_1\boldsymbol{i}_1 & h_2\boldsymbol{i}_2 & h_3\boldsymbol{i}_3 \\ \dfrac{\partial}{\partial u} & \dfrac{\partial}{\partial v} & \dfrac{\partial}{\partial w} \\ h_1Q_1 & h_2Q_2 & h_3Q_3 \end{vmatrix} \\ &= \frac{\boldsymbol{i}_1}{h_2h_3}\left[\frac{\partial(h_3Q_3)}{\partial v} - \frac{\partial(h_2Q_2)}{\partial w}\right] \\ &\quad + \frac{\boldsymbol{i}_2}{h_3h_1}\left[\frac{\partial(h_1Q_1)}{\partial w} - \frac{\partial(h_3Q_3)}{\partial u}\right] + \frac{\boldsymbol{i}_3}{h_1h_2}\left[\frac{\partial(h_2Q_2)}{\partial u} - \frac{\partial(h_1Q_1)}{\partial v}\right]\end{aligned}\right\} \quad (4)$$

6.17. Cylindrical coordinates. These are r, θ, and z, where

$$\left.\begin{aligned} x = r\cos\theta, \quad y = r\sin\theta \\ \text{Here} \qquad \\ h_r = 1, \quad h_\theta = r, \quad h_z = 1 \end{aligned}\right\} \quad (1)$$

$$ds^2 = dr^2 + r^2d\theta^2 + dz^2, \quad dV = r\,dr\,d\theta\,dz \quad (2)$$

$$\nabla^2 f = \frac{1}{r}\frac{\partial}{\partial r}\left(r\frac{\partial f}{\partial r}\right) + \frac{1}{r^2}\frac{\partial^2 f}{\partial\theta^2} + \frac{\partial^2 f}{\partial z^2} \quad (3)$$

6.18. Spherical coordinates. These are r, θ, ϕ, where

$$x = r\sin\phi\cos\theta, \quad y = r\sin\phi\sin\theta, \quad z = r\cos\phi \quad (1)$$

Here

$$h_r = 1, \quad h_\phi = r, \quad h_\theta = r\sin\phi, \quad ds^2 = dr^2 + r^2d\phi^2 + r^2\sin^2\phi\,d\theta^2 \quad (2)$$

$$dV = r^2\sin\phi\,dr\,d\phi\,d\theta \quad (3)$$

$$\nabla^2 f = \frac{1}{r^2}\frac{\partial}{\partial r}\left(r^2\frac{\partial f}{\partial r}\right) + \frac{1}{r^2\sin\phi}\frac{\partial}{\partial\phi}\left(\sin\phi\frac{\partial f}{\partial\phi}\right) + \frac{1}{r^2\sin^2\phi}\frac{\partial^2 f}{\partial\theta^2} \quad (4)$$

Some authors transpose the meaning of θ and ϕ as given here.

6.19. Parabolic coordinates. These are u, v, θ, where

$$\left.\begin{aligned} x = uv\cos\theta, \quad y = uv\sin\theta, \quad z = \tfrac{1}{2}(u^2 - v^2) \\ \sqrt{x^2 + y^2 + z^2} = \tfrac{1}{2}(u^2 + v^2) \end{aligned}\right\} \quad (1)$$

Here

$$h_u = h_v = \sqrt{u^2 + v^2}, \quad h_\theta = uv \quad (2)$$

$$ds^2 = (u^2 + v^2)(du^2 + dv^2) + u^2v^2d\theta^2, \quad dV = uv(u^2 + v^2)du\,dv\,d\theta \quad (3)$$

$$\nabla^2 f = \frac{1}{u^2 + v^2}\frac{1}{u}\frac{\partial}{\partial u}\left(u\frac{\partial f}{\partial u}\right) + \frac{1}{v}\frac{\partial}{\partial v}\left(v\frac{\partial f}{\partial v}\right) + \left(\frac{1}{u^2} + \frac{1}{v^2}\right)\frac{\partial^2 f}{\partial\theta^2} \quad (4)$$

7. Tensors

7.1. Tensors of the second rank. A tensor of the second rank, or second order, has nine components in each coordinate system. Let these be A_{pq} in one system and $\bar{A}_{pq}$ in a second system, where $p = 1, 2$, or 3 and $q = 1, 2$, or 3. If two arbitrary vectors have components U_p and V_p in the first system and $\bar{U}_p$ and $\bar{V}_p$ in the second system,

$$\sum_{p,q} U_p A_{pq} V_q = \sum_{p,q} \bar{U}_p \bar{A}_{pq} \bar{V}_q \tag{1}$$

is a scalar invariant, the same in all coordinate systems.

7.2. Summation convention. In equations involving tensors, like that in §7.1 the summation signs are often omitted in view of the convention that summation occurs on any index that appears twice.

7.3. Transformation of components. If, for the vector components,

$$\bar{U}_r = \sum_p m_{pr} U_p, \quad \bar{U}_r \bar{V}_s = \sum_{p,q} m_{pr} m_{qs} U_p V_q \tag{1}$$

the tensor components A_{pq} transform like the product $U_p V_q$ and

$$\bar{A}_{rs} = \sum_{p,q} m_{pr} m_{qs} A_{pq}$$

7.4. Matrix notation. We may write the nine A_{pq} as a square array

$$\begin{Vmatrix} A_{11} & A_{12} & A_{13} \\ A_{21} & A_{22} & A_{23} \\ A_{31} & A_{32} & A_{33} \end{Vmatrix} \tag{1}$$

This is a matrix, abbreviated as $\| A_{pq} \|$ or A. Sometimes the double vertical bars are replaced by parentheses, or omitted. (See pages 85-89.)

A vector U_p may be written as a 1 by 3 row matrix $\vec{U}_p$, or a 3 by 1 column matrix $U_p\!\uparrow$. Thus

$$\vec{U}_p = \| U_1 \quad U_2 \quad U_3 \| \quad U_p\!\uparrow = \begin{Vmatrix} U_1 \\ U_2 \\ U_3 \end{Vmatrix} \tag{2}$$

7.5. Matrix products. Let the number of columns of the matrix A_{pq} be the same as the number of rows of the matrix B_{qr}. Then the product matrix, in the order indicated, $C = AB$,

$$\| C_{pr} \| = \| A_{pq} \| \, \| B_{qr} \| \quad \text{has} \quad C_{pr} = \sum_q A_{pq} B_{qr} \tag{1}$$

as its elements, and a number of rows the same as A_{pq}, a number of columns the same as B_{qr}. Matrix multiplication is associative, $(AB)C = A(BC)$ may be written ABC. But the multiplication is not commutative. The dimensions may not allow both AB and BA to be formed, and even if they both exist they will in general be different.

7.6. Linear vector operator. From any vector U, a tensor A_{pq} may be used to generate a new vector by multiplication on the right,

$$V_p = \sum_q A_{pq} U_q \quad \text{or} \quad V_p\uparrow = \| A_{pq} \| \, U_q\uparrow \tag{1}$$

Similarly by multiplication on the left

$$W_q = \sum_p U_p A_{pq} \quad \text{or} \quad \vec{W}_q = \vec{U}_p \, \| A_{pq} \| \tag{2}$$

7.7. Combined operators. If the first process of § 7.5 is applied with a tensor A_{pq}, and then repeated on V_p with a tensor B_{pq}, the result is equivalent to a single operation with C_{pq}, where

$$C_{pq} = \sum_r B_{pr} A_{rq} \quad \text{or} \quad \| C_{pq} \| = \| B_{pr} \| \, \| A_{rq} \| \tag{1}$$

Similarly for the second operation of §7.5, where the result is equivalent to a single operation with D_{pq}, where

$$D_{pq} = \sum_r A_{pr} B_{rq} \quad \text{or} \quad \| D_{pq} \| = \| A_{pr} \| \, \| B_{rq} \| \tag{2}$$

7.8. Tensors from vectors. The products of components of two vectors, or sums of such products form a tensor. Thus

$$A_{pq} = B_p L_q + C_p M_q + D_p N_q \tag{1}$$

Any tensor may be built up of not more than three such products. For A_{pq} any tensor and $\boldsymbol{B}$, $\boldsymbol{C}$, $\boldsymbol{D}$ three noncoplanar or linearly independent vectors, values of $\boldsymbol{L}$, $\boldsymbol{M}$, and $\boldsymbol{N}$ can be found to satisfy the equation just written.

7.9. Dyadics. Write $\boldsymbol{i}_1$, $\boldsymbol{i}_2$, $\boldsymbol{i}_3$ for $\boldsymbol{i}$, $\boldsymbol{j}$, $\boldsymbol{k}$. Then

$$\boldsymbol{B} = \sum_p B_p \boldsymbol{i}_p, \quad \boldsymbol{L} = \sum_q L_q \boldsymbol{i}_q \quad \text{and} \quad \boldsymbol{BL} = \sum_{p,q} B_p L_q \boldsymbol{i}_p \boldsymbol{i}_q \tag{1}$$

The tensor of § 7.8 may be generated from

$$\sum_{p,q} A_{pq}\boldsymbol{i}_p\boldsymbol{i}_q = \boldsymbol{BL} + \boldsymbol{CM} + \boldsymbol{DN} \tag{2}$$

The first operation of § 7.7 may be effected by dot products using the right-hand factors,

$$V = \sum_p V_p\boldsymbol{i}_p = \sum_{p,q} A_{pq}\boldsymbol{i}_p(\boldsymbol{i}_q \cdot \boldsymbol{U}) = \boldsymbol{B}(\boldsymbol{L} \cdot \boldsymbol{U}) + \boldsymbol{C}(\boldsymbol{M} \cdot \boldsymbol{U}) + \boldsymbol{D}(\boldsymbol{N} \cdot \boldsymbol{U}) \tag{3}$$

Similarly the second operation of § 7.7 may be effected by dot products using the left-hand factors,

$$W = \sum_q W_q\boldsymbol{i}_q = \sum_{p,q} A_{pq}(\boldsymbol{U} \cdot \boldsymbol{i}_p)\boldsymbol{i}_q = (\boldsymbol{U} \cdot \boldsymbol{B})\boldsymbol{L} + (\boldsymbol{U} \cdot \boldsymbol{C})\boldsymbol{M} + (\boldsymbol{U} \cdot \boldsymbol{D})\boldsymbol{N} \tag{4}$$

Products of vectors, or sums of such products, used as above to generate tensors and form linear vector functions are called dyadics.

7.10. Conjugate tensor. Symmetry. If A_{pq} is a tensor, then the conjugate $A'_{pq} = A_{qp}$ is also a tensor. A tensor is symmetric if it is equal to its conjugate,

$$A_{pq} = A_{qp} \tag{1}$$

A tensor is alternating, or skew-symmetric if it is equal to the negative of its conjugate,

$$A_{pq} = -A_{qp} \tag{2}$$

Any tensor may be represented as the sum of two tensors one of which is symmetric and the other alternating by means of the identity

$$A_{pq} = \tfrac{1}{2}(A_{pq} + A'_{pq}) + \tfrac{1}{2}(A_{pq} - A'_{pq}) \tag{1}$$

7.11. Unit, orthogonal, unitary. The unit matrix has

$$A_{pq} = \delta_{pq} = 0 \quad \text{if} \quad p \neq q, \quad \text{and} \quad \delta_{pq} = 1 \quad \text{if} \quad p = q \tag{2}$$

A square matrix is singular if its determinant is zero. Each nonsingular square matrix has a reciprocal matrix $A^{-1}{}_{pq}$ such that

$$\| A_{pq} \| \, \| A^{-1}{}_{qr} \| = \| \delta_{pr} \| \quad \text{and} \quad \| A^{-1}{}_{pq} \| \, \| A_{qr} \| = \delta_{pr} \tag{3}$$

A matrix is orthogonal if its conjugate and reciprocal are equal,

$$A'_{pq} = A^{-1}{}_{pq} \quad \sum_q A_{pq}A_{qr} = \sum_q A_{qp}A_{rq} = \delta_{pr} \tag{4}$$

For a matrix with complex elements A_{pq} obtain its Hermitian conjugate $\tilde{A}_{pq}$ from A'_{pq} by replacing each element $a + bi$ by its complex conjugate $a - bi$. A matrix is unitary, or Hermitian orthogonal, if its Hermitian conjugate equals its reciprocal, $\tilde{A}_{pq} = A^{-1}{}_{pq}$.

For Cartesian coordinates the matrix m_{pq} of §7.3 is orthogonal.

7.12. Principal axes of a symmetric tensor. For any symmetric tensor A_{pq} there are three mutually perpendicular directions such that when these are taken as a new coordinate system, the new components of the tensor have the form $\bar{A}_{pq} = R_p \delta_{pq}$. The R_p are the roots of

$$\begin{vmatrix} A_{11} - R & A_{12} & A_{13} \\ A_{21} & A_{22} - R & A_{23} \\ A_{31} & A_{32} & A_{33} - R \end{vmatrix} = -R^3 + S_1 R^2 - S_2 R + S_3 = 0 \tag{1}$$

When this has three distinct roots, the directions of the new axes in the old coordinate system are found by solving

$$\left.\begin{aligned} A_{11}x_p + A_{12}y_p + A_{13}z_p &= R_p x_p \\ A_{21}x_p + A_{22}y_p + A_{23}z_p &= R_p y_p \\ A_{31}x_p + A_{32}y_p + A_{33}z_p &= R_p z_p \end{aligned}\right\} \tag{2}$$

for the ratio of x_p, y_p, z_p where p is 1, 2, or 3.

The trace

$$S_1 = A_{11} + A_{22} + A_{33} \tag{3}$$

S_3 is the determinant $| A_{pq} |$ and

$$S_2 = A_{22}A_{33} + A_{33}A_{11} + A_{11}A_{22} - A_{23}{}^2 - A_{31}{}^2 - A_{12}{}^2 \tag{4}$$

S_1, S_2, S_3 known as the first, second, and third scalar invariants of the symmetric tensor A_{pq} have the same value in all coordinate systems.

7.13. Tensors in n-dimensions. In two systems of coordinates, let a point P have coordinates $(x^1,x^2,\ldots,x^n)$ in the first system and $(\bar{x}^1,\bar{x}^2,\ldots,\bar{x}^n)$ in the second system, with relations

$$\bar{x}^p = f^p(x^1,x^2,\ldots,x^n), \quad x^p = F^p(\bar{x}^1,\bar{x}^2,\ldots,\bar{x}^n), \quad p = 1, 2, \ldots, n \tag{1}$$

If $(A^1,A^2,\ldots,A^n)$ are related to $(\bar{A}^1, \bar{A}^2,\ldots,\bar{A}^n)$ by the equations

$$\bar{A}^p = \sum_{i=1}^{n} \frac{\partial \bar{x}^p}{\partial x^i} A^i, \quad \text{or} \quad \bar{A}^p = \frac{\partial \bar{x}^p}{\partial x^i} A^i \tag{2}$$

by § 7.2, the A's are the components of a contravariant vector, or a tensor of rank one.

Similarly the relation

$$\bar{A}_p = \left(\frac{\partial x^i}{\partial \bar{x}^p}\right) A_i \tag{3}$$

holds for the components of a covariant vector, or a covariant tensor of rank one.

The expressions $d\bar{x}^m$ and dx^i are the components of a contravariant vector, and if $\phi(x^i) = \bar{\phi}(\bar{x}^m)$ is any scalar point function, the gradient with components $\partial\bar{\phi}/\partial\bar{x}^m$ and $\partial\phi/\partial x^i$ is a covariant vector.

7.14. Tensors of any rank. The scalar ϕ is a tensor of rank zero. The transformation $\bar{A}^{pq} = (\partial\bar{x}^p/\partial x^i)(\partial\bar{x}^q/\partial x^j)A^{ij}$ defines a contravariant tensor of rank two. The transformation $\bar{A}_{pq} = (\partial x^i/\partial\bar{x}^p)(\partial x^j/\partial\bar{x}^q)A_{ij}$ defines a covariant tensor of rank two. And the transformation $\bar{A}_q{}^p = (\partial\bar{x}^p/\partial x^i)(\partial x^j/\partial\bar{x}^q)A_j{}^i$ defines a mixed tensor of rank two.

Tensors of higher order are defined similarly.

7.15. The fundamental tensor. This is the symmetric covariant tensor g_{mn} such that in the Riemannian space the element of arc length is $ds^2 = g_{ij}dx^i\,dx^j$. In the determinant of the g_{ij}, let G_{ij} be the cofactor of g_{ij}, § 1.8, and let g be the value of the determinant. Then $g^{ij} = G_{ij}/g$ is a contravariant tensor. Then $g_{im}g^{jm} = \delta_i{}^j$ is a mixed tensor, where $\delta_i{}^j$ is the δ_{ij} of § 7.11.

The fundamental tensors g_{ij}, g^{ij} may be used to change covariant indices to contravariant ones and conversely. For example,

$$A^p = g^{pq}A_q, \quad A_p = g_{pq}A^q, \quad A_p{}^q = g^{iq}A_{pi} \tag{1}$$

7.16. Christoffel three-index symbols. These symbols (themselves not tensors) are defined in terms of the fundamental tensor by

$$\Gamma_{i \cdot rs} = \begin{bmatrix} r\,s \\ i \end{bmatrix} = \frac{1}{2}\left(\frac{\partial g_{ir}}{\partial x^s} + \frac{\partial g_{is}}{\partial x^r} - \frac{\partial g_{rs}}{\partial x^i}\right) \tag{1}$$

$$\Gamma^i{}_{rs} = \left\{\begin{matrix} r\,s \\ i \end{matrix}\right\} = g^{ij}\Gamma_{j \cdot rs} \tag{2}$$

They are used to form tensors by covariant differentiation, as

$$A^i{}_{;p} = \frac{\partial A^i}{\partial x^p} + \Gamma^i{}_{pj}A^j \tag{3}$$

and

$$A^t{}_{s;r} = \frac{\partial A_s{}^t}{\partial x^r} - \Gamma^i{}_{sr} A_i{}^t + \Gamma^t{}_{jr} A_s{}^j \tag{4}$$

The equations of the geodesic lines in the Riemannian space are

$$\frac{d^2x^i}{ds^2} + \Gamma^i{}_{pq} \frac{dx^p}{ds} \frac{dx^q}{ds} = 0 \tag{5}$$

7.17. Curvature tensor. The Riemann-Christoffel curvature tensor is

$$R^i{}_{jhk} = \frac{\partial}{\partial x^h} \Gamma^i{}_{jk} - \frac{\partial}{\partial x^k} \Gamma^i{}_{jh} + \Gamma^i{}_{rh} \Gamma^r{}_{jk} - \Gamma^i{}_{rk} \Gamma^r{}_{jh} \tag{1}$$

This leads to R_{ik}, the second-rank curvature tensor,

$$R_{ik} = \frac{\partial}{\partial x^r} \Gamma^r{}_{ik} - \frac{\partial}{\partial x^k} \Gamma^r{}_{ir} + \Gamma^s{}_{rs} \Gamma^r{}_{ik} - \Gamma^s{}_{ri} \Gamma^r{}_{ks} \tag{2}$$

and to R, the curvature scalar,

$$R = g^{ik} R_{ik}. \tag{3}$$

All the components of each of these curvature tensors are zero for a Euclidean, or flat space.

8. Spherical Harmonics

8.1. Zonal harmonics. In spherical coordinates, for functions independent of θ, by § 7.1 Laplace's equation is

$$\frac{\partial}{\partial r}\left(r^2 \frac{\partial f}{\partial r}\right) + \frac{1}{\sin\phi} \frac{\partial}{\partial \phi}\left(\sin\phi \frac{\partial f}{\partial \phi}\right) = 0 \tag{1}$$

This admits as particular solutions the solid zonal harmonic functions

$$r^n P_n(\cos\phi) \quad \text{and} \quad r^{-(n+1)} P_n(\cos\phi) \tag{2}$$

if $P_n(\cos\phi)$ satisfies a certain differential equation. With $\cos\phi = x$, and $P_n = y$, the equation takes the form

$$(1 - x^2) \frac{d^2y}{dx^2} - 2x \frac{dy}{dx} + n(n+1)y = 0 \tag{3}$$

which is Legendre's differential equation.

The factor $P_n(\cos\phi)$ is called a zonal harmonic.

8.2. Legendre polynomials. For n zero or a positive integer, the only solution of Legendre's differential equation of § 8.1, which is regular at 1 and -1 reduces to a polynomial. This polynomial, multiplied by the factor which makes it 1 when $x = 1$ is called the nth Legendre polynomial and is denoted by $P_n(x)$. The first few Legendre polynomials are

$$\left.\begin{aligned} &P_0(x) = 1, \quad P_1(x) = x, \quad P_2(x) = \tfrac{1}{2}(3x^2 - 1) \\ &P_3(x) = \tfrac{1}{2}(5x^3 - 3x), \quad P_4(x) = \tfrac{1}{8}(35x^4 - 30x^2 + 3) \end{aligned}\right\} \quad (1)$$

The general expression is

$$\left.\begin{aligned} P_n(x) = \frac{(2n-1)(2n-3)\ldots 3 \cdot 1}{n!} &\left[x^n - \frac{n(n-1)}{2(2n-1)} x^{n-2}\right. \\ &\left. + \frac{n(n-1)(n-2)(n-3)}{2 \cdot 4(2n-1)(2n-3)} x^{n-4} - \ldots\right] \end{aligned}\right\} \quad (2)$$

8.3. Rodrigues's formula

$$P_n(x) = \frac{1}{2^n n!} \frac{d^n(x^2-1)^n}{dx^n} \quad (1)$$

8.4. Particular values

$$P_n(1) = 1, \quad P_n(-x) = (-1)^n P_n(x) \quad (1)$$

$$P_{2n+1}(0) = 0, \quad P_{2n}(0) = (-1)^n \frac{1 \cdot 3 \cdot 5 \ldots (2n-1)}{2 \cdot 4 \cdot 6 \ldots 2n} \quad (2)$$

$$\left[\frac{dP_{2n+1}(x)}{dx}\right]_{x=0} = (-1)^n \frac{3 \cdot 5 \cdot 7 \ldots (2n+1)}{2 \cdot 4 \cdot 6 \ldots 2n} \quad (3)$$

8.5. Trigonometric polynomials. In terms of ϕ and multiples of ϕ, where as in § 8.1 we put $\cos\phi = x$, the first few polynomials are

$$\left.\begin{aligned} &P_0(\cos\phi) = 1, \quad P_1(\cos\phi) = \cos\phi, \quad P_2(\cos\phi) = \tfrac{1}{4}(3\cos 2\phi + 1) \\ &P_3(\cos\phi) = \tfrac{1}{8}(5\cos 3\phi + 3\cos\phi) \\ &P_4(\cos\phi) = \tfrac{1}{64}(35\cos 4\phi + 20\cos 2\phi + 9) \end{aligned}\right\} \quad (1)$$

The general expression is

$$\left.\begin{aligned} P_n(\cos\phi) = 2\,\frac{1 \cdot 3 \cdot 5 \ldots (2n-1)}{2^n n!} &\left[\cos n\phi + \frac{1}{1}\frac{n}{2n-1}\cos(n-2)\phi\right. \\ &+ \frac{1 \cdot 3}{1 \cdot 2}\frac{n(n-1)}{(2n-1)(2n-3)}\cos(n-4)\phi \\ &\left. + \frac{1 \cdot 3 \cdot 5}{1 \cdot 2 \cdot 3}\frac{n(n-1)(n-2)}{(2n-1)(2n-3)(2n-5)}\cos(n-6)\phi + \ldots\right] \end{aligned}\right\} \quad (2)$$

8.6. Generating functions. If $r < 1$ and $x < 1$,

$$(1 - 2rx + r^2)^{-1/2} = P_0(x) + rP_1(x) + r^2P_2(x) + r^3P_3(x) + \ldots \quad (1)$$

and if $r > 1$ and $x < 1$,

$$(1 - 2rx + r^2)^{-1/2} = \frac{1}{r} P_0(x) + \frac{1}{r^2} P_1(x) + \frac{1}{r^3} P_2(x) + \ldots \quad (2)$$

8.7. Recursion formula and orthogonality

$$nP_n(x) + (n - 1)P_{n-2}(x) - (2n - 1)xP_{n-1}(x) = 0 \quad (1)$$

$$\int_{-1}^{1} P_m(x)P_n(x)\, dx = 0 \quad \text{for } m \neq n, \quad \int_{-1}^{1} [P_n(x)]^2\, dx = \frac{2}{2n + 1} \quad (2)$$

8.8. Laplace's integral

$$P_n(x) = \frac{1}{\pi} \int_0^{\pi} \frac{du}{(x + \sqrt{x^2 - 1} \cos u)^{n+1}} = \frac{1}{\pi} \int_0^{\pi} (x + \sqrt{x^2 - 1} \cos u)^n\, du \quad (1)$$

8.9. Asymptotic expression. If $\epsilon > 0$ and $\epsilon < \phi < \pi - \epsilon$,

$$P_n(\cos\phi) = \sqrt{\frac{2}{\pi n \sin\phi}} \sin\left[\left(n + \frac{1}{2}\right)\phi + \frac{\pi}{4}\right] + O\left(\frac{1}{n^{3/2}}\right) \quad (1)$$

where O means " of the order of " for large n.

8.10. Tesseral harmonics. In spherical coordinates, by § 6.18 Laplace's equation is

$$\frac{\partial}{\partial r}\left(r^2 \frac{\partial f}{\partial r}\right) + \frac{1}{\sin\phi} \frac{\partial}{\partial\phi}\left(\sin\phi \frac{\partial f}{\partial\phi}\right) + \frac{1}{\sin^2\phi} \frac{\partial^2 f}{\partial\theta^2} = 0 \quad (1)$$

This admits as particular solutions the solid spherical harmonic functions

$$\left.\begin{array}{ll} r^n \sin m\theta\, P_n^m(\cos\phi), & r^n \cos m\theta\, P_n^m(\cos\phi) \\ r^{-(n+1)} \sin m\theta\, P_n^m(\cos\phi), & \text{and} \quad r^{-(n+1)} \cos m\theta\, P_n^m(\cos\phi) \end{array}\right\} \quad (2)$$

if $P_n^m(\cos\phi)$ satisfies a certain differential equation. With $\cos\phi = x$, and $P_n^m = y$, the equation takes the form

$$(1 - x^2) \frac{d^2y}{dx^2} - 2x \frac{dy}{dx} + \left[n(n + 1) - \frac{m^2}{1 - x^2}\right] y = 0 \quad (3)$$

which is the associated Legendre equation.

The factor $(\sin m\theta)P_n^m(\cos\phi)$ or $(\cos m\theta)P_n^m(\cos\phi)$ is called a tesseral harmonic. When $m = n$, it reduces to a sectorial harmonic. And when

$m = 0$, that involving $\cos m\theta$ reduces to $P_n^0(\cos\phi = P_n(\cos\phi)$, a zonal harmonic.

Solid harmonics are often combined in series or integrals which for fixed r and ϕ reduce to Fourier series or integrals of known functions of θ. (See §§ 14.10 to 14.16.)

8.11. Legendre's associated functions. Let m and n be each zero or a positive integer with $n \geqq m$. Then there is only one solution of the equation of § 8.10 which does not have logarithmic singularities at 1 or -1. With proper scale factor this solution is called the associated Legendre function and is denoted by $P_n^m(x)$.

$$P_n^m(x) = (1 - x^2)^{m/2} \frac{d^m P_n(x)}{dx^m} = \frac{1}{2^n n!}(1 - x^2)^{m/2} \frac{d^{n+m}}{dx^{n+m}}(x^2 - 1)^n \qquad (1)$$

$$\left.\begin{aligned} P_n^m(x) = \frac{(2n)!}{2^n n!(n-m)!}(1-x^2)^{m/2}\Bigg[x^{n-m} - \frac{(n-m)(n-m-1)}{2(2n-1)}x^{n-m-2} \\ + \frac{(n-m)(n-m-1)(n-m-2)(n-m-3)}{2\cdot 4(2n-1)(2n-3)}x^{n-m-4} - \ldots\Bigg] \end{aligned}\right\} \qquad (2)$$

8.12. Particular values

$$\left.\begin{aligned} &P_n^0(x) = P_n(x), \quad P_n^m(-x) = (-1)^n P_n^m(x) \\ &P_n^n(x) = \frac{(2n)!}{2^n n!}(1-x^2)^{n/2} = 1\cdot 3\cdot 5 \ldots (2n-1)(1-x^2)^{n/2} \end{aligned}\right\} \qquad (1)$$

With $x = \cos\phi$, the sectorial harmonic

$$P_n^n(\cos\phi) = 1\cdot 3\cdot 5\cdot \ldots (2n-1)\sin^n\phi \qquad (2)$$

8.13. Recursion formulas

$$(n-m)P_n^m(x) + (n+m-1)P_{n-2}^m(x) - (2n-1)xP_{n-1}^m(x) = 0 \qquad (1)$$

$$\left.\begin{aligned} P_n^{m+2}(x) + (n-m)(n+m+1)P_n^m(x) \\ - 2(m+1)\frac{x}{\sqrt{1-x^2}}P_n^{m+1}(x) = 0 \end{aligned}\right\} \qquad (2)$$

8.14. Asymptotic expression. If $\epsilon > 0$ and $\epsilon < \phi < \pi - \epsilon$,

$$\left.\begin{aligned} &P_n^m(\cos\phi) \\ &= (-n)^m \sqrt{\frac{2}{n\pi\sin\phi}}\sin\left[\left(n + \frac{1}{2}\right)\phi + \frac{\pi}{4} + \frac{m\pi}{2}\right] + O(n^{m-3/2}) \end{aligned}\right\} \qquad (1)$$

where O means " of the order of " for n large compared with m.

8.15. Addition theorem. Let γ be defined by

$$\cos\gamma = \cos\phi\cos\phi' + \sin\phi\sin\phi'\cos(\theta - \theta') \tag{1}$$

Thus γ is the distance on the unit sphere between the points with spherical polar coordinates $(1,\theta,\phi)$ and $(1,\theta',\phi')$. Then

$$P_n(\cos\gamma) = P_n(\cos\phi)P_n(\cos\phi') + 2\sum_{m=1}^{n}\frac{(n-m)!}{(n+m)!}P_n^{\ m}(\cos\phi)\,P_n^{\ m}(\cos\phi')\cos m(\phi - \phi')$$

8.16. Orthogonality

$$\int_{-1}^{1} P_n^{\ m}(x)\,P_k^{\ m}(x)\,dx = 0 \quad \text{for } k \neq n,$$
$$\int_{-1}^{1} P_n^{\ m}(x)^2\,dx = \frac{2}{2n+1}\,\frac{(n+m)!}{(n-m)!} \tag{1}$$

9. Bessel Functions

9.1. Cylindrical harmonics. In cylindrical coordinates, by § 6.17 Laplace's equation is

$$\frac{1}{r}\frac{\partial}{\partial r}\left(r\frac{\partial f}{\partial r}\right) + \frac{1}{r^2}\frac{\partial^2 f}{\partial\theta^2} + \frac{\partial^2 f}{\partial z^2} = 0 \tag{1}$$

This admits as particular solutions the harmonic functions

$$\left.\begin{aligned} &e^{-az}\sin n\theta\, J_n(ar), \quad e^{-az}\cos n\theta\, J_n(ar)\\ &e^{az}\sin n\theta\, J_n(ar), \quad e^{az}\cos n\theta\, J_n(ar)\end{aligned}\right\} \tag{2}$$

if $J_n(ar)$ satisfies a certain differential equation. With $ar = x$ and $J_n = y$, the equation takes the form

$$x^2\frac{d^2y}{dx^2} + x\frac{dy}{dx} + (x^2 - n^2)y = 0 \tag{3}$$

which is Bessel's differential equation.

9.2. Bessel functions of the first kind. Only one solution of the equation (3) of § 9.1 is finite for $x = 0$. With proper scale factor it is called the Bessel function of the first kind and is denoted by $J_n(x)$, where $n \geqq 0$. For n zero or a positive integer

$$J_n(x) = \sum_{k=0}^{\infty}\frac{(-1)^k(x/2)^{n+2k}}{k!(n+k)!} \tag{1}$$

In particular,

$$J_0(x) = 1 - \frac{x^2}{2^2} + \frac{x^4}{2^4(2!)^2} - \frac{x^6}{2^6(3!)^2} + \frac{x^8}{2^8(4!)^2} - \cdots \tag{2}$$

$$J_1(x) = \frac{x}{2} - \frac{x^3}{2^3 2!} + \frac{x^5}{2^5 2!3!} - \frac{x^7}{2^7 3!4!} + \frac{x^9}{2^9 4!5!} - \cdots \tag{3}$$

The expression involving the gamma function of § 13.1.

$$J_p(x) = \sum_{k=0}^{\infty} \frac{(-1)^k x^{p+2k}}{2^{p+2k} k!\, \Gamma(p+k+1)} \tag{4}$$

reduces to the above when $p = n$, zero or a positive integer, and for other positive p defines the Bessel function of order p.

9.3. Bessel functions of the second kind. When p is positive and not an integer

$$J_{-p}(x) = \sum_{k=0}^{\infty} (-1)^k \frac{x^{-p+2k}}{2^{-p+2k} k!\, \Gamma(-p+k+1)} \tag{1}$$

is a second solution of Bessel's equation of order p. Or we may use

$$Y_p(x) = N_p(x) = \frac{1}{\sin p\pi} [\cos p\pi\, J_p(x) - J_{-p}(x)] \tag{2}$$

For $n = 0$ or a positive integer, $J_{-n}(x) = (-1)^n J_n(x)$, but as $p \to n$, the limiting value of $Y_p(x)$ is $Y_n(x)$, where

$$\left.\begin{aligned} Y_n(x) = \frac{1}{\pi} \Bigg[2J_n(x) \ln \frac{x}{2} - \sum_{k=0}^{\infty} \frac{(-1)^k x^{n+2k}}{2^{n+2k} k!(n+k)!} [\psi(k+n) + \psi(k)] \\ - \sum_{r=0}^{n-1} \frac{(n-r-1)! x^{-n+2r}}{2^{-n+2r} r!} \Bigg] \end{aligned}\right\} \tag{3}$$

$$\psi(k) = -0.5772157 + 1 + \frac{1}{2} + \frac{1}{3} + \ldots + \frac{1}{k} \tag{4}$$

9.4. Hankel functions. The Hankel functions, sometimes called Bessel functions of the third kind, are defined by

$$H_p^{(1)}(x) = J_p(x) + iY_p(x) = \frac{i}{\sin p\pi} [\bar{e}^{p\pi i} J_p(x) - J_{-p}(x)] \tag{1}$$

$$H_p^{(2)}(x) = J_p(x) - iY_p(x) = -\frac{i}{\sin p\pi} [e^{p\pi i} J_p(x) - J_{-p}(x)] \tag{2}$$

In terms of the Hankel functions

$$J_p(x) = \frac{1}{2}\,[H_p^{(2)}(x) + H_p^{(1)}(x)], \quad Y_p(x) = \frac{1}{2i}\,[H_p^{(1)}(x) - H_p^{(2)}(x)] \tag{3}$$

Let A be a path in the complex plane made up of a small circuit of the point 1 in the positive direction and two lines parallel to the imaginary axis in the upper half plane extending to infinity. Thus A starts at infinity with an angle $-3\pi/2$, and ends at infinity with an angle $\pi/2$ (see § 20.3.) Then for z with positive real part

$$H_p^{(1)}(z) = \frac{\Gamma(\frac{1}{2} - p)}{2^p \pi^{3/2} i} \int_A e^{izu}(u^2 - 1)^{p-1/2}\,du \tag{4}$$

Similarly for B a path made up of a small negative circuit of the point -1 and two lines parallel to the imaginary axis in the upper half plane to infinity, starting with an angle $\pi/2$ and ending with an angle $-3\pi/2$,

$$H_p^{(2)}(z) = \frac{\Gamma(\frac{1}{2} - p)}{2^p \pi^{3/2} i} \int_B e^{izu}(u^2 - 1)^{p-1/2}\,du \tag{5}$$

The Hankel functions are complex for real values of x. But

$$i^{p+1} H_p^{(1)}(iy) \quad \text{and} \quad i^{-(p+1)} H_p^{(2)}(-iy) \tag{6}$$

are both real when y is real and positive.

When the imaginary part of z goes to plus infinity, $H_p^{(1)}(z)$ approaches zero. And when the imaginary part of z goes to minus infinity, $H_p^{(2)}(z)$ approaches zero.

9.5. Bessel's differential equation. Let $Z_p(x)$ be a solution of

$$x^2 \frac{d^2y}{dx^2} + x\frac{dy}{dx} + (x^2 - p^2)y = 0 \tag{1}$$

For p nonintegral,

$$Z_p(x) = c_1 J_p(x) + c_2 J_{-p}(x) \tag{2}$$

and for any positive or zero value of p,

$$Z_p(x) = c_1 J_p(x) + c_2 Y_p(x) \tag{3}$$

or

$$Z_p(x) = c_1 H_p^{(1)}(x) + c_2 H_p^{(2)}(x) \tag{4}$$

give general solutions with c_1, c_2 arbitrary constants.

9.6. Equation reducible to Bessel's. The equation

$$\left.\begin{aligned} &x^2 \frac{d^2y}{dx^2} + [(1 - 2A)x - 2BCx^{C+1}]\frac{dy}{dx} \\ &\quad + [(A^2 - E^2p^2) + BC(2A - C)x^C + B^2C^2x^{2C} + E^2D^2x^{2E}]y = 0 \end{aligned}\right\} \tag{1}$$

becomes Bessel's equation of order p, in Z and X if

$$y = x^A e^{Bx^C} Z, \quad X = Dx^E \tag{2}$$

and so has as its solution $y = x^A e^{Bx^C} Z_p(Dx^E)$.

And in particular $y = x^A Z_p(Dx^E)$ is the solution of

$$x^2 \frac{d^2y}{dx^2} + (1 - 2A)x \frac{dy}{dx} + [(A^2 - E^2p^2) + E^2D^2x^{2E}]y = 0 \tag{3}$$

The solution of $d^2y/dx^2 + Bxy = 0$ is

$$y = \sqrt{x}\, Z_{1/3}\left(\frac{2}{3}\sqrt{B}\, x^{3/2}\right) \tag{4}$$

9.7. Asymptotic expressions. For x large compared with p,

$$H_p^{(1)}(z) = i^{-p-1/2} 2^{1/2} (\pi z)^{-1/2} e^{iz}\, S_p(-2iz) \tag{1}$$

$$H_p^{(2)}(z) = i^{p+1/2} 2^{1/2} (\pi z)^{-1/2} e^{-iz}\, S_p(2iz) \tag{2}$$

where the asymptotic series for S_p is given by

$$\left.\begin{aligned} S_p(t) = 1 + \frac{4p^2 - 1}{1!\,4t} + \frac{(4p^2 - 1)(4p^2 - 9)}{2!\,(4t)^2} \\ + \frac{(4p^2 - 1)(4p^2 - 9)(4p^2 - 25)}{3!\,(4t)^3} + \cdots \end{aligned}\right\} \tag{3}$$

If

$$\phi_p = x - \frac{2p + 1}{4}\pi \quad \text{and} \quad S_p(2ix) = P_p(x) - iQ_p(x) \tag{4}$$

with the series P_p and Q_p real,

$$J_p(x) = 2^{1/2}(\pi x)^{-1/2}(P_p \cos\phi_p - Q_p \sin\phi_p) \tag{5}$$

$$Y_p(x) = 2^{1/2}(\pi x)^{-1/2}(P_p \sin\phi_p + Q_p \cos\phi_p) \tag{6}$$

Using the leading terms only, and O meaning " of the order of,"

$$\left.\begin{aligned} J_p(x) = 2^{1/2}(\pi x)^{-1/2} \cos\phi_p + O\left(\frac{1}{x^{3/2}}\right) \\ Y_p(x) = 2^{1/2}(\pi x)^{-1/2} \sin\phi_p + O\left(\frac{1}{x^{3/2}}\right) \end{aligned}\right\} \tag{7}$$

9.8. Order half an odd integer. Let p have the form

$$p = \frac{2n + 1}{2} = n + \frac{1}{2} \tag{1}$$

n an integer or zero.

The expressions of § 9.7 assume a closed form. The first few are

$$\left.\begin{aligned} J_{1/2}(x) &= 2^{1/2}(\pi x)^{-1/2} \sin x \\ J_{-1/2}(x) &= -Y_{-1/2}(x) = 2^{1/2}(\pi x)^{-1/2} \cos x \end{aligned}\right\} \tag{2}$$

$$J_{3/2}(x) = 2^{1/2}(\pi x)^{-1/2}\left(-\cos x + \frac{\sin x}{x}\right) \tag{3}$$

$$J_{-3/2}(x) = Y_{3/2}(x) = 2^{1/2}(\pi x)^{-1/2}\left(-\sin x - \frac{\cos x}{x}\right) \tag{4}$$

$$Y_p(x) = (-1)^{n+1} J_{-p}(x) \quad \text{if} \quad p = n + \frac{1}{2} \tag{5}$$

9.9. Integral representation

$$J_n(x) = \frac{1}{\pi}\int_0^{\pi} \cos(x \sin u - nu)\,du \tag{1}$$

9.10. Recursion formula

$$J_{p-1}(x) + J_{p+1}(x) = \frac{2p}{x} J_p(x), \quad Y_{p-1}(x) + Y_{p+1}(x) = \frac{2p}{x} Y_p(x) \tag{1}$$

9.11. Derivatives

$$\frac{dJ_p}{dx} = -\frac{p}{x} J_p + J_{p-1} = \frac{p}{x} J_p - J_{p+1} = \frac{1}{2} J_{p-1} - \frac{1}{2} J_{p+1} \tag{1}$$

$$\frac{d}{dx}[x^p J_p(ax)] = a x^p J_{p-1}(ax), \quad \frac{d}{dx}[x^{-p} J_p(ax)] = -a x^{-p} J_{p+1}(ax) \tag{2}$$

and $Y_p(x)$ satisfies similar relations.

9.12. Generating function

$$e^{(x/2)(t-1/t)} = \sum_{n=-\infty}^{\infty} J_n(x) t^n \tag{1}$$

9.13. Indefinite integrals

$$\int x J_p(ax) J_p(bx)\,dx = \frac{bx J_p(ax) J_{p-1}(bx) - ax J_{p-1}(ax) J_p(bx)}{a^2 - b^2} \tag{1}$$

$$\int x [J_p(ax)]^2\,dx = \frac{x^2}{2}\left\{[J_p(ax)]^2 - J_{p-1}(ax) J_{p+1}(ax)\right\} \tag{2}$$

9.14. Modified Bessel functions. We define

$$I_p(x) = i^{-p} J_p(ix), \quad \text{where} \quad i^{-p} = e^{-pi\pi/2} = \cos\frac{p\pi}{2} - i \sin\frac{p\pi}{2} \tag{1}$$

$$\frac{d}{dx}[x^p I_p(x)] = x^p I_{p-1}(x), \quad \frac{d}{dx}[x^{-p} I_p(x)] = x^{-p} I_{p+1}(x) \tag{2}$$

$$\frac{dI_p}{dx} = \frac{1}{2}(I_{p-1} + I_{p+1}), \quad I_{p-1} - I_{p+1} = \frac{2p}{x} I_p \tag{3}$$

For p not zero or an integer we define

$$K_p(x) = \frac{\pi}{2 \sin p\pi}[I_{-p}(x) - I_p(x)] \tag{4}$$

For $p = n$, zero, or an integer,

$$\left.\begin{aligned} K_n(x) = (-1)^{n+1} I_n(x) \ln \frac{x}{2} + \frac{1}{2}\sum_{r=0}^{n-1} \frac{(n-r-1)!\,(-1)^r}{2^{-n+2r} r!} x^{-n+2r} \\ + (-1)^n \frac{1}{2}\sum_{k=0}^{\infty} \frac{x^{n+2k}}{2^{n+2k} k!(n+k)!}[\psi(k+n) - \psi(k)] \end{aligned}\right\} \tag{5}$$

where $\psi(k)$ is that of (4) of § 9.3. For all values of p,

$$\frac{dK_p}{dx} = -\frac{1}{2}(K_{p-1} + K_{p+1}), \quad K_{p-1} - K_{p+1} = -\frac{2p}{x}K_p, \quad K_{-p} = K_p \tag{6}$$

The general solution of the differential equation

$$x^2 \frac{d^2y}{dx^2} + x\frac{dy}{dx} - (x^2 + p^2)y = 0$$

is

$$y = c_1 I_p(x) + c_2 K_p(x) \tag{7}$$

and the general solution of the differential equation

$$x^4 \frac{d^4y}{dx^4} + 2x^3 \frac{d^3y}{dx^3} - (2p^2+1)x^2 \frac{d^2y}{dx^2} + (2p^2+1)x\frac{dy}{dx} + (p^4 - 4p^2 - x^4)y = 0$$

is

$$y = c_1 J_p(x) + c_2 Y_p(x) + c_3 I_p(x) + c_4 K_p(x) \tag{8}$$

10. The Hypergeometric Function

10.1. The hypergeometric equation. The differential equation

$$x(1-x)\frac{d^2y}{dx^2} + [c - (a+b+1)x]\frac{dy}{dx} - aby = 0 \tag{1}$$

has one solution regular at the origin. With a scale factor making it 1 when $x = 0$, this solution is called the hypergeometric function and is denoted by $F(a,b;c;x)$.

10.2. The hypergeometric series. For $|x| < 1$, $F(a,b;c;x)$ equals

$$\left.\begin{aligned} 1 + \frac{a \cdot b}{1 \cdot c}x + \frac{a(a+1)b(b+1)}{1 \cdot 2 \cdot c(c+1)}x^2 \\ + \frac{a(a+1)(a+2)b(b+1)(b+2)}{1 \cdot 2 \cdot 3 \cdot c(c+1)(c+2)}x^3 + \dots \end{aligned}\right\} \tag{1}$$

With $\Gamma(c)$ as in § 13.1,

$$\left.\begin{aligned}\lim_{c\to -n}\frac{F(a,b;c;x)}{\Gamma(c)} = \frac{a(a+1)\dots(a+n)\,(b+1)\dots(b+n)}{(n+1)!}\,x^{n+1}\\ F(a+n+1,\,b+n+1;\,n+2;\,x)\end{aligned}\right\} \quad (2)$$

10.3. Contiguous functions. In general, a second solution of the differential equation of § 10.1 is given by

$$x^{1-c}\,F(a+1-c,\,b+1-c;\,2-c;\,x) \quad (1)$$

There are other functions related to $F(a,b;c;x)$ making in all a set of 24, and a number of linear relations connecting them. They form six groups of four, each set of four being equal except perhaps for sign.

For example,

$$\left.\begin{aligned}F(a,b;c;x) &= (1-x)^{c-a-b}F(c-a,\,c-b;\,c;\,x)\\ &= (1-x)^{-a}F\left(a,\,c-b;\,c;\,\frac{x}{x-1}\right)\\ &= (1-x)^{-b}F\left(b,\,c-a;\,c;\,\frac{x}{x-1}\right)\end{aligned}\right\} \quad (2)$$

is one set of four. As six distinct ones we may take

$$\left.\begin{gathered}F(a,b;c;x)\\ x^{1-c}F(a+1-c,\,b+1-c;\,2-c;\,x)\\ F(a,\,b;\,a+b+1-c;\,1-x)\\ (1-x)^{c-a-b}F(c-a,\,c-b;\,c+1-a-b;\,1-x)\\ x^{-a}F\left(a,\,a+1-c;\,a+1-b;\,\frac{1}{x}\right)\\ x^{-b}F\left(b,\,b+1-c;\,b+1-a;\,\frac{1}{x}\right)\end{gathered}\right\} \quad (3)$$

$$\left.\begin{aligned}F(a,b;c;x) = \frac{\Gamma(c)\Gamma(c-a-b)}{\Gamma(c-a)\Gamma(c-b)}F(a,\,b;\,a+b+1-c;\,1-x)\\ +\,(1-x)^{c-a-b}\frac{\Gamma(c)\Gamma(a+b-c)}{\Gamma(a)\Gamma(b)}F(c-a,\,c-b;\,c+1-a-b;\,1-x)\end{aligned}\right\} \quad (4)$$

$$F(a,b;c;x) = \frac{\Gamma(c)\Gamma(b-a)}{\Gamma(b)\Gamma(c-a)}(-x)^{-a}F\left(a,\,a+1-c;\,a+1-b;\,\frac{1}{x}\right) \quad (5)$$

10.4. Elementary functions

$$(1 + x)^n = F(-n,b\,;b\,;-x) \tag{1}$$

$$\ln(1 + x) = xF(1,1\,;2\,;-x) \tag{2}$$

$$e^x = \lim_{b\to\infty} F(1,b\,;1\,;x/b) \tag{3}$$

$$\cos nx = F(\tfrac{1}{2}n, -\tfrac{1}{2}n\,;\tfrac{1}{2}\,;\sin^2 x) \tag{4}$$

$$\sin nx = n(\sin x)F\left(\frac{1+n}{2}, \frac{1-n}{2}\,;\frac{3}{2}\,;\sin^2 x\right) \tag{5}$$

$$\ln\left(\frac{1+x}{1-x}\right) = 2xF\left(\frac{1}{2}, 1\,;\frac{3}{2}\,;x^2\right) \tag{6}$$

$$\sin^{-1} x = xF\left(\frac{1}{2}, \frac{1}{2}\,;\frac{3}{2}\,;x^2\right) \tag{7}$$

$$\tan^{-1} x = xF\left(\frac{1}{2}, 1\,;\frac{3}{2}\,;-x^2\right) \tag{8}$$

10.5. Other functions. For polynomials, see § 10.7. For the $P_n{}^m(x)$ of § 8.11:

$$P_n{}^m(x) = \frac{(x+1)^{m/2}}{(x-1)^{m/2}}\frac{1}{\Gamma(1-m)}F\left(-n, n+1\,;1-m\,;\frac{1-x}{2}\right) \tag{1}$$

$$K(k) = \int_0^{\pi/2}\frac{d\phi}{\sqrt{1-k^2\sin^2\phi}} = \frac{\pi}{2}F\left(\frac{1}{2}, \frac{1}{2}\,;1\,;k^2\right) \tag{2}$$

$$E(k) = \int_0^{\pi/2}\sqrt{1-k^2\sin^2\phi}\,d\phi = \frac{\pi}{2}F\left(-\frac{1}{2}, \frac{1}{2}\,;1\,;k^2\right) \tag{3}$$

10.6. Special relations

$$\frac{d}{dx}F(a,b\,;c\,;x) = \frac{ab}{c}F(a+1, b+1\,;c+1\,;x) \tag{1}$$

$$F(a,b\,;c\,;1) = \frac{\Gamma(c)\Gamma(c-a-b)}{\Gamma(c-a)\Gamma(c-b)} \tag{2}$$

$$F(a,b\,;c\,;x) = \frac{\Gamma(c)}{\Gamma(b)\Gamma(c-b)}\int_0^1 u^{b-1}(1-u)^{c-b-1}(1-ux)^{-a}\,du \tag{3}$$

$$\left.\begin{aligned}&F(-n+m+1, 2m+2+k\,;2m+2\,;x)\\ &\quad= \frac{(2m+1)!}{(2m+k+1)!}\frac{1}{x^{2m+1}}\frac{d^k}{dx^k}[x^{2m+k+1}(1-x)^{n-m-1}]\end{aligned}\right\} \tag{4}$$

10.7. Jacobi polynomials or hypergeometric polynomials

$$\left.\begin{aligned} J_n(p,q;x) &= F(-n, p+n; q; x) \\ &= \frac{x^{1-q}(1-x)^{q-p}}{q(q+1)\dots(q+n-1)} \frac{d^n}{dx^n}[x^{q+n-1}(1-x)^{p+n-q}] \end{aligned}\right\} \quad (1)$$

For $q > 0$, $p > q - 1$, these form a set of polynomials orthogonal with a weight function $w(x) = x^{q-1}(1-x)^{p-q}$ as in § 14.19 on 0,1. They satisfy

$$x(1-x)\frac{d^2y}{dx^2} + [q - (p+1)x]\frac{dy}{dx} + n(p+n)y = 0 \quad (2)$$

For

$$P_n(x) = J_n\left(1, 1; \frac{1-x}{2}\right) = F\left(-n, n+1; 1; \frac{1-x}{2}\right), \quad w(x) = 1 \quad (3)$$

and the Legendre polynomials of § 8.2 are orthogonal on $-1,1$. For

$$\left.\begin{aligned} T_n(x) &= \cos(n \cos^{-1} x) = J_n\left(0, \frac{1}{2}; \frac{1-x}{2}\right) \\ &= F\left(-n, n; \frac{1}{2}; \frac{1-x}{2}\right), \quad w(x) = \frac{2}{\sqrt{1-x^2}} \end{aligned}\right\} \quad (4)$$

and the Tschebycheff poylnomials $T_n(x)$ are orthogonal with weight function $1/\sqrt{1-x^2}$ on $-1, 1$.

10.8. Generalized hypergeometric functions. Let $(a)_0 = 1$, $(a)_n = a(a+1)(a+2)\dots(a+n-1)$. And define

$$\left.\begin{aligned} &{}_pF_q(a_1, a_2, \dots, a_p; b_1, b_2, \dots, b_q; x) \\ &= \sum_{n=0}^{\infty} \frac{(a_1)_n(a_2)_n \dots (a_p)_n}{(b_1)_n(b_2)_n \dots (b_q)_n} \cdot \frac{x^n}{n!} \\ &= 1 + \frac{a_1a_2\dots a_p}{b_1b_2\dots b_q}x + \frac{a_1(a_1+1)a_2(a_2+1)\dots a_p(a_p+1)}{b_1(b_1+1)b_2(b_2+1)\dots b_q(b_q+1)} \cdot \frac{x^2}{2!} + \dots \end{aligned}\right\} \quad (1)$$

For example,

$$e^x = {}_1F_1(b; b; x) \cdot {}_2F_1(a, b; c; x) = F(a, b; c; x) \quad (2)$$

of § 10.1.

The Bessel function of § 9.2,

$$J_p(x) = \frac{e^{-ix}(x/2)^p}{(p+1)} {}_1F_1\left(n + \frac{1}{2}; 2n+1; 2ix\right) \quad (3)$$

$$\left.\begin{aligned}&{}_3F_2(a, b, c\,; a+1-b, a+1-c\,; x)\\&\quad = (1-x)^{-a}\,{}_3F_2\left(\frac{a}{2}, \frac{a+1}{2}, a+1-b-c\,;\right.\\&\qquad\qquad\qquad\left. a+1-b, a+1-c\,; \frac{-4x}{(1-x)^2}\right)\end{aligned}\right\} \quad (4)$$

$$\left.\begin{aligned}&{}_3F_2(a, b, c\,; a+1-b, a+1-c\,; 1)\\&\quad = \frac{\Gamma(a/2+1)\Gamma(a+1-b)\Gamma(a+1-c)\Gamma(a/2+1-b-c)}{\Gamma(a+1)\Gamma(a/2+1-b)\Gamma(a/2+1-c)\Gamma(a+1-b-c)}\end{aligned}\right\} \quad (5)$$

10.9. The confluent hypergeometric function. $M(a,c,x) = {}_1F_1(a\,;c\,;x)$ is a solution of

$$x\frac{d^2y}{dx^2} + (c-x)\frac{dy}{dx} - ay = 0 \quad (1)$$

11. Laguerre Functions

11.1. Laguerre polynomials. The differential equation

$$x\frac{d^2y}{dx^2} + (1-x)\frac{dy}{dx} + ny = 0 \quad (1)$$

has polynomial solutions known as Laguerre polynomials and denoted by $L_n(x)$. The first few are

$$\left.\begin{aligned}&L_0(x) = 1, \quad L_1(x) = -x+1, \quad L_2(x) = x^2 - 4x + 2\\&L_3(x) = -x^3 + 9x^2 - 18x + 6\end{aligned}\right\} \quad (2)$$

In general

$$L_n(x) = (-1)^n\left(x^n - \frac{n^2}{1!}x^{n-1} + \frac{n^2(n-1)^2}{2!}x^{n-2} + \ldots + (-1)^n n!\right) \quad (3)$$

$$L_n(x) = e^x\frac{d^n(x^ne^{-x})}{dx^n} = n!\,{}_1F_1(-n\,;1\,;x) \quad (4)$$

11.2. Generating function

$$\frac{e^{-xt/(1-t)}}{1-t} = \sum_{n=0}^{\infty} L_n(x)\frac{t^n}{n!} \quad (1)$$

11.3. Recursion formula

$$L_{n+1}(x) - (2n+1-x)L_n(x) + n^2L_{n-1}(x) = 0 \quad (1)$$

11.4. Laguerre functions. The Laguerre functions $e^{-x/2}L_n(x)$ satisfy the differential equation

$$x\frac{d^2y}{dx^2} + \frac{dy}{dx} + \left(\frac{1}{2} - \frac{x}{4} + n\right)y = 0 \tag{1}$$

These functions are orthogonal on the range 0, ∞, and

$$\int_0^\infty e^{-x}L_n(x)L_m(x)dx = 0 \quad \text{if} \quad m \neq n; \quad \int_0^\infty e^{-x}[L_n(x)]^2 dx = (n!)^2 \tag{2}$$

11.5. Associated Laguerre polynomials. These are the derivatives

$$L_r{}^s(x) = \frac{d^s}{dx^s}[L_r(x)] = \frac{\Gamma(r+s+1)}{\Gamma(s+1)}\,{}_1F_1(-r\,; s+1\,; x) \tag{1}$$

They satisfy the differential equation

$$x\frac{d^2y}{dx^2} + (s+1-x)\frac{dy}{dx} + (r-s)y = 0 \tag{2}$$

11.6. Generating function

$$(-1)^s\frac{e^{-xt/(1-t)}}{(1-t)^{s+1}}t^s = \sum_{r=s}^{\infty} L_r{}^s(x)\frac{t^r}{r!} \tag{1}$$

11.7. Associated Laguerre functions. In § 11.5 put $r = n + k$ and $s = 2k + 1$. Then the associated Laguerre functions are defined as

$$e^{-x/2}x^k L_{n+k}{}^{2k+1}(x) \tag{1}$$

For x, the r of spherical coordinates, the volume element involves $x^2\,dx$, and with this element the functions, each times $x^{-1/2}$, are orthogonal.

$$\int_0^\infty e^{-x}x^k L_n{}^k(x)L_m{}^k(x)dx = 0 \quad \text{if} \quad m \neq n \tag{2}$$

$$\int_0^\infty e^{-x}x^k[L_n{}^k(x)]^2 dx = \frac{(n!)^3}{(n-k)!} \tag{3}$$

12. Hermite Functions

12.1. Hermite polynomials. The differential equation

$$\frac{d^2y}{dx^2} - 2x\frac{dy}{dx} + 2ny = 0 \tag{1}$$

has polynomial solutions known as Hermite polynomials and denoted by $H_n(x)$.

The first few are

$$\left.\begin{aligned}&H_0(x) = 1, \quad H_1(x) = 2x, \quad H_2(x) = 4x^2 - 2\\&H_3(x) = 8x^3 - 12x, \quad H_4(x) = 16x^4 - 48x^2 + 12\end{aligned}\right\} \tag{2}$$

In general, with K equal to the biggest integer in $n/2$,

$$H_n(x) = \sum_{k=0}^{K} (-1)^k \frac{1}{k!} n(n-1) \dots (n-2k+1)(2x)^{n-2k} \tag{3}$$

$$H_n(x) = (-1)^n e^{x^2} \frac{d^n e^{-x^2}}{dx^n} \tag{4}$$

$$H_{2n} = (-1)^n 2^n (2n-1)(2n-3) \dots 3 \cdot 1 \, {}_1F_1\left(-n\,; \frac{1}{2}\,; x^2\right) \tag{5}$$

$$H_{2n+1} = (-1)^n 2^{n+1} (2n+1)(2n-1) \dots 3 \cdot 1 \, x \, {}_1F_1\left(-n\,; \frac{3}{2}\,; x^2\right) \tag{6}$$

12.2. Generating function

$$e^{-t^2+2tx} = e^{x^2} e^{-(t-x)^2} = \sum_{n=0}^{\infty} H_n(x) \frac{t^n}{n!} \tag{1}$$

12.3. Recursion formula

$$H_{n+1}(x) - 2xH_n(x) + 2nH_{n-1}(x) = 0 \tag{1}$$

12.4. Hermite functions. The Hermite functions $e^{-x^2/2}H_n(x)$ satisfy the differential equation

$$\frac{d^2y}{dx^2} + (2n + 1 - x^2)y = 0 \tag{1}$$

These functions are orthogonal on the range $-\infty$, ∞, and

$$\left.\begin{aligned}&\int_{-\infty}^{\infty} e^{-x^2} H_n(x) H_m(x) dx = 0 \quad \text{if} \quad m \neq n\\&\int_{-\infty}^{\infty} e^{-x^2} [H_n(x)]^2 \, dx = 2^n n! \sqrt{\pi}\end{aligned}\right\} \tag{2}$$

13. Miscellaneous Functions

13.1. The gamma function. For p positive the integral

$$\Gamma(p) = \int_0^{\infty} x^{p-1} e^{-x} \, dx \tag{1}$$

defines the gamma function.

The infinite products

$$\frac{1}{\Gamma(z)} = e^{\gamma(z-1)} \prod_{n=1}^{\infty} \left(1 + \frac{z-1}{n}\right) e^{-(z-1)/n} = z e^{\gamma z} \prod_{n=1}^{\infty} \left(1 + \frac{z}{n}\right) e^{-z/n} \tag{2}$$

define the function for all complex values of z.

$$\gamma = 0.577216 \tag{3}$$

$$\Gamma(z) = \lim_{n\to\infty} \frac{(1 \cdot 2 \cdot 3 \ldots n)n^z}{z(z+1)(z+2)\ldots(z+n+1)} \tag{4}$$

13.2. Functional equations

$$\Gamma(z+1) = z\Gamma(z), \quad \Gamma(z)\Gamma(1-z) = \frac{\pi}{\sin \pi z} \tag{1}$$

$$\Gamma\left(\frac{z}{n}\right)\Gamma\left(\frac{z+1}{n}\right) \ldots \Gamma\left(\frac{z+n-1}{n}\right) n^{z-1/2} = (2\pi)^{(n-1)/2}\Gamma(z) \tag{2}$$

$$\Gamma\left(\frac{z}{2}\right)\Gamma\left(\frac{z+1}{2}\right)2^{z-1} = \sqrt{\pi}\,\Gamma(z) \tag{3}$$

13.3. Special values

$$\Gamma(1) = 0! = 1, \quad \Gamma(n) = (n-1)! = 1 \cdot 2 \cdot 3 \ldots (n-1) \tag{1}$$

for n a positive integer.

$$\Gamma\left(\frac{1}{2}\right) = \sqrt{\pi} \tag{2}$$

13.4. Logarithmic derivative

$$\psi(z) = \frac{d \ln \Gamma(z)}{dz} = -\gamma + \sum_{n=1}^{\infty} \left(\frac{1}{n} - \frac{1}{z+n-1}\right) \tag{1}$$

$$\psi'(z) = \frac{d\psi(z)}{dz} = \sum_{n=0}^{\infty} \frac{1}{(z+n)^2} \tag{2}$$

If the terms of a convergent series are rational functions of n, by a partial fraction decomposition the series may be summed in terms of ψ and its derivatives by the series of this section.

13.5. Asymptotic expressions. If O means "of the order of,"

$$\ln \Gamma(x) = \ln\left[\sqrt{2\pi}\, x^{x-1/2} e^{-x}\right] + \frac{1}{12x} + O\left(\frac{1}{x^3}\right) \tag{1}$$

$$x! = \Gamma(x+1) = \sqrt{2\pi}\, x^{x+1/2} e^{-x}\left[1 + \frac{1}{12x} + O\left(\frac{1}{x^2}\right)\right] \tag{2}$$

The first term of this, $\sqrt{2\pi}\, n^{n+1/2}\, e^{-n}$, is Stirling's formula for factorial n, and for large n may be used to evaluate ratios involving factorials.

13.6. Stirling's formula

$$\lim_{n\to\infty} \frac{n!}{\sqrt{2\pi}\, n^{n+1/2}\, e^{-n}} = 1 \tag{1}$$

13.7. The beta function. For p and q positive

$$B(p,q) = \int_0^1 x^{p-1}(1-x)^{q-1}\, dx = 2\int_0^{\pi/2} \cos^{2q-1}\theta \sin^{2p-1}\theta\, d\theta \tag{1}$$

In terms of the gamma function

$$B(p,q) = \frac{\Gamma(p)\Gamma(q)}{\Gamma(p+q)} \tag{2}$$

$$B(p+1, q) = \frac{p}{p+q} B(p, q), \quad B(p, 1-p) = \frac{\pi}{\sin \pi p} \tag{3}$$

13.8. Integrals. In the following integrals, the constants are such as to make all arguments of gamma functions positive.

$$\int_0^1 x^a\left(\ln\frac{1}{x}\right)^b dx = \frac{\Gamma(b+1)}{(a+1)^{b+1}}, \quad \int_0^\infty x^a e^{-bx^c} dx = \frac{1}{c}\, b^{-(a+1)/c}\,\Gamma\left(\frac{a+1}{c}\right) \tag{1}$$

The double integral of $x^P y^Q$ over the first quadrant of $(x/a)^A + (y/b)^B = 1$ is

$$\int_0^a dx \int_0^{b[1-(x/a)^A]^{1/B}} x^P y^Q\, dy = \frac{a^{P+1} b^{Q+1}}{AB} \frac{\Gamma\left(\frac{P+1}{A}\right)\Gamma\left(\frac{Q+1}{B}\right)}{\Gamma\left(\frac{P+1}{A} + \frac{Q+1}{B} + 1\right)} \tag{2}$$

The triple integral of $x^P y^Q z^R$ over the first octant of

$$\left(\frac{x}{a}\right)^A + \left(\frac{y}{b}\right)^B + \left(\frac{z}{c}\right)^C = 1$$

is

$$\frac{a^{P+1} b^{Q+1} c^{R+1}}{ABC} \frac{\Gamma\left(\frac{P+1}{A}\right)\Gamma\left(\frac{Q+1}{B}\right)\Gamma\left(\frac{R+1}{C}\right)}{\Gamma\left(\frac{P+1}{A} + \frac{Q+1}{B} + \frac{R+1}{C} + 1\right)} \tag{3}$$

13.9. The error integral

$$\mathrm{Erf}\, x = \frac{2}{\sqrt{\pi}} \int_0^x e^{-t^2} dt \tag{1}$$

The factor outside makes Erf (∞) = 1. For small x,

$$\mathrm{Erf}\, x = \frac{2}{\sqrt{\pi}} \left(x - \frac{x^3}{1!3} + \frac{x^5}{2!5} - \dots \right) \tag{2}$$

For large x the asymptotic expression is

$$\mathrm{Erf}\, x = 1 - \frac{1}{\sqrt{\pi}\, x} e^{-x^2} \left(1 - \frac{1}{2x^2} + \frac{1 \cdot 3}{(2x^2)^2} - \frac{1 \cdot 3 \cdot 5}{(2x^2)^3} + \dots \right) \tag{3}$$

$$\int_0^\infty e^{-h^2x^2} dx = \frac{\sqrt{\pi}}{2h}, \quad \int_0^\infty x e^{-h^2x^2} dx = \frac{1}{2h^2} \tag{4}$$

$$\int_0^\infty x^2 e^{-h^2x^2} dx = \frac{\sqrt{\pi}}{4h^3}, \quad \int_0^\infty x^n e^{-h^2x^2} dx = \frac{1}{2h^{n+1}} \Gamma\left(\frac{n+1}{2}\right) \tag{5}$$

13.10. The Riemann zeta function. For $z = x + iy$, $x > 1$,

$$\zeta(z) = \sum_{n=1}^{\infty} \frac{1}{n^z} = \prod_{\text{prime}} (1 - p^{-z})^{-1} \tag{1}$$

where the product extends over all prime integers p.

For $x > 1$,

$$\zeta(z) = \frac{1}{\Gamma(z)} \int_0^\infty \frac{t^{z-1}}{e^t - 1} dt \tag{2}$$

For $x > 0$,

$$\zeta(z) = \frac{1}{(1 - 2^{1-z})} \frac{1}{\Gamma(z)} \int_0^\infty \frac{t^{z-1}}{e^t + 1} dt \tag{3}$$

14. Series

14.1. Bernoulli numbers. These are generated by

$$\frac{x}{e^x - 1} = \sum_{k=0}^{\infty} B_k \frac{x^k}{k!} \tag{1}$$

$$\left.\begin{aligned} &B_0 = 1, \quad B_1 = \tfrac{1}{2}, \quad B_2 = \tfrac{1}{6}, \quad B_3 = 0, \quad B_4 = -\tfrac{1}{30}, \quad B_5 = 0, \quad B_6 = \tfrac{1}{42} \\ &B_7 = 0, \quad B_8 = -\tfrac{1}{30}, \quad B_9 = 0, \quad B_{10} = \tfrac{5}{66}, \quad B_{11} = 0, \quad B_{12} = -\tfrac{691}{2730} \end{aligned}\right\} \tag{2}$$

14.2. Positive powers. For m any positive integer, the finite sum

$$1^m + 2^m + 3^m + \dots + n^m = \frac{(B+n)^{m+1} - B^{m+1}}{m+1} \tag{1}$$

where on the right is meant the result of expanding by the binomial theorem § 1.3 and then replacing powers B^k by the Bernoulli numbers B_k of § 14.1. In particular

$$1 + 2 + 3 + \dots + n = \tfrac{1}{2} n(n+1) \tag{2}$$

$$1^2 + 2^2 + 3^2 + \dots + n^2 = \tfrac{1}{6} n(n+1)(2n+1) \tag{3}$$

$$1^3 + 2^3 + 3^3 + \dots + n^3 = \tfrac{1}{4} n^2(n+1)^2 \tag{4}$$

14.3. Negative powers. The sum of reciprocal even powers

$$1 + \frac{1}{2^{2m}} + \frac{1}{3^{2m}} + \frac{1}{4^{2m}} + \dots = \frac{(-1)^m (2\pi)^{2m} B_{2m}}{2(2m)!} = \zeta(2m) \tag{1}$$

In particular,

$$\zeta(2) = \frac{\pi^2}{6}, \quad \zeta(4) = \frac{\pi^4}{90}, \quad \zeta(6) = \frac{\pi^6}{945} \tag{2}$$

This shows that for large m, the numerical value B_{2m} becomes infinite like

$$\frac{2(2m)!}{(2\pi)^{2m}} \tag{3}$$

which exceeds $(m/10)^m$.

14.4. Euler-Maclaurin sum formula. For m and n positive integers,

$$\left.\begin{aligned} \sum_{k=0}^{m} f(k) &= \int_0^m f(t)\,dt + \frac{1}{2}[f(0) + f(m)] \\ &+ \sum_{k=1}^{n} \frac{B_{2k}}{(2k)!}[f^{(2k-1)}(m) - f^{(2k-1)}(0)] + \frac{mB_{2n+2}}{(2n+2)!} f^{(2n+2)}(\theta m) \end{aligned}\right\} \tag{1}$$

where θ is a suitable number in the interval $0 < \theta < 1$.

14.5. Power series. For simple functions power series may be obtained by use of Taylor's theorem of § 3.25. For the series listed, the expression following a series indicates a region in which the series converges. No expression is added if the series converges for all values of x.

14.6. Elementary functions

$$\left.\begin{aligned}(1+x)^n = 1 + nx + \frac{n(n-1)}{2!}x^2 + \frac{n(n-1)(n-2)}{3!}x^3 + \ldots \\ + \frac{n!}{(n-k)!k!}x^k + \ldots, \quad |x| < 1\end{aligned}\right\} \quad (1)$$

$$e^x = 1 + x + \frac{x^2}{2!} + \frac{x^3}{3!} + \ldots \quad (2)$$

$$\ln(1+x) = x - \frac{x^2}{2} + \frac{x^3}{3} - \frac{x^4}{4} + \ldots, \quad |x| \le 1 \quad (3)$$

$$\sin x = x - \frac{x^3}{3!} + \frac{x^5}{5!} - \frac{x^7}{7!} + \ldots \quad (4)$$

$$\cos x = 1 - \frac{x^2}{2!} + \frac{x^4}{4!} - \frac{x^6}{6!} + \ldots \quad (5)$$

$$\tan x = x + \frac{x^3}{3} + \frac{2x^5}{15} + \frac{17x^7}{315} + \frac{62x^9}{2835} + \ldots, \quad |x| < \frac{\pi}{2} \quad (6)$$

$$\sec x = 1 + \frac{x^2}{2!} + \frac{5x^4}{4!} + \frac{61x^6}{6!} + \frac{1385x^8}{8!} + \ldots, \quad |x| < \frac{\pi}{2} \quad (7)$$

$$\sin^{-1} x = x + \frac{x^3}{6} + \frac{1\cdot 3}{2\cdot 4}\cdot\frac{x^5}{5} + \frac{1\cdot 3\cdot 5}{2\cdot 4\cdot 6}\cdot\frac{x^7}{7} + \ldots, \quad |x| < 1 \quad (8)$$

$$\tan^{-1} x = x - \frac{x^3}{3} + \frac{x^5}{5} - \frac{x^7}{7} + \ldots, \quad |x| < 1 \quad (9)$$

$$\sqrt{1-x^2}\,\sin^{-1} x = x - \frac{x^3}{3} - \frac{2x^5}{3\cdot 5} - \frac{2\cdot 4x^7}{3\cdot 5\cdot 7} - \ldots, \quad |x| < 1 \quad (10)$$

$$\frac{\sin^{-1} x}{\sqrt{1-x^2}} = x + \frac{2x^3}{3} + \frac{2\cdot 4x^5}{3\cdot 5} + \frac{2\cdot 4\cdot 6x^7}{3\cdot 5\cdot 7} + \ldots, \quad |x| < 1 \quad (11)$$

$$\sinh x = x + \frac{x^3}{3!} + \frac{x^5}{5!} + \frac{x^7}{7!} + \ldots \quad (12)$$

$$\cosh x = 1 + \frac{x^2}{2!} + \frac{x^4}{4!} + \frac{x^6}{6!} + \ldots \quad (13)$$

$$\tanh x = x - \frac{x^3}{3} + \frac{2x^5}{15} - \frac{17x^7}{315} + \frac{62x^9}{2835} - \ldots, \quad |x| < \frac{\pi}{2} \quad (14)$$

$$\sinh^{-1} x = x - \frac{1}{2}\cdot\frac{x^3}{3} + \frac{1\cdot 3}{2\cdot 4}\cdot\frac{x^5}{5} - \frac{1\cdot 3\cdot 5}{2\cdot 4\cdot 6}\cdot\frac{x^7}{7} + \ldots, \quad |x| < 1 \tag{15}$$

$$\sinh^{-1} x = \ln 2x + \frac{1}{2}\cdot\frac{1}{2x^2} - \frac{1\cdot 3}{2\cdot 4}\cdot\frac{1}{4x^4} + \frac{1\cdot 3\cdot 5}{2\cdot 4\cdot 6}\cdot\frac{1}{6x^6} - \ldots, \quad |x| > 1 \tag{16}$$

$$\cosh^{-1} x = \ln 2x - \frac{1}{2}\cdot\frac{1}{2x^2} - \frac{1\cdot 3}{2\cdot 4}\cdot\frac{1}{4x^4} - \frac{1\cdot 3\cdot 5}{2\cdot 4\cdot 6}\cdot\frac{1}{6x^6} - \ldots, \quad |x| > 1 \tag{17}$$

$$\tanh^{-1} x = x + \frac{x^3}{3} + \frac{x^5}{5} + \frac{x^7}{7} + \ldots, \quad |x| < 1 \tag{18}$$

14.7. Integrals

$$\int_0^x e^{-x^2}\,dx = x - \frac{x^3}{3} + \frac{x^5}{5\cdot 2!} - \frac{x^7}{7\cdot 3!} + \ldots \tag{1}$$

$$\int_0^x \cos(x^2)dx = x - \frac{x^5}{5\cdot 2!} + \frac{x^9}{9\cdot 4!} - \frac{x^{13}}{13\cdot 6!} + \ldots \tag{2}$$

$$\int_0^x \sin(x^2)dx = \frac{x^3}{3} - \frac{x^7}{7\cdot 3!} + \frac{x^{11}}{11\cdot 5!} - \frac{x^{15}}{15\cdot 7!} + \ldots \tag{3}$$

$$\int_0^x \frac{\sin x}{x}\,dx = x - \frac{x^3}{3\cdot 3!} + \frac{x^5}{5\cdot 5!} - \ldots \tag{4}$$

$$-\int_x^\infty \frac{\cos t}{t}\,dt = 0.577216 + \ln x - \frac{x^2}{2\cdot 2!} + \frac{x^4}{4\cdot 4!} - \ldots, \quad x > 0 \tag{5}$$

$$\int_x^\infty e^{-t}\frac{dt}{t} = -0.577216 - \ln x + x - \frac{x^2}{2\cdot 2!} + \frac{x^3}{3\cdot 3!} - \ldots, \quad x > 0 \tag{6}$$

14.8. Expansions in rational fractions. In this section the prime on a summation means that the term for $n = 0$ is to be omitted

$$\cot z = \frac{1}{z} + \sum_{n=-\infty}^{\infty}{}' \left[\frac{1}{z - n\pi} + \frac{1}{n\pi}\right] = \frac{1}{z} + \sum_{n=1}^{\infty} \frac{2z}{z^2 - n^2\pi^2} \tag{1}$$

$$\csc z = \frac{1}{z} + \sum_{n=-\infty}^{\infty}{}' (-1)^n \left[\frac{1}{z - n\pi} + \frac{1}{n\pi}\right] = \frac{1}{z} + \sum_{n=1}^{\infty} (-1)^n \frac{2z}{z^2 - n^2\pi^2} \tag{2}$$

$$\csc^2 z = \sum_{n=-\infty}^{\infty} \frac{1}{(z - n\pi)^2}, \quad \sec z = \sum_{n=1}^{\infty} (-1)^{n-1} \frac{4(2n-1)\pi}{(2n-1)^2\pi^2 - 4x^2} \tag{3}$$

14.9. Infinite products for the sine and cosine. The prime means omit $n = 0$.

$$\sin z = z \prod_{n=-\infty}^{\infty}{}' \left(1 - \frac{z}{n\pi}\right) e^{z/n} = z \prod_{n=1}^{\infty} \left(1 - \frac{z^2}{n^2\pi^2}\right) \tag{1}$$

$$\cos z = \prod_{n=1}^{\infty} \left[1 - \frac{4z^2}{(2n-1)^2\pi^2}\right] \tag{2}$$

14.10. Fourier's theorem for periodic functions. The function $f(x)$ is periodic of period p if

$$f(x + p) = f(x) \tag{1}$$

The period p and frequence ω are related by the equations

$$\omega = \frac{2\pi}{p} \quad \text{and} \quad p = \frac{2\pi}{\omega} \tag{2}$$

Then for any constant c, the Fourier coefficients are

$$\left.\begin{aligned} a &= \frac{1}{p} \int_c^{c+p} f(x)\,dx, \quad a_n = \frac{2}{p} \int_c^{c+p} f(x) \cos n\omega x \, dx \\ b_n &= \frac{2}{p} \int_c^{c+p} f(x) \sin n\omega x \, dx \end{aligned}\right\} \tag{3}$$

It is often convenient to take $c = 0$, or $c = -p/2$.

The Fourier series for $f(x)$ is

$$f(x) = a + \sum_{n=1}^{\infty} (a_n \cos n\omega x + b_n \sin n\omega x) \tag{4}$$

A regular arc is a continuous curve with finite arc length. The function $f(x)$ is piecewise regular if its graph on any finite interval is made up of a finite number of pieces, each of which is a regular arc or an isolated point. For any piecewise regular periodic function the Fourier series will converge to $f(x)$ at all points of continuity, and to $\frac{1}{2}[f(x+) + f(x-)]$ at all points of discontinuity. Here $f(x+)$ is the limit approached from the right, and $f(x-)$ is the limit approached from the left, at x.

14.11. Fourier series on an interval. For any function $f(x)$, defined for $c < x < c + p$, but not necessarily periodic, the relations of § 14.10 may be used to find a Fourier series of period p which represents $f(x)$ on the interval $c, c + p$.

14.12. Half-range Fourier series. For any function $f(x)$, piecewise regular on $0,L$ the Fourier cosine series of period $2L$ which represents $f(x)$ for $0 < x < L$ is

$$f(x) = a + \sum_{n=1}^{\infty} a_n \cos n\omega x \tag{1}$$

where

$$a = \frac{1}{L}\int_0^L f(x)dx, \quad a_n = \frac{2}{L}\int_0^L f(x) \cos n\omega x \, dx, \quad \omega = \frac{\pi}{L} \tag{2}$$

The Fourier sine series of period $2L$ for $f(x)$ is

$$f(x) = \sum_{n=1}^{\infty} b_n \sin n\omega x \tag{3}$$

where

$$b_n = \frac{2}{L}\int_0^L f(x) \sin n\omega x \, dx, \quad \omega = \frac{\pi}{L} \tag{4}$$

14.13. Particular Fourier series. For $0 < x < L$,

$$1 = \frac{4}{\pi}\left(\sin\frac{\pi x}{L} + \frac{1}{3}\sin\frac{3\pi x}{L} + \frac{1}{5}\sin\frac{5\pi x}{L} + \ldots\right) \tag{1}$$

$$x = \frac{2L}{\pi}\left(\sin\frac{\pi x}{L} - \frac{1}{2}\sin\frac{2\pi x}{L} + \frac{1}{3}\sin\frac{3\pi x}{L} - \ldots\right) \tag{2}$$

$$x = \frac{L}{2} - \frac{4L}{\pi^2}\left(\cos\frac{\pi x}{L} + \frac{1}{3^2}\cos\frac{3\pi x}{L} + \frac{1}{5^2}\cos\frac{5\pi x}{L} + \ldots\right) \tag{3}$$

$$\left.\begin{aligned} Ax + B = \frac{1}{\pi}\Big[&(4B + 2LA)\sin\frac{\pi x}{L} - \frac{2LA}{2}\sin\frac{2\pi x}{L} \\ &+ \frac{4B + 2LA}{3}\sin\frac{3\pi x}{L} - \frac{2LA}{4}\sin\frac{4\pi x}{L} + \ldots\Big]\end{aligned}\right\} \tag{4}$$

If

$$\left.\begin{aligned} &f(x) = H \text{ for } c < x < c + w, \\ &f(x) = 0 \text{ for } c + w < x < c + 2L, \\ \text{and}\quad &f(x + 2L) = f(x);\end{aligned}\right\} \tag{5}$$

then for this periodic square pulse,

$$f(x) = \frac{Hw}{2L} + \frac{2H}{\pi}\sum_{n=1}^{\infty}\frac{1}{n}\sin\frac{n\pi w}{2L}\cos\frac{n\pi}{L}\left(x - c - \frac{w}{2}\right) \tag{6}$$

For $0 < x < \pi$,

$$\frac{\pi}{4} = \sin x + \frac{\sin 3x}{3} + \frac{\sin 5x}{5} + \frac{\sin 7x}{7} + \dots \tag{7}$$

$$-\frac{x}{2} + \frac{\pi}{4} = \frac{\sin 2x}{2} + \frac{\sin 4x}{4} + \frac{\sin 6x}{6} + \dots \tag{8}$$

$$-\frac{\pi x}{4} + \frac{\pi^2}{8} = \cos x + \frac{\cos 3x}{3^2} + \frac{\cos 5x}{5^2} + \frac{\cos 7x}{7^2} + \dots \tag{9}$$

$$\frac{x^2}{4} - \frac{\pi x}{4} + \frac{\pi^2}{24} = \frac{\cos 2x}{2^2} + \frac{\cos 4x}{4^2} + \frac{\cos 6x}{6^2} + \dots \tag{10}$$

$$-\frac{\pi x^2}{8} + \frac{\pi^2 x}{8} = \sin x + \frac{\sin 3x}{3^3} + \frac{\sin 5x}{5^3} + \frac{\sin 7x}{7^3} + \dots \tag{11}$$

$$\frac{x^3}{12} - \frac{\pi x^2}{8} + \frac{\pi^2 x}{24} = \frac{\sin 2x}{2^3} + \frac{\sin 4x}{4^3} + \frac{\sin 6x}{6^3} + \dots \tag{12}$$

$$\ln \left| \sin \frac{x}{2} \right| = -\ln 2 - \frac{\cos x}{1} - \frac{\cos 2x}{2} - \frac{\cos 3x}{3} - \dots \tag{13}$$

$$\ln \left| \cos \frac{x}{2} \right| = -\ln 2 + \frac{\cos x}{1} - \frac{\cos 2x}{2} + \frac{\cos 3x}{3} - \dots \tag{14}$$

$$\int_0^x \ln \left| \tan \frac{u}{2} \right| du = -2\left(\frac{\sin x}{1^2} + \frac{\sin 3x}{3^2} + \frac{\sin 5x}{5^2} + \dots\right) \tag{15}$$

$$\int_0^x \ln |\, 2 \sin u \,| \, du = -2\left(\frac{\sin 2x}{2^2} + \frac{\sin 4x}{4^2} + \frac{\sin 6x}{6^2} + \dots\right) \tag{15}$$

$$\frac{1}{2} \tan^{-1} \frac{2r \sin x}{1 - r^2} = \sum_{n=1}^{\infty} \frac{r^{2n-1}}{2n-1} \sin (2n-1)x, \quad |\, r \,| < 1 \tag{17}$$

$$\frac{1}{2} \tan^{-1} \frac{2r \cos x}{1 - r^2} = \sum_{n=1}^{\infty} (-1)^{n-1} \frac{r^{2n-1}}{2n-1} \cos (2n-1)x, \quad |\, r \,| < 1 \tag{18}$$

For $-\pi < x < \pi$,

$$e^{ax} = \frac{2 \sinh a\pi}{\pi} \left[\frac{1}{2a} + \sum_{n=1}^{\infty} (-1)^n \frac{a \cos nx - n \sin nx}{a^2 + n^2}\right] \tag{19}$$

$$\sin ax = \frac{2 \sin a\pi}{\pi} \left[\frac{\sin x}{1^2 - a^2} - \frac{2 \sin 2x}{2^2 - a^2} + \frac{3 \sin 3x}{3^2 - a^2} - \dots\right] \tag{20}$$

$$\cos ax = \frac{2a \sin a\pi}{\pi}\left[\frac{1}{2a^2} + \frac{\cos x}{1^2 - a^2} - \frac{\cos 2x}{2^2 - a^2} + \frac{\cos 3x}{3^2 - a^2} - \dots\right] \tag{21}$$

$$\sinh ax = \frac{2 \sinh a\pi}{\pi}\left[\frac{\sin x}{1^2 + a^2} - \frac{2 \sin 2x}{2^2 + a^2} + \frac{3 \sin 3x}{3^2 + a^2} - \dots\right] \tag{22}$$

$$\cosh ax = \frac{2a \sinh a\pi}{\pi}\left[\frac{1}{2a^2} - \frac{\cos x}{1^2 + a^2} + \frac{\cos 2x}{2^2 + a^2} - \frac{\cos 3x}{3^2 + a^2} + \dots\right] \tag{23}$$

14.14. Complex Fourier series. The Fourier series of § 14.10 may be written in complex form as

$$f(x) = \sum_{n=-\infty}^{\infty} C_n e^{in\omega x} \tag{1}$$

where

$$C_n = \frac{1}{p}\int_c^{c+p} f(x)e^{-in\omega x}\, dx, \quad \omega = \frac{2\pi}{p} \tag{2}$$

The complex C_n are related to the coefficients of § 14.10 by

$$C_0 = a, \quad C_n = \frac{a_n - ib_n}{2}, \quad C_{-n} = \frac{a_n + ib_n}{2} \tag{3}$$

$$\frac{(1 - r \cos \omega x) + ir \sin \omega x}{1 + 2r \cos \omega x + r^2} = \sum_{n=0}^{\infty} r^n e^{in\omega x}, \quad |r| < 1 \tag{4}$$

$$\left.\begin{aligned} &\ln \frac{1}{\sqrt{1 - 2r \cos \omega x + r^2}} + i \tan^{-1} \frac{r \sin \omega x}{1 - r \cos \omega x} = \sum_{n=1}^{\infty} \frac{1}{n} r^n e^{in\omega x} \\ &\qquad\qquad |r| < 1 \end{aligned}\right\} \tag{5}$$

14.15. The Fourier integral theorem. Let $f(x)$ be piecewise regular, so redefined at points of discontinuity that

$$f(x) = \tfrac{1}{2}[f(x+) + f(x-)] \tag{1}$$

and such that $\int_{-\infty}^{\infty} |f(x)|\, dx$ is finite. Then

$$f(x) = \frac{1}{\pi}\int_0^{\infty} du \int_{-\infty}^{\infty} \cos u(x - t)\, f(t)\, dt \tag{2}$$

In complex form,

$$f(x) = \lim_{A\to\infty} \frac{1}{2}\int_{-A}^{A} du \int_{-\infty}^{\infty} e^{iu(x-t)} f(t)\, dt \tag{3}$$

14.16. Fourier transforms. The Fourier transform of $f(x)$ is

$$F(u) = \int_{-\infty}^{\infty} e^{-iut} f(t)\, dt \tag{1}$$

and by § 14.15,

$$f(x) = \lim_{A\to\infty} \frac{1}{2\pi} \int_{-A}^{A} e^{iux} F(u)\, du \tag{2}$$

In a linear system, if $R(x,u)$ is the response to e^{iux}, the response to $f(x)$ is

$$R(x) = \lim_{A\to\infty} \frac{1}{2\pi} \int_{-A}^{A} R(x,u)F(u)\, du \tag{3}$$

14.17. Laplace transforms. If $f(t) = 0$ for $t < 0$, its Laplace transform is

$$F(p) = \int_{0}^{\infty} e^{-pt} f(t)\, dt \tag{1}$$

We write $F(p) = \text{Lap } f(t)$.
For derivatives,

$$\text{Lap } f'(t) = -f(0+) + p \text{ Lap } f(t) \tag{2}$$

$$\text{Lap } f''(t) = -f'(0+) - pf(0+) + p^2 \text{ Lap } f(t) \tag{3}$$

and so on. These relations reduce a linear differential equation with constant coefficients to an algebraic equation in the transform. We have

$$\text{Lap } \frac{t^n}{n!} = \frac{1}{p^{n+1}}, \quad \text{Lap } e^{-at} \sin kt = \frac{k}{(p+a)^2 + k^2} \tag{5}$$

$$\text{Lap } e^{-at} \cos kt = \frac{p+a}{(p+a)^2 + k^2}, \quad \text{Lap } 1 = \frac{1}{p} \tag{6}$$

If

$$\left.\begin{aligned} h(t) &= \int_0^t f(u)g(t-u)\, du, \\ \text{Lap } h(t) &= [\text{Lap } f(t)]\,[\text{Lap } g(t)] \end{aligned}\right\} \tag{7}$$

For examples of the use of Laplace transforms to solve systems of differential equations, see Franklin, P., *Fourier Methods*, McGraw-Hill Book Company, Inc., New York, 1949, Chap. 5.

14.18. Poisson's formula. For small x, the right member converges rapidly and may be used to compute the value of the left member in

$$\frac{1}{2} + \sum_{n=1}^{\infty} e^{-(nx)^2} = \frac{\sqrt{\pi}}{x}\left(\frac{1}{2} + \sum_{n=1}^{\infty} e^{-(n\pi/x)^2}\right) \tag{1}$$

14.19. Orthogonal functions. Many eigenvalue problems arising from differential systems or integral equations have eigenfunctions which form a complete orthogonal set. For these functions $\phi_n(x)$

$$\int \phi_n(x)\phi_m(x)dx = 0 \quad \text{if} \quad m \neq n; \qquad \int [\phi_n(x)]^2\, dx = N_n \tag{1}$$

and for sufficiently regular functions $f(x)$,

$$f(x) = \sum_{n=1}^{\infty} c_n \phi_n(x) \tag{2}$$

where

$$c_n = \frac{1}{N_n} \int f(x)\phi_n(x)\, dx \tag{3}$$

The integrals are all taken over the fundamental interval for the problem, and the expansion holds in this interval. Except for special conditions sometimes required at the ends of the interval, or at infinity when the interval is infinite or semi-infinite, the regularity conditions are usually similar to those for Fourier series.

It is possible to create a set of orthogonal functions $\phi_n(x)$ from any linearly independent infinite set $f_n(x)$ by setting

$$\phi_1(x) = f_1(x), \quad \phi_2(x) = f_2(x) - \left[\frac{1}{N_1} \int \phi_1(x) f_2(x) dx\right] \phi_1(x) \tag{4}$$

$$\phi_n(x) = f_n(x) - \sum_{i=1}^{n-1} \left[\frac{1}{N_i} \int \phi_i(x) f_n(x) dx\right] \phi_i(x) \tag{5}$$

Examples are $P_n(x)$ on $-1,1$ of § 8.8 ; $P_n{}^m(x)$ for fixed m on $-1,1$ of § 8.16 ; $x^{1/2}J_p(a_{pn}x)$ where $J_p(a_{pn}) = 0$ for fixed positive p on 0,1 of § 9.13 ; $e^{-x/2}L_n(x)$ on $0, \infty$ of § 11.4 ; $e^{-x/2}x^{k+1}L_{n+k}{}^{2k+1}(x)$ for fixed k on $0, \infty$ of § 11.7; $e^{-x^2/2}H_n(x)$ on $-\infty, \infty$ of § 12.4; $\sqrt{x^{q-1}(1-x)^{p-q}}\,J_n(p,q\,;x)$ on 0,1 of § 10.7; $1/(\sqrt[4]{1-x^2})T_n(x)$ on $-1,1$ of § 10.7.

14.20. Weight functions. The functions $g_n(x)$ are orthogonal with a weighting factor $w(x)$ when

$$\int w(x)g_n(x)g_m(x)dx = 0 \quad \text{if} \quad m \neq n \tag{1}$$

For such functions the expansion is

$$F(x) = \sum_{n=1}^{\infty} C_n g_n(x) \tag{2}$$

where

$$C_n = \frac{1}{M_n} \int w(x)F(x)g_n(x)dx, \quad M_n = \int w(x)[g_n(x)]^2\, dx \tag{3}$$

Illustrations related to some of the examples mentioned at the end of § 14.19 are $J_p(a_{pn}x)$ with $w(x) = x$; $L_n(x)$ with $w(x) = e^{-x}$; $H_n(x)$ with $w(x) = e^{-x^2}$; $J_n(p,q\,;x)$ with $w(x) = x^{q-1}(1-x)^{p-q}$; $T_n(x)$ with $w(x) = 1/\sqrt{1-x^2}$.

15. Asymptotic Expansions

15.1. Asymptotic expansion. Let the expression

$$a_0 + \frac{a_1}{x} + \frac{a_2}{x^2} + \frac{a_3}{x^3} + \dots \tag{1}$$

be related to a function $F(x)$ in such a way that for any fixed n

$$\lim_{x \to \infty} x^n \left[F(x) - a_0 - \frac{a_1}{x} - \dots - \frac{a_n}{x^n} \right] = 0 \tag{2}$$

The limit holds in some sector of the complex plane. The expression in general is a divergent series, since it can converge only if $F(z)$ is analytic at infinity. In any case it is called an asymptotic expansion for the function. The terms asymptotic or semiconvergent series are also used.

The error committed when we employ a finite number of terms in place of $f(x)$ is frequently of the same order of magnitude as the numerical value of the next following term. Thus asymptotic expressions can be used in computation like convergent alternating series as long as the terms remain small. For examples see §§ 8.9, 8.14, 9.7, 13.5, and 13.9.

If the function $F(x)$ is an infinite series of terms $\Sigma\, u_k(x)$, an asymptotic series may sometimes be found by taking $u_k(x)$ as the $f(k)$ in the Euler-Maclaurin sum formula of § 14.4.

15.2. Borel's expansion. Let

$$\phi(t) = \sum_{n=0}^{\infty} A_n t^n$$

be any function for which the integral

$$I(x) = \int_0^\infty e^{-tx} t^p \phi(t)\,dt \tag{1}$$

converges. Then the expansion

$$I(x) = \frac{\Gamma(p+1)}{x^{p+1}} \left[A_0 + (p+1)\frac{A_1}{x} + (p+1)(p+2)\frac{A_2}{x^2} + \dots \right] \tag{2}$$

is usually an asymptotic series for $I(x)$. One extension is

$$m \int_0^\infty e^{-(xt)^m}\, t^{m-1} \phi(t)\, dt = \sum_{n=0}^{\infty} A_n\, \Gamma\left(\frac{n}{m} + 1\right) \frac{1}{x^{n+m}} \tag{3}$$

15.3. Steepest descent. If the path C is suitably chosen in the complex plane (see § 20.3) the integral

$$F(x) = \int_C e^{-xg(t)}h(t)dt \tag{1}$$

may be transformed into integrals whose expansion can be found by § 15.2. The path C is one along which $g(t)$ remains real and changes most rapidly. It will pass through a saddle point of the surface $R(u,v)$ where $t = u + iv$ and $g(t) = R(u,v) + iI(u,v)$. At this saddle point t_0, $g(t_0) = 0$ and $g'(t_0) = 0$, and expanding about it

$$s = g(t) = (t - t_0)^2[g_0 + g_1(t - t_0) + g_2(t - t_0)^2 + \ldots] \tag{2}$$

The path C consists of two parts L_1 and L_2. On L_1, we find $t_1(s)$ by integrating dt_1 between 0 and s to obtain $t_1 - t_0$. And on L_2, we get $t_2(s)$ by integrating dt_2 between 0 and s to obtain $t_2 - t_0$. Here

$$\frac{dt_1}{ds} = \frac{1}{2} s^{-1/2} \sum_{n=0}^{\infty} a_n s^{n/2}, \qquad \frac{dt_2}{ds} = \frac{1}{2} s^{-1/2} \sum_{n=0}^{\infty} (-1)^{n+1} a_n s^{n/2}$$

where in terms of the coefficients g_k,

$$a_0 = g_0^{-1/2}, \quad a_1 = g_0^{-1}\left(-\frac{g_1}{g_0}\right), \quad a_2 = g_0^{-3/2}\left(-\frac{3g_2}{2g_0} + \frac{3 \cdot 5g_1^2}{8g_2^2}\right)$$

$$a_3 = g_0^{-2}\left(-\frac{2g_3}{g_0} + \frac{6g_1g_2}{g_0^2} - \frac{4g_1^3}{g_0^3}\right)$$

Then in terms of s,

$$F(x) = \int_{L_1} e^{-xs}h(t_1)\frac{dt_1}{ds}\,ds + \int_{L_2} e^{-xs}h(t_2)\frac{dt_2}{ds}\,ds$$

Since s is real, if s runs from 0 to infinity on the path $C = L_1 + L_2$, the sum of these integrals in general has an asymptotic expansion which follows from Borel's expansion (2) of § 15.2.

16. Least Squares

16.1. Principle of least squares. Let $n > r$, so that the n observation equations

$$f_k(a_1, a_2, \ldots, a_r) = s_k, \quad k = 1, 2, \ldots n \tag{1}$$

form an overdetermined system for the determination of the r unknown constants a_q. Then when the n observed quantities s_k have comparable accuracy, the kth residual is taken as

$$v_k = f_k(a_1, a_2, \ldots, a_r) - s_k \tag{2}$$

The principle of least square asserts that the best approximation to the a_q is the set for which the sum of the squares of the residuals is a minimum. Necessary conditions for

$$S = \sum_{k=1}^{n} v_k^2 \tag{3}$$

to be a minimum are

$$\frac{\partial S}{\partial a_1} = 0, \quad \frac{\partial S}{\partial a_2} = 0, \quad \ldots, \quad \frac{\partial S}{\partial a_r} = 0 \tag{4}$$

These constitute the normal equations. In general they determine a unique solution for a_q which gives the desired best approximation.

16.2. Weights. When the relative accuracy of the s_k in § 16.1 is known to be different, we assign weights w_k such that the quantities $w_k s_k$ have comparable accuracy. Then the residual v_k is taken as

$$v_k = w_k f_k(a_1, a_2, \ldots, a_r) - w_k s_k \tag{5}$$

and otherwise the principle is applied as in § 16.1.

16.3. Direct observations. When the s_k are direct observations of a quantity a_1, the residual v_k is $a_1 - s_k$, and the principle gives the average

$$a_1 = \frac{1}{n}(s_1 + s_2 + \cdots + s_n) \tag{1}$$

as the best approximation for a_1.

16.4. Linear equations. When linear in the a_q, the n observation equations may be written

$$\sum_{q=1}^{r} A_{kq} a_q = s_k, \quad k = 1, 2, \ldots, n \tag{1}$$

In this case the normal equations are

$$\sum_{q=1}^{r} \sum_{k=1}^{n} A_{kp} A_{kq} a_q = \sum_{k=1}^{n} A_{kp} s_k, \quad p = 1, 2, \ldots, r \tag{2}$$

16.5. Curve fitting. Suppose the n points (x_k, y_k) follow approximately a straight line. To find the best line when we regard y_k as the observed value corresponding to x_k, we write

$$a_1 + a_2 x_k = y_k \tag{1}$$

as the observation equation. And we determined a_1 and a_2 by solving the two simultaneous normal equations

$$\left.\begin{aligned} na_1 + \left(\sum_{k=1}^{n} x_k\right) a_2 &= \sum_{k=1}^{n} y_k \\ \left(\sum_{k=1}^{n} x_k\right) a_1 + \left(\sum_{k=1}^{n} x_k^2\right) a_2 &= \sum_{k=1}^{n} x_k y_k \end{aligned}\right\} \quad (2)$$

Similarly, to find the best parabola we use

$$a_1 + a_2 x_k + a_3 x_k^2 = y_k \quad (3)$$

as the observation equations, and proceed as in § 16.4, or we may use a polynomial of the mth degree, with the observation equations

$$a_1 + a_2 x_k + a_3 x_k^2 + \dots + a_{m+1} x_k^m = y_k \quad (4)$$

16.6. Nonlinear equations. When the functions f_k are not linear in the a_q, we get a first approximation $\bar{a}_q$ by graphical means when $r = 2$, or by solving some set of r of the observation equations. Then we expand in Taylor's series about $\bar{a}_q$, and neglect second-order terms. The result is

$$f_k(\bar{a}_1,\bar{a}_2,\dots,\bar{a}_r) + \sum_{q=1}^{r} \frac{\partial f_k}{\partial \bar{a}_q} \Delta a_k = s_k \quad (1)$$

We treat these as observation equations in the unknown Δa_k. With

$$\partial f_k/\partial \bar{a}_q \text{ for } A_{kq}, \quad \Delta a_q \text{ for } a_q, \quad s_k - f_k(\bar{a}_1,\bar{a}_2,\dots,\bar{a}_r) \text{ for } s_k,$$

they have the form (1) discussed in § 16.4.

17. Statistics

17.1. Average. The average of a finite set of observed values $s_1, s_2, \dots, s_n$ is

$$\mu_1(s) = \bar{s} = \frac{1}{n}(s_1 + s_2 + \dots + s_n) \quad (1)$$

17.2. Median. When a sequence is arranged in order of magnitude the number in the central position, or the average of the two nearest to a central position, is called the median. Essentially, there are as many numbers in the sequence larger than the median as there are smaller than the median.

17.3. Derived averages. The kth moment is the average of the kth powers so that

$$\mu_k(s) = \overline{s^k} = \frac{1}{n}(s_1{}^k + s_2{}^k + \ldots + s_n{}^k) \tag{1}$$

17.4. Deviations. To the set of observed values $s_1, s_2, \ldots, s_n$ corresponds an average $\bar{s}$, § 17.1, and a set of deviations

$$d_1 = s_1 - \bar{s}, \quad d_2 = s_2 - \bar{s}, \quad \ldots, \quad d_n = s_n - \bar{s} \tag{1}$$

The average of a set of deviations is zero, $\mu_1(d) = \bar{d} = 0$.

For the mean-square deviation § 17.3,

$$\mu_2(d) = \mu_2(s) - [\mu_1(s)]^2 \quad \text{or} \quad \overline{d^2} = \overline{s^2} - (\bar{s})^2 \tag{2}$$

The square root of $\mu_2(d)$, or root-mean-square value of the deviation is called the standard deviation σ.

$$\sigma = \sqrt{\mu_2(d)} = \sqrt{\frac{1}{n}\Sigma d_r{}^2} \tag{3}$$

17.5. Normal law. The deviations are samples of a continuous variable x, whose distribution function may frequently be taken as the normal law of error

$$y = \frac{h}{\sqrt{\pi}} e^{-h^2x^2} \tag{1}$$

That is $P_{a,b}$ the probability that an error is between a and b, is

$$P_{a,b} = \frac{h}{\sqrt{\pi}} \int_a^b e^{-h^2x^2} dx \tag{2}$$

In terms of the error function of § 13.9.

$$P_{a,b} = \tfrac{1}{2}(\text{Erf } hb - \text{Erf } ha) \quad \text{and} \quad P_{-b,b} = \text{Erf } hb \tag{3}$$

For a given normal distribution, h is called the measure of precision.

17.6. Standard deviation. For the normal law of § 17.5 the second moment is

$$\mu_2(x) = \int_{-\infty}^{\infty} x^2 \frac{h}{\sqrt{\pi}} e^{-h^2x^2} dx = \frac{1}{2h^2} \tag{1}$$

Thus the standard deviation is

$$\sigma = \sqrt{\mu_2(x)} = \frac{1}{\sqrt{2}h} \quad \text{and} \quad h = \frac{1}{\sqrt{2}\sigma} = \frac{0.7071}{\sigma} \tag{2}$$

All the odd moments, $\mu_{2k+1}(x) = 0$, and for the higher even moments

$$\mu_{2k}(x) = 1 \cdot 3 \cdot 5 \cdot \ldots (2k-1)\frac{1}{2^k h^{2k}} = 1 \cdot 3 \cdot 5 \cdot \ldots (2k-1)\sigma^{2k} \quad (3)$$

17.7. Mean absolute error. For the normal law of § 17.5 the mean absolute error is

$$\mu_1(|\, x \,|) = 2\int_0^\infty x \frac{h}{\sqrt{\pi}} e^{-h^2x^2} dx = \frac{1}{h\sqrt{\pi}} \quad (1)$$

Thus the mean absolute error

$$\mu = \frac{1}{h\sqrt{\pi}} \quad \text{and} \quad h = \frac{1}{\mu\sqrt{\pi}} = \frac{0.5643}{\mu} \quad (2)$$

17.8. Probable error. For the normal law of § 17.5 the particular error which is just as likely to be exceeded as not is called the probable error, ϵ. Thus

$$\tfrac{1}{2} = P_{-\varepsilon,\varepsilon} = \text{Erf}\, h\epsilon \quad \text{and} \quad h\epsilon = 0.4769 \quad (1)$$

And the probable error

$$\epsilon = \frac{0.4769}{h} = 0.6745\sigma = 0.8453\mu \quad (2)$$

17.9. Measure of dispersion. To find a measure of precision h which makes the normal law fit a set of deviations, we identify the standard deviation of § 17.6 with $\sqrt{n/(n-1)}$ times that of § 17.4. Thus

$$\frac{1}{\sqrt{2}h} = \sigma = \sqrt{\frac{1}{n-1}\Sigma d_r^2} \quad \text{and} \quad \epsilon = \frac{0.6745}{\sqrt{n-1}}\sqrt{\Sigma d_r^2} \quad (1)$$

Or, we may identify the mean absolute error of § 17.7 with $\sqrt{n/(n-1)}$ times that formed from the d_r, $\mu_1(|\, d_r \,|)$. Thus

$$\frac{1}{h\sqrt{\pi}} = \mu = \frac{1}{\sqrt{n(n-1)}}\Sigma|\, d_r \,| \quad \text{and} \quad \epsilon = \frac{0.8453}{\sqrt{n(n-1)}}\Sigma|\, d_r \,| \quad (2)$$

These are easier to calculate, but slightly less accurate, than the values found from the standard deviation. In some cases both values are found, and a rough agreement is considered a check on the normal character of the distribution. In either case the probable error of the mean is taken as $1/\sqrt{-}$ times that for a single observation, $\epsilon_m = \epsilon/\sqrt{n}$.

17.10. Poisson's distribution. A variable taking on integral values only, $k = 1, 2, 3, \ldots$ has Poisson's distribution law if the probability that the variable equals k is

$$P_k = \frac{a^k e^{-a}}{k!} \tag{1}$$

For this distribution the average value of the variable

$$\mu_1(k) = a \tag{2}$$

and the standard deviation

$$\sigma = \sqrt{\mu_2(k)} = \sqrt{a}, \quad a = \sigma^2 \tag{3}$$

These relations may be used to find the parameter a from the average or standard deviation of a sample set of observations which follow this law.

17.11. Correlation coefficient. Let x and y be two variables for each of which the deviations from the average have a normal distribution. Then for a set of corresponding pairs $x_1,y_1; x_2,y_2; \ldots; x_n,y_n$ the correlation coefficient of the sample is

$$r = \frac{\Sigma(x_i - \bar{x})(y_i - \bar{y})}{\sqrt{\Sigma(x_i - \bar{x})^2 \cdot \Sigma(y_i - \bar{y})^2}} = \frac{\Sigma d_i D_i}{\sqrt{\Sigma d_i^2 \cdot \Sigma D_i^2}} \tag{1}$$

where

$$\bar{x} = \mu_1(x) = \frac{1}{n}\Sigma x_i, \quad d_i = x_i - \bar{x} \tag{2}$$

$$\bar{y} = \mu_1(y) = \frac{1}{n}\Sigma y_i, \quad D_i = y_i - \bar{y} \tag{3}$$

For $r = 0$ there is no correlation. For $r = 1$ there is strict proportionality $y = Cx$ with a positive C, and for $r = -1$ with a negative C. For $|r|$ near 1, the ellipses of equal probability for the distribution of points x_i, y_i are long and thin, approximating a straight line. The constant c determined from

$$\tan 2\phi = \frac{2\Sigma d_i D_i}{\Sigma d_i^2 - \Sigma D_i^2}, \quad c = \tan\phi \tag{4}$$

leads to the straight line $y = cx$ such that the sum of the squares of the distances to all the observed points x_i, y_i is a minimum.

18. Matrices

18.1. Matrix. The elements of a matrix are a system of mn numbers $a_{ik}(i = 1,2,\ldots,m; k = 1,2,\ldots,n)$. These form a matrix when written as a rectangular array of m rows and n columns

$$\left\| \begin{matrix} a_{11} & a_{12} & \cdots & a_{1n} \\ a_{21} & a_{22} & \cdots & a_{2n} \\ \cdots & \cdots & \cdots & \cdots \\ a_{m1} & a_{m2} & \cdots & a_{mn} \end{matrix} \right\| \tag{1}$$

We use the abbreviations $\| a_{ik} \|$ or a single letter A. The double bars are sometimes replaced by parentheses. (See §§ 7.4, 7.5.)

18.2. Addition. The relation

$$A = B \quad \text{means} \quad a_{ik} = b_{ik} \tag{1}$$

$$A + B = C \quad \text{means} \quad a_{ik} + b_{ik} = c_{ik} \tag{2}$$

$$sA = A \quad \text{means} \quad sa_{ik} = b_{ik}, \text{ where } s \text{ is a scalar} \tag{3}$$

In each of these relations all the matrices involved must have the same dimensions. If these are m by n the matrix equation is equivalent to mn scalar equations.

18.3. Multiplication. Two matrices A and B, taken in this order, are said to be conformable if the number of columns of A equals the number of rows of B. Let A be m by N and B be N by n. Then the product $C = AB$ is an m by n matrix with

$$c_{ik} = \sum_{j=1}^{N} a_{ij} b_{jk} \tag{1}$$

If $n \neq m$, then B and A will not be conformable, and there is no product BA.

If A is m by n and B is n by m, then AB is an m by m matrix and BA is an n by n matrix, of different dimensions if $m \neq n$.

When A and B are each n by n, or square matrices of order n, the product AB and BA will each be n by n, but in general will not be the same. Matrix multiplication is not commutative.

Matrix multiplication is associative,

$$A(BC) = (AB)C = ABC \tag{2}$$

18.4. Linear transformations. Let $X\!\uparrow\, = \| x_k \|$ be a 1 by n column matrix, $Y\!\uparrow\, = \| y_j \|$ be a 1 by N column matrix, and $Z\!\uparrow\, = \| z_i \|$ be a 1 by m column matrix. Then if A is m by N, and B is N by n, in matrix form the transformation

$$y_j = \sum_{k=1}^{n} b_{jk} x_k \quad \text{is} \quad Y\!\uparrow\, = BX\!\uparrow \tag{1}$$

and the transformation

$$z_i = \sum_{j=1}^{N} a_{ij} y_j \quad \text{is} \quad Z\!\uparrow\, = AY\!\uparrow \tag{2}$$

If the transformation which takes the x_k into the z_i directly is

$$z_i = \sum_{k=1}^{n} c_{ik} x_k \quad \text{is} \quad Z{\uparrow} = CX{\uparrow}, \quad \text{then} \quad C = AB \quad \text{and} \quad Z{\uparrow} = ABX{\uparrow} \tag{3}$$

18.5. Transposed matrix. If $A = \| a_{ij} \|$ is m by N, the transposed matrix $A' = \| a_{ji} \|$ is N by m. For a product $C = AB$, the transposed matrix is $C' = B'A'$.

For $X{\uparrow}$, $Y{\uparrow}$, $Z{\uparrow}$ the column matrices of § 18.4 with a single column, the transposed matrices will be row matrices with a single row. We denote them by $\vec{X}$, $\vec{Y}$, $\vec{Z}$. Then the transformations of § 18.4 may be written in terms of row matrices as

$$\vec{Y} = \vec{X}B', \quad \vec{Z} = \vec{Y}A', \quad \vec{Z} = \vec{X}C', \quad \text{or} \quad \vec{Z} = \vec{X}B'A', \quad \text{since } C' = B'A'$$

18.6. Inverse matrix. The unit matrix is a square matrix with ones on the main diagonal and the remaining elements zero. Its elements are $\delta_{ik} = 0$ if $i \neq k$, and $= 1$ if $i = k$.

A square matrix is singular if its determinant is zero. Each nonsingular square matrix has a reciprocal matrix A^{-1} such that

$$AA^{-1} = \| \delta_{ik} \| \quad \text{and} \quad A^{-1}A = \| \delta_{ik} \| \tag{1}$$

Let $| A |$ denote the determinant $| a_{ik} |$, and A_{ik} the cofactor of a_{ik} or product of $(-1)^{i+k}$ by the determinant obtained from $| a_{ik} |$ by striking out the ith row and kth column. (See § 1.8.) Then explicitly A^{-1} has as elements

$$a^{-1}{}_{ik} = \frac{A_{ki}}{| A |} \tag{2}$$

18.7. Symmetry. For square matrices symmetry and skew-symmetry are defined as for tensors in § 7.10. Orthogonal and unitary, or Hermitian orthogonal matrices are defined in § 7.11.

18.8. Linear equations. The set of n linear equations in n unknowns x_k,

$$\sum_{k=1}^{n} a_{ik} x_k = b_i, \quad i = 1,2,\ldots,n \tag{1}$$

determine a square matrix $A = \| a_{ik} \|$ and two column matrices $B{\uparrow} = \| B_i \|$ and $X{\uparrow} = \| x_i \|$. In matrix form we may write

$$AX{\uparrow} = B{\uparrow} \tag{2}$$

If A is nonsingular, find its solution by premultiplying by the inverse matrix A^{-1} of § 18.6.

$$X\uparrow = A^{-1}B\uparrow \quad \text{or} \quad x_i = \frac{1}{|A|}\sum_{k=1}^{n} A_{ki}b_k \tag{3}$$

18.9. Rank. For a rectangular matrix we get a subdeterminant of the qth order by selecting those elements in some set of q rows and some set of q columns. If at least one subdeterminant of order r is not zero, but all the subdeterminants of order $r + 1$ and hence those of higher order vanish, the matrix is of rank r.

With any set of m linear equations in n unknowns x_k,

$$\sum_{k=1}^{n} a_{ik}x_k = b_i, \quad i = 1,2,\ldots,m \tag{1}$$

or $$AX\uparrow = B\uparrow \quad \text{with} \quad A = \| a_{ik} \|, \quad X\uparrow = \| x_k \|, \quad B\uparrow = \| b_i \| \tag{2}$$

we associate two matrices. The matrix of the system is the m by n matrix A, and we obtain from A the augmented matrix, which is an m by $(n + 1)$ matrix, by adding an $(m + 1)$st column whose elements are those of $B\uparrow$. Then the condition that the system of equations is consistent in the sense of having one or more sets of solutions for the unknown x_k is that the matrix and augmented matrix have the same rank.

18.10. Diagonalization of matrices. Let $A = \| a_{ij} \|$ be a square n by n matrix, and $\| \delta_{ij} \|$ be the unit matrix of § 18.6. Then $\| \lambda\delta_{ij} - a_{ij} \|$ is the characteristic matrix of A. The roots of the characteristic equation, $| \lambda\delta_{ij} - a_{ij} | = 0$ are the eigenvalues of A. Suppose that they are all distinct, $\lambda_1, \lambda_2, \ldots, \lambda_n$. Find solutions of

$$\sum_{k=1}^{n} a_{ik}x_{kj} = \lambda_j x_{ij}.$$

The ratios of the x_{ij}, for each j, are determined, § 18.9. They may be scaled to make the vector $x_{1j}, x_{2j}, \ldots, x_{nj}$ have unit length. For distinct λ_j, any two of these vectors are orthogonal. The orthogonal (§ 7.11) matrix $X = \| x_{ij} \|$ is such that $X^{-1}AX = \| \lambda_i\delta_{ij} \|$, a diagonal matrix.

When the eigenvalues are not all distinct, a reduction to diagonal form may not be possible, and we are led into the theory of elementary divisors.

Whether all eigenvalues are distinct or not, whenever A is symmetric, § 7.10, unitary, § 7.11, or Hermitian, $\tilde{H} = H$, a unitary matrix U can be found such that $U^{-1}AU = \| \lambda_i \delta_{ij} \|$, a diagonal matrix. Here $U = \| u_{ij} \|$, where if λ_j is a root of $| \lambda\delta_{ij} - a_{ij} |$ of multiplicity m, so that $\lambda_j = \lambda_{j+1} = \dots = \lambda_{j+m-1}$, the columns of u_{ij} from j to $j + m - 1$, u_{is}, are found from the m linearly independent vector solutions of

$$\sum_{k=1}^{n} a_{ij} x_{kt} = \lambda_j x_{it}$$

We take $v_{i1} = x_{i1}$,

$$v_{i2} = x_{i2} - \frac{1}{V_1} \left(\sum_{p=1}^{n} \bar{v}_{p1} x_{p2} \right) v_{i1}$$

where the bar means complex conjugate

$$v_{it} = x_{it} - \sum_{k=1}^{t-1} \frac{1}{V_k} \left(\sum_{p=1}^{n} \bar{v}_{pk} x_{pt} \right) v_{ik}$$

Here

$$V_k = \sum_{p=1}^{n} \bar{v}_{pk} v_{pk}$$

Then $u_{i,j+t-1} = (1/\sqrt{V_t}) v_{it}$.

The eigenvalues of a real symmetric or of a Hermitian matrix, $\tilde{H} = H$, are all real. The eigenvalues of a real orthogonal or of a unitary matrix are all of absolute value 1, ± 1 or $e^{i\phi}$.

19. Group Theory

19.1. Group. Let a rule of combination be given which determines a third mathematical object C from two given objects A and B, taken in that order. We call the rule " multiplication " ; call C the " product," and write $C = AB$. Then a system composed of a set of elements $A, B, \dots,$ and this one rule of combination is called a group if the following conditions are satisfied.

I. If A and B are any elements of the set, whether distinct or not, the product $C = AB$ is also an element of the set.

II. The associative law holds ; that is if A, B, C are any elements of the set, $(AB)C = A(BC)$ and may be written ABC.

III. The set contains an identity or unit element I which is such that every element is unchanged when combined with it,

$$IA = AI = A \tag{1}$$

IV. If A is any element, the set contains an inverse element A^{-1}, such that

$$A^{-1}A = AA^{-1} = I \tag{2}$$

19.2. Quotients. For any two elements A and B there is a left-hand quotient of B by A, such that $AX = B$. This $X = A^{-1}B$. There is also a right-hand quotient of B by A, such that $YA = B$. This $Y = BA^{-1}$. Let F be a fixed element of the set, and V be a variable element taking on all possible values successively. Then V^{-1}, VF, FV, $V^{-1}F$, and FV^{-1} each runs through all possible values.

19.3. Order. If there is an infinite number of elements in the set, the group is an infinite or group of infinite order. If there is a finite number, g, of elements in the set, the group is a finite group of order g.

19.4. Abelian group. If the rule of combination is commutative so that all cases $AB = BA$, the group is an Abelian group. In particular a finite group whose elements are all powers of a single element A, such as A, A^2, A^3, ..., $A^g = I$ is necessarily Abelian and is called a cyclic group.

19.5. Isomorphy. We use a single capital letter to indicate either a group or its set of elements. Two groups G and G' are isomorphic if it is possible to establish a one-to-one correspondence between their elements G and G$'$ of such a sort that if A, B are elements of G and A', B', are the corresponding elements of G', then AB corresponds to $A'B'$.

If it is possible to establish an m-to-one correspondence between the elements G and G' of this sort, the group G is multiply isomorphic to the group G'.

19.6. Subgroup. Let G be a finite group of order g. Let the elements H be a subset of the elements G such that if A_H and B_H are any two elements in H, their product A_HB_H is also in H. Then the system composed of the elements H and the rule of combination for G is a group in the sense of § 19.1. We call this group H a subgroup of the group G. The order of the subgroup H, h, is a divisor of g. And the integral quotient $j = g/h$ is called the index of H with respect to G.

19.7. Normal divisor. The elements $A^{-1}HA$, where A is any fixed element of G and H takes on all the elements of some subgroup H, themselves form a subgroup of G, which is said to be conjugate to the subgroup H. If, as A takes on all possible values as an element of G, the resulting conjugate subgroups of H are all identical, then the subgroup H is called a normal divisor of the group G.

19.8. Representation. Let M be a set of h-rowed square matrices, each of which is nonsingular, which form a group when the rule of combination is matrix multiplication as defined in § 18.3. If the group M is simply or multiply isomorphic to a group G, the group M is called a representation of G in terms of matrices or linear transformations, § 18.4. The number of rows in the matrix, or of variables in the linear transformations, h, is called the degree or dimension of the representation.

Two representations M' and M are equivalent if for some fixed matrix A, $M' = A^{-1}MA$. If for some $p + q = h$, each matrix M consists of two square matrices of p and q rows, respectively, on the main diagonal, surrounded by a p by q and a q by p rectangle of zeros, the representation M is reducible. Otherwise it is irreducible. For each finite group G of order g there are a finite number c of irreducible representations not equivalent to one another. If these have dimensions $h_1, h_2, \ldots, h_c$, each of the h_n is a divisor of g, and

$$h_1^2 + h_2^2 + \ldots + h_c^2 = g$$

19.9. Three-dimensional rotation group. The elements of this infinite group are real three by three orthogonal matrices, § 7.11, $B = \| b_{ij} \|$. As in § 18.4, $Y = BX$, where $x_1, x_2, x_3 = x, y, z$ of a first coordinate system and $y_1, y_2, y_3 = x', y', z'$ of a second rotated coordinate system. The particular matrices

$$R_z(\alpha) = \begin{Vmatrix} \cos\alpha & \sin\alpha & 0 \\ -\sin\alpha & \cos\alpha & 0 \\ 0 & 0 & 1 \end{Vmatrix}, \quad R_y(\beta) = \begin{Vmatrix} \cos\beta & 0 & -\sin\beta \\ 0 & 1 & 0 \\ \sin\beta & 0 & \cos\beta \end{Vmatrix} \tag{1}$$

correspond to rotations about $0Z$ through α and about $0Y$ through β. For a suitable choice of the Euler angles α, β, and γ any matrix of the rotation group $B = R(\alpha,\beta,\gamma) = R_z(\gamma)\, R_y(\beta)\, R_z(\alpha)$.

Consider next the unimodular matrices of order two, or unitary matrices with determinant unity,

$$U = \begin{Vmatrix} a + bi & c + di \\ -c + di & a - bi \end{Vmatrix} \tag{2}$$

where $a^2 + b^2 + c^2 + d^2 = 1$. As in § 7.11, $U_{pq} = U^{-1}{}_{pq}$. These make up the special unitary group.

The Pauli spin matrices are defined by

$$P_1 = \begin{Vmatrix} 0 & 1 \\ 1 & 0 \end{Vmatrix}, \quad P_2 = \begin{Vmatrix} 0 & i \\ -i & 0 \end{Vmatrix}, \quad P_3 = \begin{Vmatrix} 1 & 0 \\ 0 & -1 \end{Vmatrix} \tag{3}$$

The Hermitian matrix with $x^2 + y^2 + z^2 = 1$,

$$\begin{Vmatrix} z & x + iy \\ x - iy & -z \end{Vmatrix} = H(x,y,z) = xP_1 + yP_2 + zP_3 \tag{4}$$

has $\tilde{H} = H$. And for any unimodular matrix U, if

$$U^{-1}HU = x'P_1 + y'P_2 + z'P_3,$$

the transformation from x,y,z to x',y',z' is a three-dimensional rotation. The rotations $R_z(\alpha)$ and $R_y(\beta)$ correspond to $U_1(\alpha)$ and $U_2(\beta)$ where

$$U_1(\alpha) = \begin{Vmatrix} e^{i\alpha/2} & 0 \\ 0 & e^{-i\alpha/2} \end{Vmatrix}, \quad U_2(\beta) = \begin{Vmatrix} \cos\dfrac{\beta}{2} & -\sin\dfrac{\beta}{2} \\ \sin\dfrac{\beta}{2} & \cos\dfrac{\beta}{2} \end{Vmatrix} \tag{5}$$

Thus $R(\alpha,\beta,\gamma)$ correspond to $U(\alpha,\beta,\gamma) = U_1(\gamma)U_2(\beta)U_1(\alpha)$, or

$$U(\alpha,\beta,\gamma) = \begin{Vmatrix} e^{i(\alpha+\gamma)/2}\cos\dfrac{\beta}{2} & -e^{-i(\alpha-\gamma)/2}\sin\dfrac{\beta}{2} \\ e^{i(\alpha-\gamma)/2}\sin\dfrac{\beta}{2} & e^{-i(\alpha+\gamma)/2}\cos\dfrac{\beta}{2} \end{Vmatrix} \tag{6}$$

The irreducible representations, § 19.8, by matrices of order $(2j + 1)$, of the special unitary U group are given by

$$\left.\begin{aligned} U_{pq}{}^{(j)}(\alpha,\beta,\gamma) = \sum_m \frac{(-1)^{m-q-p}\sqrt{(j+p)!\,(j-p)!\,(j+q)!\,(j-q)!}}{(j-p-m)!\,(j+q-m)!\,(m+p-q)!\,m!} \\ \times e^{iq\alpha}\cos^{2j+q-p-2m}\frac{\beta}{2}\sin^{p+2m-q}\frac{\beta}{2}\,e^{ip\gamma} \end{aligned}\right\} \tag{7}$$

Here $p,q = -j, -j+1, \ldots, j-1, j$, and $m = 0, 1, 2, \ldots$ where we stop the summation by putting $1/(-N)! = 0$; and $j = 0, \frac{1}{2}, 1, \frac{3}{2}, 2, \ldots$.

The group characters, or traces, § 7.12, of these matrices are

$$\chi^{(j)}(\alpha) = 1 + 2\cos\alpha + \ldots + 2\cos j\alpha = \frac{\sin(j+\frac{1}{2})\alpha}{\sin\alpha/2} \tag{8}$$

for $j = 0, 1, 2, 3, \ldots$.

For $j = \frac{1}{2}, \frac{3}{2}, \frac{5}{2}, \ldots,$

$$\chi^{(j)}(\alpha) = 2\cos\frac{\alpha}{2} + 2\cos\frac{3\alpha}{2} + \ldots + 2\cos j\alpha = \frac{\sin(j+\frac{1}{2})\alpha}{\sin\alpha/2} \tag{9}$$

Each matrix $R(\alpha,\beta,\gamma)$ corresponds to two U matrices, namely, $U(\alpha,\beta,\gamma)$ and $U(\alpha + 2\pi, \beta, \gamma)$. The group $R(\alpha,\beta,\gamma)U(\alpha,\beta,\gamma)$ is simply isomorphic with $U(\alpha,\beta,\gamma)$ and has all the matrices $U_{pq}{}^{(j)}$ given above as representations.

But for the group $R(\alpha,\beta,\gamma)$ the matrices $U_{pq}{}^{(j)}$ and characters $\chi^{(j)}(\alpha)$ with $j = 0, 1, 2, 3, \ldots$ give a complete representation.

20. Analytic Functions

20.1. Definitions. Consider the complex variable $z = x + iy$ where $i^2 = -1$. We associate $z = x + iy$ with the point (x,y). The single-valued function $f(z)$ has a derivative $f'(z)$ at z if

$$f'(z) = \lim_{\Delta z \to 0} \frac{[f(z + \Delta z) - f(z)]}{z}$$

where $\Delta z \to 0$ through any complex values.

Let z be a variable and z_0 a fixed point in R, any open (boundary excluded) simply connected region. The function $f(z)$ is analytic in R if and only if any of the following four conditions hold.

a. $f(z)$ has a derivative $f'(z)$ at each point of R.

b. $f(z)$ may be integrated in R in the sense that the integral $\oint f(z)\,dz = 0$ about every closed path in R. Thus

$$F(z) = \int_{z_0}^{z} f(z)\,dz$$

is a single-valued analytic function of z independent of the path in R.

c. $f(z)$ has a Taylor's series expansion, §§ 3.25, 20.4, in powers of $(z - z_0)$ about each point z_0 in R.

d. $f(z) = u(x,y) + iv(x,y)$, where both $u(x,y)$ and $v(x,y)$ have continuous partial derivatives that satisfy the Cauchy-Riemann differential equations,

$$\frac{\partial u}{\partial x} = \frac{\partial v}{\partial y}, \quad \frac{\partial u}{\partial y} = -\frac{\partial v}{\partial x} \tag{1}$$

The functions u and v are conjugate potential functions, and each satisfies Laplace's equation

$$\frac{\partial^2 u}{\partial x^2} + \frac{\partial^2 u}{\partial y^2} = 0 \tag{2}$$

20.2. Properties. If $f(z)$ is analytic at all points of some circle with center S, but not at S, then $z_0 = S$ is an isolated singular point. It is a pole of order n if n is the smallest positive integer for which $(z - z_0)^n f(z)$ is bounded. If there is no such n, then z_0 is an essential singularity.

The function $f(x + iy) = u(x,y) + iv(x,y)$ effects a mapping from the x,y to the u,v plane. At any point where $f'(z) \neq 0$, the mapping is conformal, preserving angles of image curves and shapes and ratios of distances for infinitesimal figures.

20.3. Integrals. The integral

$$\left.\begin{aligned} \int_a^b f(z)\, dz &= \int_a^b (u + iv)\,(dx + i\, dy) \\ &= \int_a^b (u\, dx - v\, dy) + i \int_a^b (v\, dx + u\, dy) \end{aligned}\right\} \quad (1)$$

Each of the real differentials $(u\, dx - v\, dy)$ and $(v\, dx + u\, dy)$ is exact. The integral

$$\int_a^b f(z) dz$$

is not changed when the path from a to b is continuously varied without crossing any singular point. And

$$\oint f(z) dz = \int_a^a f(z) dz = 0$$

over any closed path not enclosing any singular point. For the same type of path enclosing t, traversed in the positive sense, Cauchy's integral formula asserts that

$$f(t) = \frac{1}{2\pi i} \oint \frac{f(z)}{z - t}\, dz \quad (2)$$

20.4. Laurent expansion. Let $f(z)$ be analytic for $0 < | z - z_0 | < r$. Then for z in this range,

$$\left.\begin{aligned} f(z) &= \sum_{m=-\infty}^{\infty} a_m (z - z_0)^m \\ &= \dots + \frac{a_{-2}}{(z - z_0)^2} + \frac{a_{-1}}{z - z_0} + a_0 + a_1(z - z_0) + a_2(z - z_0)^2 + \dots \end{aligned}\right\} \quad (1)$$

$$a_m = \frac{1}{2\pi i} \oint_C (z - z_0)^{-m-1} f(z) dz \quad (2)$$

where C is a circle $| z - z_0 | = a$ with $a < r$ traversed in the positive direction. When $f(z)$ is analytic at z_0, all the a_m with negative subscripts are zero, and the expansion reduces to Taylor's series, § 3.25, with $a_n = (1/n!)f^{(n)}(z)$.

When $f(z)$ has a pole of order n at z_0, § 20.2, the a_m are zero for $m < -n$. In this case the residue a_{-1} may be found from

$$a_{-1} = \frac{1}{(n-1)!} \frac{d^{n-1}}{dz^{n-1}} [(z-z_0)^n f(z)] \Big|_{z=z_0} \tag{3}$$

In particular for a pole of the first order, $n = 1$, the residue a_{-1} may be found from

$$a_{-1} = \lim_{z \to z_0} [(z-z_0)f(z)] \tag{4}$$

For an isolated essential singularity, § 20.2, there are an infinite number of negative powers in the expansion, and we still call a_{-1} the residue.

20.5. Laurent expansion about infinity. Similarly let $f(z)$ be analytic for $|z| > r$. Then for z in this range,

$$f(z) = \sum_{m=-\infty}^{\infty} a_m z^m, \quad a_m = \frac{1}{2\pi i} \oint z^{-m-1} f(z) dz \tag{1}$$

where C is a circle $|z| = a$ with $a > r$ traversed in the positive direction. In this case for $z = \infty$, $f(z)$ is analytic if there are no positive powers, $f(z)$ has a pole of order n if all the a_m are zero for $m > n$, and $f(z)$ has an isolated essential singularity if there are an infinite number of positive powers.

20.6. Residues. About a circle C, $|z - z_0| = a$ as in 20.4,

$$\oint_C (z-z_0)^m dz = 0 \tag{1}$$

when $m = 0, 1, \pm 2, \pm 3, \ldots$, except for $m = 1$, when

$$\oint_C \frac{dz}{z - z_0} = 2\pi i \tag{2}$$

Termwise integration of the Laurent series of § 20.5 over C gives $\oint_C f(z)\, dz = 2\pi i a_{-1}$. And about any closed contour passing through regular points only, but enclosing singular points z_k, $\oint f(z)\, dz = 2\pi i[\Sigma R(z_k)]$, where $R(z_k)$ is the residue a_{-1} or coefficient of $1/(z - z_k)$ in the Laurent expansion appropriate to z_k. For poles the residues may be found as indicated in § 20.4.

For a number of examples of the use of this residue theorem to compute real integrals see Franklin, *Methods of Advanced Calculus*, Chap 5.

21. Integral Equations

21.1. Fredholm integral equations. Fredholm's integral equation (F) is

$$U(x) = f(x) + \lambda \int_a^b K(x,t)U(t)dt \tag{1}$$

The kernel $K(x,t)$ and $f(x)$ are given functions, and $U(x)$ is to be found. We assume that

$$\int_a^b \int_a^b [K(x,t)]^2 \, dx \, dt = W$$

is finite. With (F) we associate the homogeneous equation (F_h), which is

$$U(x) = \lambda \int_a^b K(x,t)U(t)dt$$

and the transposed equation (F_t), which is

$$\tilde{U}(x) = f(x) + \lambda \int_a^b K(t,x)\tilde{U}(t)dt.$$

The homogeneous transposed equation (F_{ht}) is

$$\tilde{U}(x) = \int_a^b K(t,x)\tilde{U}(t)dt.$$

The number ρ of linearly independent solutions $u_1(x)$, $u_2(x)$, ..., $u_\varrho(x)$ of (F_h) is finite and equal to the number of linearly independent solutions $u_1(x)$, $u_2(x)$, ..., $u_\varrho(x)$ of (F_{ht}). The number ρ is the defect of the kernel $K(x,t)$ for the value λ. The general solution of (F_h) is

$$c_1u_1(x) + c_2u_2(x) + \ldots + c_\varrho u_\varrho(x).$$

If $\rho = 0$, then (F) and (F_t) each has a unique solution for any $f(x)$. There is a solving kernel, or resolvent, $\Gamma(x,t;\lambda)$ such that

$$U(x) = f(x) + \lambda \int_a^b \Gamma(x,t;\lambda)f(t)dt \quad \text{for } (F),$$

and

$$\tilde{U}(x) = f(x) + \lambda \int_a^b \Gamma(t,x;\lambda)f(t)dt \quad \text{for } (F_t).$$

Fredholm's form of the resolvent is

$$\Gamma(x,t;\lambda) = \frac{D(x,t;\lambda)}{D(\lambda)} \tag{2}$$

where

$$D(x,t;\lambda) = K(x,t) + \sum_{n=1}^{\infty} (-1)^n D_n(x,t) \frac{\lambda^n}{n!}$$

and

$$D(\lambda) = \sum_{n=0}^{\infty} (-1)^n D_n \frac{\lambda^n}{n!}$$

The D's are found in succession from

$$D_m = \int_a^b D_{m-1}(x,x)dx, \quad D_m(x,t) = D_m K(x,t) - m \int_a^b K(x,s)D_{m-1}(s,t)ds.$$

These series converge for all λ.

For sufficiently small λ, we have the Neumann series

$$\Gamma(x,t;\lambda) = K(x,t) + \lambda K_2(x,t) + \lambda^2 K_3(x,t) + \dots + \lambda^n K_{n+1}(x,t) + \dots \tag{3}$$

The iterated kernels $K_m(x,t)$ are found in succession from

$$K_1(x,t) = K(x,t) \quad \text{and} \quad K_m(x,t) = \int_a^b K(x,s)K_{m-1}(s,t)ds \tag{4}$$

If $\rho > 0$, then (F) has a solution only if

$$\int_a^b u_i(t)f(t)dt = 0$$

for $i = 1, 2, \dots, \rho$. The general solution of (F) is one solution plus the general solution of (F_h).

21.2. Symmetric kernel. In § 21.1 let $K(x,t) = K(t,x)$. Then (F_h) has nonzero solutions for certain determined discrete values $\lambda_1, \lambda_2, \dots, \lambda_n, \dots$, called the eigenvalues, for which the defect $\rho > 0$. The corresponding solutions $u_i(x)$ of (F_h) are the eigenfunctions. For any λ, $\rho \leqq \lambda^2 W$. For W see § 21.1.

Every symmetric kernel, not identically zero, has at least one and at most a denumerable infinity of eigenvalues, with no finite limit point. These eigenvalues are all real if $K(x,t)$ is real and symmetric. At most A^2W eigenvalues have numerical values not exceeding A, and $\lambda_n^2 \geqq 1/W$. Also

$$\sum_{n=1}^{\infty} \frac{1}{\lambda_n^2} \leqq W$$

where each eigenvalue of defect ρ is counted ρ times.

The ρ functions for a λ of defect ρ may be taken as of norm, the N_n of § 14.19, unity, and orthogonal to one another. Then if the functions for a λ

with $\rho = 1$ are taken as of norm 1, the totality of eigenvalues will form a real normal and orthogonal set as in § 14.19. Thus

$$\int_a^b [u_n(x)]^2\, dx = 1, \quad \int_a^b u_n(x)u_m(x)dx = 0, \quad m \neq n.$$

The bilinear series

$$K(x,t) = \sum_{n=1}^{\infty} \frac{u_n(x)u_n(t)}{\lambda_n} \tag{5}$$

converges for a Mercer kernel with only a finite number of positive, or only a finite number of negative eigenvalues. Otherwise it converges in the mean,

$$\lim_{n\to\infty} \int (K - S_n)^2 dx = 0$$

where S_n is the sum to n terms.

If (F) has a solution, it is given by the Schmidt series

$$U(x) = f(x) + \lambda \sum_{n=1}^{\infty} \frac{f_n}{\lambda_n - \lambda} u_n(x) \tag{6}$$

where

$$f_n = \int_a^b f(t)u_n(t)dt.$$

Any function which is sourcewise representable,

$$F(x) = \int_a^b K(x,t)\phi(t)dt \tag{7}$$

has an absolutely and uniformly convergent series development

$$F(x) = \sum_{n=1}^{\infty} F_n u_n(x) \quad \text{where} \quad F_n = \int_a^b F(t)u_n(t)dt. \tag{8}$$

21.3. Volterra integral equations. Volterra's integral equation

$$(V) \quad \text{is} \quad U(x) = f(x) + \lambda \int_0^x K(x,t)U(t)dt \tag{1}$$

This always has a solution

$$U(x) = f(x) + \lambda \int_0^x \Gamma(x,t;\lambda)f(t)dt \tag{2}$$

where the resolvent

$$\Gamma(x,t;\lambda) = K(x,t) + \lambda K_2(x,t) + \lambda^2 K_3(x,t) + \ldots + \lambda^n K_{n+1}(x,t) + \ldots \tag{3}$$

The iterated kernels $K_m(x,t)$ are found in succession from

$$K_1(x,t) = K(x,t), \quad K_m(x,t) = \int_0^x K(x,s)K_{m-1}(s,t)ds \tag{4}$$

This Neumann series converges for all values of x.

If $H(x,x) \neq 0$, we may reduce the solution of the equation of the first kind,

$$g(x) = \lambda \int_0^x H(x,t)U(t)dt \tag{5}$$

to the solution of an equation of type (V). Differentiation with respect to x leads to

$$g'(x) = \lambda \int_0^x \frac{\partial H}{\partial x} U(t)dt + \lambda H(x,x)U(x) \tag{6}$$

This is of type (V) with

$$f(x) = \frac{g'(x)}{\lambda H(x,x)}, \quad K(x,t) = \frac{-1}{\lambda H(x,x)} \frac{\partial H}{\partial x} \tag{7}$$

21.4. The Abel integral equation. Abel's equation is

$$g(x) = \int_a^x \frac{U(t)}{(x-t)^q} dt, \quad \text{with} \quad 0 < q < 1 \quad \text{and} \quad g(a) = 0 \tag{1}$$

For this kernel, singular at $t = x$, the unique continuous solution is

$$U(t) = \frac{\sin q\pi}{\pi} \frac{d}{dt} \int_a^t \left[\frac{f(x)}{(t-x)^{1-q}} dx \right] \tag{2}$$

21.5. Green's function. Let $L(y) = E(x)$ be the general linear differential equation of the nth order (4) of § 5.13. The solution satisfying given boundary conditions may often be written as

$$y(x) = \int_a^b G(x,t)E(t)dt \tag{1}$$

where $G(x,t)$ is the Green's function for the given equation and boundary conditions. This function satisfies $L(G) = 0$ except at $x = t$. At $x = t$, $G(x,t)$ as a function of x is continuous, together with its first $(n-2)$ derivatives, but

$$\left. \frac{d^{n-1}G}{dx^{n-1}} \right|_{x=t-}^{x=t+} = -\frac{1}{A_n}$$

where, as in (4) of § 5.13, A_n is the coefficient of d^ny/dx^n in $L(y)$. The Green's function $G(x,t)$ must also satisfy the same given boundary conditions as were imposed on $y(x)$.

The solution of the differential equation $L(y) + \lambda w(x)y = F(x)$ satisfying the given boundary conditions satisfies the Fredholm equation (F) of § 21.1

$$y(x) = \lambda \int_a^b G(x,t)w(t)y(t)dt + \int_a^b G(x,t)F(t)dt \tag{2}$$

If $[vL(u) - uL(v)]dx$ is an exact differential, $L(y)$ is self-adjoint, and the Green's function, if it exists, is necessarily symmetric, so that $G(x,t) = G(t,x)$. In this case let

$$U(x) = y(x)\sqrt{w(x)}, \quad f(x) = \sqrt{w(x)} \int_a^b G(x,t)F(t)dt$$

Then

$$U(x) = \lambda \int_a^b G(x,t)\sqrt{w(x)w(t)}\,U(t)dt + f(x) \tag{3}$$

which is an integral equation with symmetric kernel like (F) of § 21.2 with $K(x,t) = G(x,t)\sqrt{w(x)w(t)}$.

All the conclusions of § 21.2 apply, and in particular any n times differentiable function satisfying the boundary conditions can be developed in a series in terms of the $u_n(x)$. Hence there are an infinite number of eigenvalues and the eigenfunctions form a complete set.

21.6. The Sturm - Liouville differential equation. If $L(u) = (pu')' - qu$, then

$$[vL(u) - uL(v)]dx = d[p(vu' - uv')] \tag{1}$$

Thus the Sturm-Liouville equation

$$L(u) + \lambda wu = 0, \quad \text{or} \quad (pu')' - qu + \lambda wu = 0 \tag{2}$$

is self-adjoint, and the theory of §§ 21.5 and 21.2 applies to it.

We describe a number of important particular examples. We write $p(x)$ and $q(x)$ in place of p and q when p and q are used as parameters with other meanings.

a. Let the boundary conditions be $u(-1)$ and $u(1)$ finite. With $p = 1 - x^2$, $q = 0$, $w = 1$, we have the Legendre equation of § 8.1 with eigenvalues $n(n + 1)$ and eigenfunctions the polynomials $P_n(x)$ of § 8.2. With

$$p = 1 - x^2, \quad q = \frac{m^2}{1 - x^2}, \quad w = 1 \tag{3}$$

we have the associated Legendre equation of § 8.10 with eigenvalues $n(n + 1)$, $n \geqq m$, and eigenfunctions the polynomials $P_n{}^m(x)$ of § 8.11. With

$$p(x) = x^q(1 - x)^{p-q}, \quad q(x) = 0, \quad w(x) = x^{q-1}(1 - x)^{p-q} \tag{4}$$

we have the Jacobi equation of § 10.7 with eigenvalues $n(p+n)$ and eigenfunctions the polynomials $J_n(p,q;x)$. With

$$p = (1-x^2)^{1/2}, \quad q = 0, \quad w = (1-x^2)^{-1/2} \tag{5}$$

we have the Tschebycheff equation of § 10.7 with eigenvalues n^2 and eigenfunctions the polynomials $T_n(x)$.

b. Let the boundary conditions be $u(0)$ finite, and appropriate behavior at plus infinity. With

$$p = xe^{-x}, \quad q = 0, \quad w = e^{-x} \tag{6}$$

we have the Laguerre equation of § 11.1 with eigenvalues n and eigenfunctions the polynomials $L_n(x)$. With

$$p = x, \quad q = \frac{x}{4} - \frac{1}{2}, \quad w = 1 \tag{7}$$

we have the Laguerre equation of § 11.4 with eigenvalues n and eigenfunctions $e^{-x/2}L_n(x)$. With

$$p = x^{s-1}e^{-x}, \quad q = 0, \quad w = x^s e^{-x} \tag{8}$$

we have the associated Laguerre equation of § 11.5 with eigenvalues $r-s$ and eigenfunctions the polynomials $L_r^{\,s}$.

c. Let the boundary conditions be appropriate behavior at plus infinity and at minus infinity. With

$$p = e^{-x^2}, \quad q = 0, \quad w = e^{-x^2} \tag{9}$$

we have the Hermite equation of § 12.1 with eigenvalues $2n$ and eigenfunctions the polynomials $H_n(x)$. With

$$p = 1, \quad q = x^2 - 1, \quad w = 1 \tag{10}$$

we have the Hermite equation of § 12.4 with eigenvalues $2n$ and eigenfunctions $e^{-x^2/2}H_n(x)$.

d. Let the boundary conditions be $u(0)$ finite, $u(1) = 0$. With

$$p = x, \quad q = \frac{n^2}{x}, \quad w = x, \quad \lambda = a^2 \tag{11}$$

we have an equivalent of the modified Bessel equation

$$x^2u'' + xu' + (-n^2 + a^2x^2)u = 0,$$

which is satisfied by $Z_n(ax)$ by §§ 9.5 and 9.6. The eigenvalues are $a_{nr}^{\,2}$, where $J_n(a_{nr}) = 0$, and the eigenfunctions are the $J_n(a_{nr}x)$, § 9.2. Here n is fixed, and the eigenvalues and eigenfunctions correspond to $r = 1, 2, 3, \ldots$. For

$$p = x^c(1-x)^{a+b+1-c}, \quad q = 0, \quad w = x^{c-1}(1-x)^{a+b-c}, \quad \lambda = -ab \tag{12}$$

the Sturm-Liouville equation reduces to the hypergeometric equation of § 10.1.

21.7. Examples of Green's function. In each of the following examples, the Green's function $G(x,t) = G_1(x,t)$ for $x \leqq t$ and $G(x,t) = G_1(t,x)$ for $x \geqq t$. Here G_{1m} means a modified Green's function.

a. Let the boundary conditions be $u(0) = 0$, $u(1) = 0$.

For $$L(u) = u'', \quad G_1(x,t) = (1-t)x \tag{1}$$

For $$L(u) = u'' + k^2u, \quad G_1(x,t) = \frac{\sin kx \sin k(1-t)}{k \sin k} \tag{2}$$

For $$L(u) = u'' - k^2u, \quad G_1(x,t) = \frac{\sinh kx \sinh k(1-t)}{k \sinh k} \tag{3}$$

b. Let the boundary conditions be $u(-1) = u(1)$, $u'(-1) = u'(1)$.

For $$L(u) = u'', \quad G_{1m}(x,t) = \frac{1}{4}(x-t)^2 + \frac{1}{2}(x-t) + \frac{1}{6} \tag{4}$$

For $$L(u) = u'' + k^2u, \quad G_1(x,t) = -\frac{\cos k(x-t+1)}{2k \sin k} \tag{5}$$

For $$L(u) = u'' - k^2u, \quad G_1(x,t) = \frac{\cosh k(x-t+1)}{2k \sinh k} \tag{6}$$

c. Let the boundary conditions be $u(0) = 0$, $u'(1) = 0$.

For $$L(u) = u'', \quad G_1(x,t) = x \tag{7}$$

For $$L(u) = u'' + k^2u, \quad G_1(x,t) = \frac{\sin kx \cos k(1-t)}{k \cos k} \tag{8}$$

For $$L(u) = u'' - k^2u, \quad G_1(x,t) = \frac{\sinh kx \cosh k(1-t)}{k \cosh k} \tag{9}$$

d. Let the boundary conditions be $u(-1) = 0$, $u(1) = 0$.

For $$L(u) = u'', \quad G_1(x,t) = \frac{1}{2}(1+x-t-xt) \tag{10}$$

e. Let the boundary conditions be

$$u(0) = -u(1), \quad u'(0) = -u'(1). \tag{11}$$

For $\quad L(u) = u'', \quad G_1(x,t) = \frac{1}{4} + \frac{x-t}{2}$ $\hfill (12)$

f. Let the boundary conditions be $u(+\infty)$ and $u(-\infty)$ finite.

For $\quad L(u) = u'' - u, \quad G_1(x,t) = \frac{1}{2} e^{t-x}$ $\hfill (13)$

g. Let the boundary conditions be $u(0) = 0$, $u'(0) = 0$, $u(1) = 0$, $u'(1) = 0$.

For $\quad L(u) = u^{iv}, \quad G_1(x,t) = \frac{1}{6} x^2(1-t)^2(2xt + x - 3t)$ $\hfill (14)$

h. Let the boundary conditions be $u(-1)$ and $u(1)$ finite.

For $\quad L(u) = [(1-x^2)u']'$, $\hfill (15)$

$$G_1(x,t) = -\frac{1}{2} \ln (1-x)(1-t) + \ln 2 - \frac{1}{2} \tag{16}$$

With this $L(u)$, Legendre's equation of § 8.1 is

$$L(y) + n(n+1)y = 0 \tag{17}$$

$$\left.\begin{aligned} &\text{For} \quad L(u) = [(1-x^2)u']' - \frac{m^2 u}{1-x^2}, \quad m \neq 0, \\ &G_1(x,t) = \frac{1}{2m} \left[\frac{(1+x)(1-t)}{(1-x)(1+t)}\right]^{m/2} \end{aligned}\right\} \tag{18}$$

With this $L(u)$, the associated Legendre equation of § 8.10 is

$$L(y) + n(n+1)y = 0 \tag{19}$$

i. Let the boundary, conditions be $u(0)$ finite, $u(1) = 0$.

For $\quad L(u) = (xu')', \quad G_1(x,t) = -\ln x$ $\hfill (20)$

With this $L(u)$, Bessel's equation (3) of § 9.1 for $n = 0$ is

$$L(y) + xy = 0 \tag{21}$$

For $\quad L(u) = (xu')' - \frac{n^2}{x} u, \quad n \neq 0, \quad G_1(x,t) = \frac{1}{n}\left[\left(\frac{x}{t}\right)^n - (xt)\right]^n$ $\hfill (22)$

With this $L(u)$, Bessel's equation (3) of § 9.1 is

$$L(y) + xy = 0 \tag{23}$$

Bibliography

1. Algebra

FINE, H. B., *College Algebra*, Ginn & Company, Boston, 1905.

RICHARDSON, M., *College Algebra*, Prentice-Hall, Inc., New York, 1947.

2. Trigonometry and space geometry

BRENKE, W. C., *Plane and Spherical Trigonometry*, The Dryden Press, Inc., New York, 1943.

FINE, H. B. and THOMPSON, H. D., *Coordinate Geometry*, The Macmillan Company, New York, 1918.

3. and 4. Calculus

DE HAAN, B., *Nouvelles tables d'intégrales définies*, P. Engels, Leyden, 1867.

DE HAAN, B., *Tables d'intégrales définies*, P. Engels, Leyden, 1858-1864.

DWIGHT, H. B., *Tables of Integrals and Other Mathematical Data*, The Macmillan Company, New York, 1947.

FINE, H. B., *Calculus*, The Macmillan Company, New York, 1937.

FRANKLIN, P., *Differential and Integral Calculus*, McGraw-Hill Book Company, Inc., New York, 1953.

FRANKLIN, P., *Methods of Advanced Calculus*, McGraw-Hill Book Company, Inc., New York, 1944.

PIERCE, B. O., *A Short Table of Integrals*, Ginn & Company, Boston, 1929.

WIDDER, D. V., *Advanced Calculus*, Prentice-Hall, Inc., New York, 1947.

5. Differential equations

FORSYTH, A. R., *Treatise on Differential Equations*, Macmillan and Co., Ltd., London, 1885. (Dover reprint)

INCE, E. L., *Ordinary Differential Equations*, Longmans, Green & Co., London, 1927. (Dover reprint)

MORRIS, M. and BROWN, O. E., *Differential Equations*, 3d ed., Prentice-Hall, Inc., New York, 1952.

6. Vector analysis

PHILLIPS, H. B., *Vector Analysis*, John Wiley & Sons, Inc., New York, 1933.

7. Tensors

BRAND, L., *Vector and Tensor Analysis*, John Wiley & Sons, Inc., New York, 1947.

EDDINGTON, A. S., *The Mathematical Theory of Relativity*, Cambridge University Press, London, 1930.

EISENHART, L. P., *Riemannian Geometry*, rev. ed., Princeton University Press, Princeton, 1949.

KRON, G., *Short Course in Tensor Analysis for Electrical Engineers*, John Wiley & Sons, Inc., New York, 1942. (Dover reprint)

MCCONNELL, A. J., *Applications of the Absolute Differential Calculus*, Blackie & Son, Ltd., London, 1931. (Dover reprint)

WILLS, A. P., *Vector and Tensor Analysis*, Prentice-Hall, Inc., New York, 1938. (Dover reprint)

8. to 15. Special functions

CARSLAW, H. S., *Theory of Fourier Series and Integrals*, Macmillan and Co., Ltd., London, 1921. (Dover reprint)

COURANT, R. and HILBERT, D., *Methoden der mathematischen Physik*, Vol. 1, Julius Springer, Berlin, 1931. (An English translation, *Methods of Mathematical Physics*, is available from Interscience Publishers, Inc., New York, 1943.

FRANKLIN, P., *Fourier Methods*, McGraw-Hill Book Company, Inc., New York, 1949. (Dover reprint)

GRAY, A., MATHEWS, G. B. and MACROBERT, T. M., *A Treatise on Bessel Functions*, Macmillan and Co., Ltd., London, 1931.

HOBSON, E. W., *The Theory of Spherical and Ellipsoidal Harmonics*, Cambridge University Press, London, 1931.

KNOPP, K., *Theory and Application of Infinite Series*, trans. Young, R. C., Blackie & Son, Ltd., London, 1928.

McLACHLAN, N. W., *Bessel Functions for Engineers*, 2d ed., Oxford University Press, Oxford, 1955.

MARGENAU, H. and MURPHY, G. M., *The Mathematics of Physics and Chemistry*, D. Van Nostrand Company, Inc., New York, 1943.

WATSON, G. N., *Theory of Bessel Functions*, Cambridge University Press, London, 1944.

WHITTAKER, E. T. and WATSON, G. N., *A Course of Modern Analysis*, Cambridge University Press, London, 1927.

18. Matrices

BÔCHER, M., *Introduction to Higher Algebra*, The Macmillan Company, New York, 1907.

DICKSON, L. E., *Modern Algebraic Theories*, Benjamin H. Sanborn and Co., Chicago, 1926. (Dover reprint)

FRAZER, R. A., DUNCAN, W. J. and COLLAR, A. R., *Elementary Matrices*, Cambridge University Press, London, 1938.

TURNBULL, H. W. and AITKEN, A. C., *The Theory of Canonical Matrices*, Blackie & Son, Ltd., London, 1932.

19. Groups

SPEISER, A., *Theorie der Gruppen von endlicher Ordnung*, Julius Springer, Berlin, 1927.

WEYL, H., *The Classical Groups*, Princeton University Press, Princeton, 1939.

WEYL, H., *Theory of Groups and Quantum Mechanics*, Methuen & Co., Ltd., London, 1931. (Dover reprint)

WIGNER, E., *Gruppentheorie und ihre Anwendung auf Quantenmechanik der Atomspektren*, F. Vieweg und Sohn, Braunschweig, 1931.

20. Analytic functions

HURWITZ, A. and COURANT, R., *Vorlesungen über allgemeine Funktionentheorie und elliptische Funktionen*, Julius Springer, Berlin, 1925.

PIERPONT, J., *Functions of a Complex Variable*, Ginn & Company, Boston, 1914. (Dover reprint)

21. Integral Equations

FRANK, P. and VON MISES, R., *Differential- und Integralgleichungen der Mechanik und Physik*, M. S. Rosenberg, New York, 1943. (Dover reprint)

Collections of formulas

ADAMS, E. P. and HIPPISLEY, R. L., *Smithsonian Mathematical Formulae and Tables of Elliptic Functions*, Smithsonian Institution, Washington, 1922.

JAHNKE, E. and EMDE, F., *Funktionentafeln mit Formeln und Kurven*, B. G. Teubner, Leipzig, 1938. (Dover reprint)

MADELUNG, E., *Die mathematischen Hilfsmittel des Physikers*, Julius Springer, Berlin, 1936.

Chapter 2

STATISTICS

By Joseph M. Cameron

Statistical Engineering Laboratory
National Bureau of Standards

This chapter on statistics was added, at the special request of numerous physicists, to elaborate some of the formulas of mathematical statistics as given in abbreviated form in Chapter 1. Experimental physicists have become increasingly aware of the importance of assigning to their measures some parameter such as mean error, probable error, or limit of error to represent the precision of their measurements and to serve as a yardstick for judging differences. Nevertheless, the literature of physics abounds with examples where the scientist has failed to take full advantage of modern statistical methods. This chapter represents an attempt to give the major formulas of mathematical statistics in a form directly useful for the research student.

1. Introduction

1.1. Characteristics of a measurement process. A sequence of measurements on the same subject derived from the same measurement process (i.e. with the environmental conditions and procedures maintained throughout the sequence) tends to cluster about some central value for the sequence. If, as the number of measurements is increased indefinitely, the average for the sequence approaches, in the probability sense, a value μ, then μ is the limiting mean of the sequence.

An individual value x_i differs from the limiting mean μ because of uncontrolled or random fluctuations in the environmental conditions or procedures. These individual values will follow some frequency distribution about the limiting mean. The specification of this distribution is made in terms of a mathematical frequency law involving several parameters including the limiting mean μ, a measure of the dispersion about the limiting mean, and other parameters as necessary.

The dispersion of values about the mean is measured in terms of the square root of the second moment about the mean, called the standard deviation, σ. This standard deviation is the limit, in the probability sense, of the square root of the average value of the squared deviation of the individual values from the mean as the number of values is increased indefinitely.

Formulas for estimating the parameters of a distribution from a finite set of data, and methods of making statistical tests of hypothesis concerning these parameters are presented in this chapter.

1.2. Statistical estimation. The experimenter has at hand only a finite set of measurements which are considered a subset of an infinite sequence. He will, therefore, be in ignorance of the *parameters* of the process such as the limiting mean μ, the standard deviation σ, and any other parameters that may exist. The finite set of data is to be used to estimate the unknown parameters. For example, the average or the median may be used to estimate the limiting mean. The function of the data used for estimation is called an *estimator*; the numerical value obtained by using this estimator on a set of data is called an *estimate*.

Estimators of parameters may be either point estimators, such as median or average, or may be interval estimators, such as confidence limits.

A sequence of estimates, generated by use of an estimator on small sets of values randomly selected from some distribution, will have a limiting mean, a standard deviation, and follow some distribution. Just as one technique of measurement may show superiority over another, so also may one method of estimation (i.e., one estimator) show superiority over another in terms of bias and precision.

The exact form of the underlying distribution of the basic measurements cannot be settled, and upon the appropriate choice of the form of the distribution depends the correctness of the inferences drawn from the data.

1.3. Notation. Unless otherwise stated in the formulas that follow:

Individual measurements are indicated by

$$x_1 \quad x_2 \quad x_3 \quad \ldots \quad x_i \quad \ldots \quad x_n$$

where the subscript refers to the order in which the measurements were made.

Each x_i is assumed to be independent in the statistical sense from all other measurements in the set, and all x_i are assumed to follow the same distribution.

Measurements ranked in order of magnitude are indicated by

$$x_{(1)}, \quad x_{(2)}, \quad \ldots \quad x_{(i)} \quad \ldots \quad x_{(n)}$$

such that
$$x_{(1)} \leq x_{(2)} \leq \ldots \leq x_{(n)}$$

Parameters of distributions are indicated by Greek letters, μ, σ, β, etc., while estimates of these parameters are indicated by italic letters $\bar{x}$, s, b, etc.

2. Standard Distributions

2.1. The normal distribution.* A variate that follows the frequency function

$$f(x) = \frac{1}{\sigma\sqrt{2\pi}} e^{-(x-\mu)^2/2\sigma^2} \qquad -\infty < x < \infty \tag{1}$$

is said to be normally distributed with

$$\text{mean value of } x = \mu \qquad \text{variance of } x = \sigma^2$$

The normal distribution is quite often a fairly good approximation to the distribution of random fluctuations in physical phenomena. This general applicability of the normal distribution is verified by experience and supported in part by theory.

By virtue of the *central limit theorem*, which states that for certain conditions the sum of a large number of independent random variables is asymptotically normally distributed, the distribution of the random fluctuations in a measurement process tends toward normality because the variations are due to a multitude of minor random variations in the process.†

2.2. Additive property. If $x_1, x_2, \ldots, x_n$ are normally and independently distributed with means $\mu_1, \mu_2, \ldots, \mu_n$, and variances $\sigma_1^2, \sigma_2^2, \ldots, \sigma_n^2$, then

$$c_1x_1 + c_2x_2 + \ldots + c_nx_n$$

where the c_i are arbitrary constants and will be normally distributed with

$$\text{mean } \sum_{i=1}^{n} c_i\mu_i \quad \text{and} \quad \text{variance } \sum_{i=1}^{n} c_i^2\sigma_i^2$$

* Tables of the standardized normal distribution

$$\frac{1}{\sqrt{2\pi}} \int_{-x}^{x} e^{-t^2/2} dt$$

to 15 decimals are given in *Tables of Normal Probability Functions*, National Bureau of Standards, Applied Mathematics Series, No. 23, 1953.

† Cramer, H., *Mathematical Methods of Statistics*, Princeton University Press, 1946, pp. 213-220; and *Random Variables and Probability Distributions*, Cambridge Tracts in Mathematics, No. 36, 1937, p. 113.

2.3. The logarithmic-normal distribution. If $y = \ln x$ is normally distributed with mean μ, and standard deviation σ, then x will have mean ξ

$$\xi = e^{\mu+\sigma^2/2} \tag{1}$$

and variance σ_x^2.

$$\sigma_x^2 = e^{2\mu+\sigma^2}(e^{\sigma^2} - 1) = \xi^2(e^{\sigma^2} - 1) \tag{2}$$

Note: The use of the average of the x's as an estimate of ξ is inefficient when the variance of the y's is greater than 0.7. The use of s^2 as an estimate of σ_x^2 is inefficient when the variance of the y's is greater than 0.1. For efficient estimators see FINNEY, D. J., *Supplement to Journal of the Royal Statistical Society*, Vol. 7, No. 2, 1941, pp. 155-161.

2.4. Rectangular distribution. If the probability that x lies in the interval $x_0 + dx$ is the same for all values of x_0 and x is restricted to the interval (α, β), then x follows the rectangular distribution with frequency function

$$f(x) = \frac{1}{\beta - \alpha} \qquad (\alpha < x < \beta) \tag{1}$$

$$\text{mean value of } x = \frac{\beta + \alpha}{2}, \qquad \text{variance of } x = \frac{(\beta - \alpha)^2}{12}$$

2.5. The χ^2 distribution. The sum of squares of n independent normal variates having mean zero and variance unity follow the χ^2 distribution for n degrees of freedom. The χ^2 distribution enters in statistical tests of goodness of fit, homogeneity, and a variety of other purposes because the standardized variable

$$\frac{\text{estimate of mean} - \text{mean}}{\text{standard deviation of estimate of mean}}$$

tends to be asymptotically normal with mean zero and variance unity. The χ^2 distribution for n degrees of freedom has frequency function

$$f(\chi^2) = \frac{1}{2^{n/2}\Gamma(n/2)} e^{-\chi^2/2}(\chi^2)^{(n-2)/2} \qquad (0 < \chi^2 < \infty) \tag{1}$$

$$\text{mean of } \chi^2 = n, \qquad \text{variance of } \chi^2 = 2n$$

The χ^2 distribution has the *additive property* that the sum of k independent χ^2 variates based on $n_1, n_2, n_3, \ldots, n_k$ degrees of freedom respectively follows a χ^2 distribution for

$$\sum_{i=1}^{k} n_i$$

degrees of freedom.

A table of values of χ^2 exceeded with probability P is given in most texts on statistics.

2.6. Student's *t*-distribution. In sets of size $(n + 1)$ from a normal distribution the ratio of the average to standard deviation ($t = \bar{x}/s$) follows Student's t-distribution with n degrees of freedom. Frequency function for the t-distribution is

$$f(t) = \frac{\Gamma[(n+1)/2]}{\Gamma(n/2)} \cdot \frac{1}{\sqrt{n\pi}} \cdot \frac{1}{(1 + t^2/n)^{(n+1)/2}} \qquad (-\infty < t < \infty) \qquad (1)$$

$$\text{mean value of } t = 0 \qquad \text{variance of } t = \frac{n}{n-2} \qquad (n > 2)$$

Tables of the values of t exceeded in absolute value with probability α are given in most modern texts on statistics, i.e., t_0 for which

$$\int_{-t_0}^{t_0} f(t)dt = 1 - \alpha$$

2.7. The *F* distribution. If s_1^2 and s_2^2 based on m and n degrees of freedom, respectively, are two independent estimates of the same σ^2 for a normal distribution, $F = s_1^2/s_2^2$ will follow the F distribution for (m,n) degrees of freedom (the number of degrees of freedom for the numerator is always quoted first).

The frequency function of F is

$$(F) = \frac{\Gamma[(m+n)/2]}{\Gamma(m/2)\Gamma(n/2)} \left(\frac{m}{n}\right)^{m/2} \frac{F^{(m-2)/2}}{[1 + (m/n)F]^{(m+n)/2}} \qquad (0 < F < \infty) \qquad (1)$$

$$\text{mean of } F = \frac{n}{n-2} \qquad (n > 2)$$

$$\text{variance of } F = \left(\frac{n}{n-2}\right)^2 \left[\frac{2(m+n-2)}{m(n-4)}\right] \qquad (n > 4)$$

Tables of values of F exceeded with probability 0.05 and 0.01 are given in G. W. Snedecor, *Statistical Methods*, Iowa University Press, 4th edition (1946), and in a number of recent texts on statistics.

2.8. Binomial distribution. If on any single trial, the probability of occurrence of an event is P and the probability of nonoccurrence of the event is $1 - P$, the probability of $k = 0, 1, \ldots, n$ occurrences in n trials is given by the successive terms in the expansion of

$$[(1 - P) + P]^n = (1 - P)^n + n(1 - P)^{n-1}P + \ldots \qquad (1)$$

The average number of occurrences in n trials will be

$$nP$$

and the standard deviation of the number of occurrences will be

$$\sqrt{nP(1-P)}$$

Tables of individual terms and cumulative terms for $P = .01(01).50$ and $n = 1(1)49$ are given in *Tables of the Binomial Probability Distribution*, National Bureau of Standards Applied Mathematics Series, No. 6, 1950.

2.9. Poisson distribution. If the probability of the occurrence of an event in an interval (of time, space, etc.) of length dx is proportional to the length of the interval, i.e., probability of occurrence is equal to, say, $\lambda\, dx$, the probability of k independent occurrences in an interval of length x is given by (for $\mu = \lambda x$)

$$\frac{e^{-\mu}\mu^k}{k!}$$

The average number of occurrences is μ; the standard deviation of the number of occurrences is $\sqrt{\mu}$. Additive property : If $k_1, k_2, \ldots, k_n$ independently follow Poisson distributions with means $\mu_1, \mu_2, \ldots, \mu_n$, then

$$\sum_{i=1}^{n} k_i$$

also follows a Poisson distribution, with mean $\sum_{i=1}^{n} \mu_i$.

Note : For P small (less than 0.1) the Poisson may be used as an approximation to the binomial taking $\mu = nP$.

Tables of individual terms and cumulative sums are given in MOLINA, E. C., *Poissons' Exponential Binomial Limit*, D. Van Nostrand Company, Inc., 1947.

3. Estimators of the Limiting Mean

3.1. The average or arithmetic mean. Symbol $\bar{x}$.

$$\bar{x} = \frac{1}{n}\sum_{i=1}^{n} x_i$$

If the x_i have limiting mean μ and standard deviation σ,

$$\text{limiting mean of } \bar{x} = \mu$$

$$\text{s.d. of } \bar{x} = \sigma/\sqrt{n}$$

3.2. The weighted average. If each x_i has assigned to it a weight of w_i, the weighted mean is defined as

$$\frac{\sum_{i=1}^{n} w_i x_i}{\sum_{i=1}^{n} w_i} \tag{1}$$

If the x_i have same limiting mean μ and standard deviation σ,

$$\text{limiting mean of weighted average} = \mu$$

$$\text{s.d. of weighted mean} = \sigma \frac{\sqrt{\sum_{i=1}^{n} w_i^2}}{\sum_{i=1}^{n} w_i}$$

3.3. The median. The median is that value that divides the set of values $x_{(i)}$ into two equal halves. The median of n values is

$$x_{(n+1)/2} \quad \text{for } n \text{ odd}$$

$$\tfrac{1}{2}(x_{(n/2)} + x_{(n/2+1)}) \quad \text{for } n \text{ even}$$

If the x_i have limiting mean μ and standard deviation σ and if the distribution of the x's is symmetrical about μ,

$$\text{limiting mean of the median} = \mu$$

If the x_i follow a normal distribution the standard deviation of the median approaches, for large n,

$$\sqrt{\frac{\pi}{2}} \cdot \frac{\sigma}{\sqrt{n}} = 1.2533 \frac{\sigma}{\sqrt{n}} \tag{1}$$

For values of $n < 10$ the s.d. is approximately $1.2\,\sigma/\sqrt{n}$.

4. Measures of Dispersion

4.1. The Standard deviation. Symbol s.

$$s = \sqrt{\frac{\sum_{i=1}^{n}(x_i - \bar{x})^2}{n-1}} \tag{1}$$

If the x_i follow a normal distribution with limiting mean μ and standard deviation σ,

$$\left.\begin{aligned} \text{limiting mean of } s &= \left[\frac{\Gamma(n/2)}{\Gamma[(n-1)/2]}\sqrt{\frac{2}{n-1}}\right]\sigma \\ &= a\sigma \sim \sqrt{1-\frac{1}{2(n-1)}}\,\sigma \end{aligned}\right\} \qquad (2)$$

$$\text{s.d. of } s = \sqrt{1-a^2}\,\sigma \sim \frac{\sigma}{\sqrt{2(n-1)}} \qquad (3)$$

The distribution of s is approximately normal for $n > 30$.

Note: In all statistical analysis involving comparing of averages or precision of two processes, the square of the standard deviation, or variance, is used. If such analysis is intended, the estimator s is to be preferred over other estimators of standard deviation because s^2 has a limiting mean σ^2, the other estimators do not. These alternate estimates are :

a. The estimator

$$\sqrt{\frac{\sum_{i=1}^{n}(x_i-\bar{x})^2}{n}} = \sqrt{\frac{n-1}{n}}\,s \qquad (4)$$

for which, if the x_i follow a normal distribution,

$$\text{limiting mean of } \sqrt{\frac{n-1}{n}}\,s = \sqrt{\frac{n-1}{n}}\,a\sigma$$

$$\text{s.d. of } \sqrt{\frac{n-1}{n}}\,s = \sqrt{\frac{n-1}{n}}\sqrt{1-a^2}\,\sigma$$

b. The estimator

$$\frac{1}{a}\,s = \frac{\Gamma[(n-1)/2]}{\Gamma(n/2)}\sqrt{\frac{n-1}{2}}\sqrt{\frac{\sum_{i=1}^{n}(x_i-\bar{x})^2}{n-1}} \qquad (5)$$

for which, if the x_i follow a normal distribution,

$$\text{limiting mean of } \frac{1}{a}\,s = \sigma$$

$$\text{s.d. of } \frac{1}{a}\,s = \sqrt{\frac{1}{a^2}-1}\,\sigma$$

c. The estimator

$$0.67449 \sqrt{\frac{\sum_{i=1}^{n} (x_i - \bar{x})^2}{n}}$$

the so-called "probable error." For the case of a normal distribution has mean value

$$0.67449 \sqrt{\frac{n-1}{n}}\, a\sigma$$

and standard deviation

$$0.67449 \sqrt{\frac{n-1}{n}} \sqrt{1 - a^2}\, \sigma$$

For a normal distribution with mean μ and standard deviation σ, the interval $(\mu - 0.67449\,\sigma, \mu + 0.67449\,\sigma)$ will include 50 per cent of the frequency distribution. When σ is estimated from finite sets of data such an exact probability statement cannot be made (see tolerance limits).

4.2. The variance $s^2 = \sum_{i=1}^{n} (x_i - \bar{x})^2/(n-1)$. Whatever the distribution of x's, the limiting mean of $s^2 = \sigma^2$.

If the x's follow a normal distribution with standard deviation σ,

$$\text{standard deviation of } s^2 = \sqrt{\frac{2}{n-1}}\, \sigma^2 \qquad (1)$$

and $(n-1)s^2/\sigma^2$ follows a χ^2 distribution with $(n-1)$ degrees of freedom.

4.3. Average deviation, $\frac{1}{n}\sum_{i=1}^{n} |x_i - \bar{x}|$

If the x_i are normally distributed, limiting mean μ and standard deviation σ,

$$\text{limiting mean of average deviation} = \sqrt{\frac{2}{\pi}} \sqrt{\frac{n-1}{n}}\, \sigma \qquad (1)$$

$$\left.\begin{aligned} \text{s.d. of average deviation} &= \sqrt{\frac{2}{\pi}} \frac{\sqrt{n-1}}{n} \left[\frac{\pi}{2} + \sqrt{n(n-2)} - n \right. \\ &\left. + \sin^{-1} \frac{1}{n-1}\right]^{1/2} \sigma \sim \frac{\sigma}{\sqrt{n}} \sqrt{1 - \frac{2}{\pi}} \end{aligned}\right\} \qquad (2)$$

for large n. Note that in comparing two average deviations, account must be taken of any difference in the number of observations in the two estimates.

4.4. Range: difference between largest and smallest value in a set of observations, $x_{(n)} = x_{(1)}$. If the x_i follow a normal distribution with limiting mean μ and standard deviation σ,

$$\text{limiting mean of range} = (\text{constant})\ \sigma$$

Note: Tables of this constant for different values of n are given in *Biometrika*, Vol. 24, p. 416, along with values of the standard deviation of the range. With the use of these tables an estimate of the standard deviation can be obtained by dividing the range of the data by this constant. For example: for sets of 5 measurements the constant is 2.33,
10 measurements the constant is 3.08,
20 measurements the constant is 3.73,
30 measurements the constant is 4.09,
100 measurements the constant is 5.02.

5. The Fitting of Straight Lines

5.1. Introduction. When the mathematical relationship or law between two variates is known or assumed to be linear, and exact knowledge of the relationship is obscured by random errors in the measurement of one or both of the variates, the formulas of this section are applicable.

A clear distinction must be made between the case of estimating the parameters of a linear *physical law* and the case of estimating the parameters of *linear regression*. In the physical law case there are two mathematical variables, X and Y, related by a linear equation. Our inability to measure without error obscures this relationship. In the linear regression case the average values of y for given x's are related to x by a linear equation. In the linear regression case there is a bivariate distribution of x and y. There are two distinct regression lines (that of $\bar{y}$ on x and $\bar{x}$ on y) and there is a correlation between x and y. The regression lines and the correlation are properties of the underlying bivariate distribution.

5.2. The case of the underlying physical law. Let X and Y represent the variables of the physical law, and:

a. The two variables X and Y are related by a law of the form $Y = \alpha + \beta X$ or its converse $X = (-\alpha/\beta) + (1/\beta)Y$.

b. At any X_i repeated measurements y, of the variable Y, are dispersed about $Y_i = \alpha + \beta X_i$ with a standard deviation to be called σ_{y_i}, the functional form of the error distribution being the same at all X_i.

c. At any Y_i repeated measurements, x_i, of the variate X are dispersed about $X_i = -\alpha/\beta + (1/\beta)Y_i$ with a standard deviation to be called σ_{x_i}, the functional form of the distribution of errors being the same at all Y_i.

d. In practice there are obtained n paired values (x_i, y_i) which correspond to paired values (X_i, Y_i) from the linear law. The problem is to obtain the best estimate, $Y = a + bX$ of the linear law $Y = \alpha + \beta X$ from the n values (x_i, y_i).

If one variable, X, is known to be measured without error, the following tables (pages 118-119) are applicable.

If both variates are subject to error, i.e., if $\sigma_{y_i} = \sigma_y$ and $\sigma_{x_i} - \sigma_x$ and the error distribution is normal, the following technique is applicable.

Step 1. Arrange the n paired values (x_i, y_i) in order of magnitude according to one of the variates, say x.

Step 2. Divide this ranked set of n values into three nearly equal groups, so that in the two extreme groups there are the same number, say k. Thus the center has $n - 2k$.

Step 3. Compute the averages, $\bar{x}_1$ and $\bar{y}_1$ of x's and y's in the first group, the averages $\bar{x}_3$ and $\bar{y}_3$ of the x's and y's in the third group, and the grand averages $\bar{x}$ and $\bar{y}$.

Estimator of slope $$b' = \frac{\bar{y}_3 - \bar{y}_1}{\bar{x}_3 - \bar{x}_1}$$

Estimator of intercept $$a = \bar{y} - b'\bar{x}$$

Estimator of error in y's,

$$s_y = \sqrt{\left[\sum_{i=1}^{n} (y_i - \bar{y})^2 - b' \sum_{i=1}^{n} (x_i - \bar{x})(y_i - \bar{y})\right] \Big/ -1}$$

Estimator of error in x's,

$$s_x = \sqrt{\left[\sum_{i=1}^{n} (x_i - \bar{x})^2 - (1/b') \sum_{i=1}^{n} (x_i - \bar{x})(y_i - \bar{y})\right] \Big/ n-1}$$

The distributions of the estimators depend on the value of the unknown slope β. See Wald, A., *Annals of Mathematical Statistics*, Vol. 11, 1940, p. 284; Bartlett, M. S., *Biometrics*, Vol. 5, 1949, p. 207.

Error distribution of y's the same for all X, i.e., $\sigma_{y_i}^2 = \sigma^2$

Parameter	Estimator when $\alpha = 0$	Estimator when $\alpha \neq 0$
β	$b = \dfrac{\sum_{i=1}^{n} x_i y_i}{\sum_{i=1}^{n} x_i^2}$	$b = \dfrac{\sum_{i=1}^{n} (x_i - \bar{x})(y_i - \bar{y})}{\sum_{i=1}^{n} (x_i - \bar{x})^2}$
α	$a = 0$	$a = \bar{y} - b\bar{x}$
σ^2	$s = \dfrac{\sum_{i=1}^{n} y_i^2 - \dfrac{\left[\sum_{i=1}^{n} x_i y_i\right]^2}{\sum_{i=1}^{n} x_i^2}}{n - 1}$	$s = \dfrac{\sum_{i=1}^{n} (y_i - \bar{y})^2 - \dfrac{\left[\sum_{i=1}^{n} (x_i - \bar{x})(y_i - \bar{y})\right]^2}{\sum_{i=1}^{n} (x_i - \bar{x})^2}}{n - 2}$
s.d. of estimate of slope : σ_b	$s_b = \dfrac{s}{\sqrt{\sum_{i=1}^{n} x_i^2}}$	$s_b = \dfrac{s}{\sqrt{\sum_{i=1}^{n} (x_i - \bar{x})^2}}$
s.d. of estimate of intercept : σ_a	$\sigma_a = 0$	$s_a = s\sqrt{\dfrac{1}{n} + \dfrac{\bar{x}^2}{\sum_{i=1}^{n} (x_i - \bar{x})^2}}$
s.d. of Y_0, a point on the line at $X = X_0 : \sigma_{Y_0}$	$s_{Y_0} = \dfrac{sX_0}{\sqrt{\sum_{i=1}^{n} x_i^2}}$	$s_{Y_0} = s\sqrt{\dfrac{1}{n} + \dfrac{(X_0 - \bar{x})^2}{\sum_{i=1}^{n} (x_i - \bar{x})^2}}$

ERROR DISTRIBUTION DIFFERENT FOR EACH X, $\sigma_{y_i} = \sigma/w_i$

Parameter	Estimator when $\alpha = 0$	Estimator when $\alpha \neq 0$
β	$b = \dfrac{\sum_{i=1}^{n} w_i^2 x_i y_i}{\sum_{i=1}^{n} w_i^2 x_i^2}$	$b = \dfrac{\sum_{i=1}^{n} w_i^2 x_i y_i - \left[\left(\sum_{i=1}^{n} w_i^2 x_i\right)\left(\sum_{i=1}^{n} w_i^2 y_i\right) \Big/ \sum_{i=1}^{n} w_i^2\right]}{\sum_{i=1}^{n} w_i^2 x_i^2 - \left[\left(\sum_{i=1}^{n} w_i^2 x_i\right)^2 \Big/ \sum_{i=1}^{n} w_i^2\right]}$
α	$\alpha = 0$	$a = \dfrac{\sum_{i=1}^{n} w_i^2 y_i}{\sum_{i=1}^{n} w_i^2} - b\,\dfrac{\sum_{i=1}^{n} w_i^2 x_i}{\sum_{i=1}^{n} w_i^2}$
σ^2	$s^2 = \dfrac{1}{n-1}\left[\sum_{i=1}^{n} w_i^2 y_i^2 - \dfrac{\left(\sum_{i=1}^{n} w_i^2 x_i y_i\right)^2}{\sum_{i=1}^{n} w_i^2 x_i^2}\right]$	$s = \dfrac{1}{n-2}\left\{\sum_{i=1}^{n} w_i^2 y_i^2 - \dfrac{\left(\sum_{i=1}^{n} w_i^2 y_i\right)^2}{\sum_{i=1}^{n} w_i^2} - \dfrac{\left[\sum_{i=1}^{n} w_i^2 x_i y_i - \dfrac{\left(\sum_{i=1}^{n} w_i^2 x_i\right)\left(\sum_{i=1}^{n} w_i^2 y_i\right)}{\sum_{i=1}^{n} w_i^2}\right]^2}{\sum_{i=1}^{n} w_i^2 x_i^2 - \left[\left(\sum_{i=1}^{n} w_i^2 x_i\right)^2 \Big/ \sum_{i=1}^{n} w_i^2\right]}\right\}$
s.d. of estimate of slope : σ_b	$s_b = \dfrac{}{\sqrt{\sum_{i=1}^{n} w_i x^2{}_i{}^2}}$	$s_b = \dfrac{s}{\sqrt{\sum_{i=1}^{n} w_i^2 x_i^2 - \dfrac{\left(\sum_{i=1}^{n} w_i^2 x_i\right)^2}{\sum_{i=1}^{n} w_i^2}}}$

Note: One of the estimates, s_y^2 or s_x^2, can be negative. Such negative values are to be expected more often when n is small, or when the parameter (σ_x^2 or σ_y^2) being estimated is near zero. The distribution of the estimates has not been worked out.

6. Linear Regression

6.1. Linear regression. If two variates x and y vary concurrently (as for example, the height and weight of men age 30) and a linear relation exists between the average value of y and the corresponding value of x, and also a linear relation between the average value of x and the corresponding y, the formulas of this section are applicable.

Parameter	Estimator	Variance
α	$a = \bar{y} - b\bar{x}$	—
β	$b = \dfrac{\sum_{i=1}^{n} (x_i - \bar{x})(y_i - \bar{y})}{\sum_{i=1}^{n} (x_i - \bar{x})^2}$	$\dfrac{1}{n-3} \dfrac{\sigma_y^2}{\sigma_x^2} (1 - \rho^2)$
ρ	$r = \dfrac{\sum_{i=1}^{n} (x_i - \bar{x})(y_i - \bar{y})}{\sqrt{\sum_{i=1}^{n} (x_i - \bar{x})^2 \sum_{i=1}^{n} (y_i - \bar{y})^2}}$	$\dfrac{1}{n} (1 - \rho^2)^2$ for large n
α'	$a' = \bar{x} - b'\bar{y}$	—
β'	$b' = \dfrac{\sum_{i=1}^{n} (x_i - \bar{x})(y_i - \bar{y})}{\sum_{i=1}^{n} (y_i - \bar{y})^2}$	$\dfrac{1}{n-3} \dfrac{\sigma_x^2}{\sigma_y^2} (1 - \rho^2)$
σ_y^2	$s_y^2 = \dfrac{1}{n-1} \sum_{i=1}^{n} (y_i - \bar{y})^2$	—
σ_x^2	$s_x^2 = \dfrac{1}{n-1} \sum_{i=1}^{n} (x_i - \bar{x})^2$	—

For a given x_i, the y's have a variance σ_y^2, and mean value

$$\bar{y}_i = \alpha + \beta x_i$$

and for a given y_i, the x's have a variance σ_x^2, and mean value

$$\bar{x}_i = \alpha' + \beta' y_i$$

where $\bar{y}$ and $\bar{x}$ are the averages of an infinite number of values at y and x, respectively. Thus there are two lines to determine, and further there is the parameter ρ, the correlation between x and y.

7. The Fitting of Polynomials

7.1. Unequal intervals between the x's. Let n paired values (x_i, y_i) be observed. If the variate x is measured without error and the variate y is related to x by the equation

$$y = \alpha_0 + \alpha_1 x + \alpha_2 x^2 + \alpha_3 x^3 + \dots \qquad (1)$$

and the errors in y follow a distribution with mean zero and standard deviation σ, the constants are estimated by minimizing the sum of squares of the y deviations, i.e., the estimates $a_0, a_1, a_2, \dots$ of $\alpha_0, \alpha_1, \alpha_2, \dots$ are those values for which

$$\sum_{i=1}^{n} (y_i - a_0 - a_1 x - a_2 x^2 - \dots)^2 = \text{minimum} \qquad (2)$$

This minimization process leads to a set of simultaneous equations:

$$\left.\begin{aligned} \Sigma y - na_0 - a_1\Sigma x - a_2\Sigma x^2 - \dots &= 0 \\ \Sigma xy - a_0\Sigma x - a_1\Sigma x^2 - a_2\Sigma x^3 - \dots &= 0 \\ \Sigma x^2 y - a_0\Sigma x^2 - a_1\Sigma x^3 - a_2\Sigma x^4 - \dots &= 0 \quad \text{etc.} \end{aligned}\right\} \qquad (3)$$

where the summations are over $i = 1, 2, \dots, n$.

For the estimators of the parameters of linear equation $y = \alpha_0 + \alpha_1 x$ the solution is

$$a_0 = \frac{\begin{vmatrix} \Sigma y & \Sigma x \\ \Sigma xy & \Sigma x^2 \end{vmatrix}}{D_2} \qquad a_1 = \frac{\begin{vmatrix} n & \Sigma y \\ \Sigma x & \Sigma xy \end{vmatrix}}{D_2} \qquad (4)$$

where

$$D_2 = \begin{vmatrix} n & \Sigma x \\ \Sigma x & \Sigma x^2 \end{vmatrix}$$

and the sum of squares of deviation of the y's from the line is

$$S_2^2 = \Sigma y^2 - a_0\Sigma y - a_1\Sigma xy.$$

For the estimators of the parameters of a parabola, $y = \alpha_0 + \alpha_1 x + \alpha_2 x^2$,

$$\left.\begin{aligned} a_0 &= \frac{\begin{vmatrix} \Sigma y & \Sigma x & \Sigma x^2 \\ \Sigma xy & \Sigma x^2 & \Sigma x^3 \\ \Sigma x^2 y & \Sigma x^3 & \Sigma x^4 \end{vmatrix}}{D_3} \qquad a_1 = \frac{\begin{vmatrix} n & \Sigma y & \Sigma x^2 \\ \Sigma x & \Sigma xy & \Sigma x^3 \\ \Sigma x^2 & \Sigma x^2 y & \Sigma x^4 \end{vmatrix}}{D_3} \\ a_2 &= \frac{\begin{vmatrix} n & \Sigma x & \Sigma y \\ \Sigma x & \Sigma x^2 & \Sigma xy \\ \Sigma x^2 & \Sigma x^3 & \Sigma x^2 y \end{vmatrix}}{D_3} \end{aligned}\right\} \tag{5}$$

where

$$D_3 = \begin{vmatrix} n & \Sigma x & \Sigma x^2 \\ \Sigma x & \Sigma x^2 & \Sigma x^3 \\ \Sigma x^2 & \Sigma x^3 & \Sigma x^4 \end{vmatrix}$$

and the sum of squares of y deviation from the parabola is

$$S_3^2 = \Sigma y^2 - a_0 \Sigma y - a_1 \Sigma xy - a_2 \Sigma x^2 y \tag{6}$$

The extension to polynomials of higher order is obvious. The estimate of the standard deviation σ_1 of the y determination

$$\text{is} \quad s = \sqrt{\frac{S_2^2}{n-2}} \quad \text{for the line} \tag{7}$$

$$\text{is} \quad s = \sqrt{\frac{S_3^2}{n-3}} \quad \text{for the parabola} \tag{8}$$

$$\text{and} \quad s = \sqrt{\frac{S_k^2}{n-k}} \quad \text{for a polynomial of degree } k-1 \tag{9}$$

For testing the significance of any of the constants fitted, the reduction in the total sum of squares of the y's is studied as follows :

		Degrees of freedom	Mean square
Total sum of squares . .	$\sum_{i=1}^{n} (y_i - \bar{y})^2 = S_1^2$	$n-1$	—
Deviation from linear	S_2^2	$n-2$	$S_2^2/(n-2)$
Reduction of sum of squares due to linear terms	$S_1^2 - S_2^2$	1	$S_1^2 - S_2^2$
Deviations from quadratic	S_3^2	$n-3$	$S_3^2/(n-3)$
Reduction of sum of squares due to quadratic terms	$S_2^2 - S_3^2$	1	$S_2^2 - S_3^2$
etc.			

To test the significance of a coefficient, a_k, compute

$$F = \frac{S_{k-1}^2 - S_k^2}{S_k^2/(n-k)} \tag{10}$$

If the distribution of errors in the y measurement is normal, F will follow the F distribution for 1, $n-k$ degrees of freedom. If the computed F is greater than the value of F for 1, $n-k$ degrees of freedom exceeded with probability α, the reduction in sum of squares due to fitting the constant a_k (coefficient of x^{k-1}) is regarded as significant at the 100 α per cent level of significance.

Each of the a_k are seen to be a linear function of the y's, so that the standard deviation of any particular a_i can be obtained by writing

$$a_k - \sum_{i-1}^{n} c_i y_i,$$

and using the formula for the variance of a linear function, noting that the mean value of $y_i = \alpha_0 + \alpha_1 x_i + \alpha_2 x_i^2 + \dots$.

7.2. The case of equal intervals between the x's — the method of orthogonal polynomials. In order to find the best fitting polynomial

$$y = a_0 + a_1 x + a_2 x^2 + \dots \tag{1}$$

when the x values are evenly spaced, it is convenient to fit instead the polynomial

$$y = A + B\xi_1' + C\xi_2' + \dots \tag{2}$$

where ξ_1', ξ_2', ξ_3', ... are orthogonal polynomials of degree 1, 2, 3, ...

$$\xi_0 = 1$$

$$\xi_1' = (x - \bar{x})\lambda,$$

$$\xi_2' = \left[(x-\bar{x})^2 - \frac{n^2-1}{12}\right]\lambda_2$$

$$\vdots$$

$$\xi_k' = \left[\xi_1'\xi_n' - \frac{r^2(n^2-r^2)}{4(4r^2-1)}\xi_{n-1}'\right]\lambda_r$$

The λ's are chosen so that all the coefficients of these polynomials are integers. The coefficients of these orthogonal polynomials and a description of the method of analysis are given in FISHER, R. A., and YATES, F., *Statistical Tables*, 4th edition, Hafner Publishing Company, New York, 1953 for ξ_1', to ξ_5', for $n = 3(1)75$.

7.3. Fitting the coefficients of a function of several variables. Just as a distinction was made in the case of the straight line between the underlying physical law situation and the linear regression situation, so also must this distinction be preserved with the extension to this multivariate case.

If Y is related by a physical law to variates X_1, X_2, ... by the equation

$$Y = \alpha_0 + \alpha_1 X_1 + \alpha_2 X_2 + \dots \tag{1}$$

one is interested only in estimating the coefficients α_k. There is no question of the correlation between, say, X_1 and X_3. In the physical law case when X_1, X_2, and X_3 are regarded as free from error and Y as subject to errors following a distribution with mean zero and standard deviation σ, for n values $(y_i, X_{1i}, X_{2i}, \dots)$ the estimators a_k for the parameter α_k are, for the case of the law $Y = \alpha_0 + \alpha_1 X_1 + \alpha_2 X_2$,

$$\left.\begin{aligned} a_0 &= \frac{\begin{vmatrix} \Sigma y & \Sigma X_1 & \Sigma X_2 \\ \Sigma X_1 y & \Sigma X_1^2 & \Sigma X_1 X_2 \\ \Sigma X_2 y & \Sigma X_1 X_2 & \Sigma X_2^2 \end{vmatrix}}{D_3} \qquad a_1 = \frac{\begin{vmatrix} n & \Sigma y & \Sigma X_2 \\ \Sigma X_1 & \Sigma X_1 y & \Sigma X_1 X_2 \\ \Sigma X_2 & \Sigma X_2 y & \Sigma X_2^2 \end{vmatrix}}{D_3} \\ a_2 &= \frac{\begin{vmatrix} n & \Sigma X_1 & \Sigma y \\ \Sigma X_1 & \Sigma X_1^2 & \Sigma X_1 y \\ \Sigma X_2 & \Sigma X_1 X_2 & \Sigma X_2 y \end{vmatrix}}{D_3} \end{aligned}\right\} \tag{2}$$

where
$$D_3 = \begin{vmatrix} n & \Sigma X_1 & \Sigma X_2 \\ \Sigma X_1 & \Sigma X_1^2 & \Sigma X_1 X_2 \\ \Sigma X_2 & \Sigma X_1 X_2 & \Sigma X_2^2 \end{vmatrix}$$

The sum of squares of deviations of y from the fitted equation is

$$S_3^2 = \Sigma y^2 - a_0 \Sigma y - a_1 \Sigma X_1 y - a_2 \Sigma X_2 y \tag{3}$$

The extension to include cases of more variables is straightforward, and the tests of significance of the coefficients is analogous to the case of fitting a polynomial.

7.4. Multiple regression. In the multiple regression case where a number of variates are varying concurrently (as, for example, weight of males age 30, as a function of height, wrist circumference, and waist girth) is treated in R. A. FISHER, *Statistical Methods for Research Workers*, Oliver and Boyd, Section 29. In multiple regression there are partial correlations and partial regression coefficients to be estimated.

8. Enumerative Statistics

8.1. Estimator of parameter of binomial distribution. If in n trials, m occurrences of some event are observed, then the estimator for the parameter P is

$$p = \frac{m}{n}. \tag{1}$$

This estimate will have a standard deviation

$$\text{s.d. of } p = \sqrt{\frac{P(1-P)}{n}}. \tag{2}$$

$100(1-\alpha)$ per cent confidence intervals (P_1, P_2) for the parameter P of the binomial for k occurrences in a sample of n are given by solving

$$\sum_{i=k}^{n} \binom{n}{i} P^i (1-P)^{n-i} - \alpha/2 \tag{3}$$

for P, giving the lower limit P_1, and

$$\sum_{i=0}^{k} \binom{n}{i} P^i (1-P)^{n-i} = \alpha/2 \tag{4}$$

for P, giving the upper limit P_2.

The computation is simplified by using the incomplete beta function (see Mood, A. M., *Introduction to Theory of Statistics*, McGraw-Hill Book Company, Inc., 1950, pp. 233-235).

8.2. Estimator of parameter of Poisson distribution. If in an interval of x units, k occurrences are observed, the estimator for the parameter λ of the Poisson is k/x, which has a standard deviation of $\sqrt{\lambda/x}$.

If n independent estimates of λ, k_1/x_1, k_2/x_n, ..., k_n/x_n are available, the best estimate of λ is

$$\frac{\sum_{i=1}^{n} k_i}{\sum_{i=1}^{n} x_i}$$

which has standard deviation $\sqrt{\lambda \Big/ \sum_{i=1}^{n} x_i}$.

The average of the ratios has standard deviation

$$\sqrt{\frac{\lambda}{n^2}\sum_{i=1}^{n}\left(\frac{1}{x_i}\right)} \geq \sqrt{\frac{\lambda}{\sum_{i=1}^{n} x_i}}$$

To test the conformance of a series of n values of k_i to the Poisson law where the interval length is the same for all determinations, compute

$$\chi^2 = \frac{\sum_{i=1}^{n}(k_i - \bar{k})^2}{\bar{k}} \tag{5}$$

where $\bar{k} = (1/n)\sum_{i=1}^{n} k_i$.

This statistic will have approximately the χ^2 distribution for $n - 1$ degrees of freedom.

8.3. Rank correlation coefficient. N objects are ranked by two methods. Denote the ranking of the ith object by the first method by r_{1i} and its ranking by the second method by r_{2i}, and the difference $d_i = r_{1i} - r_{2i}$.

The rank correlation coefficient is defined as

$$1 - \frac{6\sum_{i=1}^{N} d_i^2}{N(N^2 - 1)}$$

For a complete treatment of rank correlation see KENDALL, M. G., *Rank Correlation Methods*, Charles Griffin & Co., Ltd., London, 1948.

9. Interval Estimation

9.1. Confidence interval for parameters. Instead of a point estimator of a parameter, an interval estimator may be used. A random interval (L_1, L_2), depending only on the observed set of data having the property that $100(1 - 2\alpha)$ per cent of such intervals computed will include the value of the parameter being estimated, is called the $100(1 - 2\alpha)$ *per cent confidence interval.*

No probability statement can be made regarding an individual confidence interval since a particular interval either includes the parameter, or it does not. It is the method of estimation that carries with it the probability of correctly bracketing the parameter.

For a given distribution of measurements it is possible to set down the distribution of estimates of a parameter derived by using a given estimator and a sequence of random sets from the original distribution. For example, the probability that $\bar{x}$ based on n measurements lies in the interval $(\mu - 1.96\,\sigma/\sqrt{n},\ \mu + 1.96\,\sigma/\sqrt{n})$ is 0.95. From this it follows that the random interval $(\bar{x} - 1.96\,\sigma/\sqrt{n},\ \bar{x} + 1.96\,\sigma/\sqrt{n})$ will include the parameter μ with probability 0.95, and the interval $(\bar{x} - 1.96\,\sigma/\sqrt{n},\ \bar{x} + 1.96\,\sigma/\sqrt{n})$ is therefore the 95 per cent confidence interval estimator of μ.

9.2. Confidence interval for the mean of a normal distribution. For a set of n measurements from a normal population the interval (L_1,L_2) where

$$L_1 = \bar{x} - t_{(2\alpha,n-1)}\frac{s}{\sqrt{n}} \tag{1}$$

and

$$L_2 = \bar{x} + t_{(2\alpha,n-1)}\frac{s}{\sqrt{n}} \tag{2}$$

where $t_{(2\alpha,n-1)}$ is the value of Student's t for $n - 1$ degrees of freedom exceeded in absolute value with probability 2α, defines the $100(1 - 2\alpha)$ per cent confidence interval for μ, the mean of a normal distribution.

9.3. Confidence interval for the standard deviation of a normal distribution. From a set of n measurement (i.e., an estimate of standard deviation based on $n - 1$ degrees of freedom) from a normal distribution the interval (L_1,L_2) where

$$L_1 = \sqrt{\frac{(n-1)s^2}{\chi^2_{(\alpha,n-1)}}} \tag{1}$$

and

$$L_2 = \sqrt{\frac{(n-1)s^2}{\chi^2_{(1-\alpha,n-1)}}} \tag{2}$$

and where $\chi^2_{(\alpha,n-1)}$ is the value of χ^2 for $df = n - 1$ exceeded with probability α, defines the $100(1 - 2\alpha)$ per cent confidence interval for σ.

Note: (L_1^2,L_2^2) defines the $100(1 - 2\alpha)$ per cent confidence interval for σ^2.

9.4. Confidence interval for slope of straight line. If the variate x is free from errors and the variate y_i,

$$y_i = (\alpha + \beta x_i) + \epsilon_i \tag{1}$$

where ϵ_i is a random variable, normally distributed with mean 0 and standard deviation σ for all i, then (L_1, L_2) is the $100(1 - 2\alpha)$ per cent confidence interval for β, the slope of the line.

$$L_1 = b - t_{(2\alpha, n-2)} s_b \tag{2}$$

$$L_2 = b + t_{(2\alpha, n-2)} s_b \tag{3}$$

where $t_{(2\alpha, n-2)}$ is the value of Student's t, exceeded in absolute value with probability 2α.

Note: If the intercept is known to be zero, replace $t_{(2\alpha, n-2)}$ by $t_{(2\alpha, n-1)}$ in the formulas for L_1 and L_2.

9.5. Confidence interval for intercept of a straight line. Under the same conditions required for a confidence interval for the slope,

$$L_1 = a - t_{(2\alpha, n-2)} s_a \tag{1}$$

$$L_2 = a + t_{(2\alpha, n-2)} s_a \tag{2}$$

define the $100(1 - 2\alpha)$ per cent confidence interval estimator for α, the intercept of a straight line.

9.6. Tolerance limits. Limits between which a given percentage, P, of the values of a distribution lie are called tolerance limits. For example, for the normal distribution with mean μ and standard deviation σ, 95 per cent of the values lie between the limits $\mu - 1.96\sigma$ and $\mu + 1.96\sigma$. If $\bar{x}$ and s are estimates of μ and σ, it cannot be stated that the limits $\bar{x} - 1.96s$ and $\bar{x} + 1.96$ include 95 per cent of the population. Here $\bar{x} + Ks$ (where K is a constant) is a random variable which is approximately normally distributed with mean $\mu + Ks$ and approximate standard deviation $\sigma\sqrt{1/n + K^2/2(n-1)}$ where n is the number of observations upon which $\bar{x}$ is based and $(n - 1)$ is the number of degrees of freedom in the estimate s.

The value of K can be adjusted so that the limits $(\bar{x} - Ks)$ and $(\bar{x} - ks)$ will in a given percentage γ of the cases include a given proportion P of the original distribution. See *Techniques of Statistical Analysis*, edited by C. Eisenhart, M. W. Hastay, and W. A. Wallis, McGraw-Hill Book Company, Inc., 1947, Chapter 2, for tables of values of K for various n, γ, and P.

10. Statistical Tests of Hypothesis

10.1. Introduction. Investigations are concerned not only with estimating certain parameters, but also with comparing the estimates obtained with certain assumed values or with other estimates. For example, an average value for a certain quantity is derived from two separate experiments. The question is asked: Can the averages be considered in agreement? In this case the hypothesis to be tested, called the *null hypothesis*, is that the limiting means of the distribution from which the averages arose are identical.

The specification of the statistical test requires a statement of the *alternate hypothesis* against which the null hypothesis is tested. The statistical test will lead to either acceptance or rejection of the null hypothesis.

Any statistical test may result in a wrong decision. Errors of two kinds may be committed: *Type I error:* rejecting the null hypothesis when it is in fact true and, *Type II error:* accepting the null hypothesis when it is in fact false. For any statistical test the risk α of making a Type I error can be made as small as desired.

A statistic T, some function of the observed value, is chosen in the light of the null and alternate hypotheses. When the null hypothesis is true this statistic will tend to fall within a certain range; when the alternate hypothesis is true the statistic T will tend to fall in some other range. These ranges will in general overlap. The value selected to serve as the boundary between these ranges determines the risks of making a wrong decision.

The null hypothesis is rejected if an observed value of a test statistic exceeds the critical value. In this section the critical value specified is such that there is a chance α of wrongly rejecting the null hypothesis. The test is said to be conducted at the 100α per cent *level of significance.*

If in conducting a test of a hypothesis at the 100α per cent level, a value is obtained that falls short of the critical value, the statement can be made that "there is no evidence of a difference." If the observed value of a test statistic exceeds the critical value, the statement can be made that "there is evidence of a difference since as divergent a set of values as found would occur by chance only 100α per cent of the time if no real difference existed."

10.2. Test of whether the mean of a normal distribution is greater than a specified value.

a. *Observed values:* $\bar{x}$ and s based on n measurements considered as a set of values from a normal distribution with mean μ, and standard deviation σ.

b. *Null hypothesis :* $\mu = \mu_0$

c. *Alternate hypothesis :* $\mu > \mu_0$

d. *Test statistic :* $$t = \frac{\bar{x} - \mu_0}{s/\sqrt{n}}$$

t follows Student's t-distribution for $n - 1$ degrees of freedom if null hypothesis is true.

e. *Rejection criterion :* Reject the null hypothesis if the observed value of t is greater than the value of t for $n - 1$ degrees of freedom, exceeded with probability α.

10.3. Test of whether the mean of a normal distribution is different from some specified value.

a. *Observed values :* $\bar{x}$ and s based on n measurements considered as a set of values from a normal distribution with mean μ, and standard deviation σ.

b. *Null hypothesis :* $\mu = \mu_0$

c. *Alternate hypothesis :* $\mu \neq \mu_0$

d. *Test statistic :* $$t = \frac{|\bar{x} - \mu_0|}{s/\sqrt{n}}$$

t follows Student's t-distribution for $n - 1$ degrees of freedom if null hypothesis is true.

e. *Rejection criterion :* Reject the null hypothesis if the observed value of t is greater than the value of t for $n - 1$ degrees of freedom, exceeded in absolute value with probability α. (This is equivalent to the value of t exceeded with probability $\alpha/2$).

10.4. Test of whether the mean of one normal distribution is greater than the mean of another normal distribution.

a. *Observed values :* $\bar{x}_1$ and s_1 based on n measurements considered as a set of values from a normal distribution with mean μ_1, and standard deviation σ, and $\bar{x}_2$ and s_2 based on m measurements considered as a set of values from a normal distribution with mean μ_2 and standard deviation σ.

b. *Null hypothesis :* $\mu_1 = \mu_2$

c. *Alternate hypothesis :* $\mu_1 > \mu_2$

d. *Test statistic :*

$$t = \frac{\bar{x}_1 - \bar{x}_2}{\sqrt{(m+n)/mn}\ \sqrt{[(n-1)s_1^2 + (m-1)s_2^2]/(m+n-2)}}$$

t follows Student's t distribution for $(n + m - 2)$ degrees of freedom when null hypothesis is true.

e. *Rejection criterion :* Reject the null hypothesis if the observed value of t is greater than the value of t for $(n + m - 2)$ degrees of freedom exceeded with probability α.

10.5. Test of whether the means of two normal distributions differ.

a. *Observed values :* $\bar{x}_1$ and s_1 based on n measurements considered as a set of values from a normal distribution with mean μ_1, and standard deviation σ, and $\bar{x}_2$ and s_2 based on m measurements considered as a set of values from a normal distribution with mean μ_2 and standard deviation σ.

b. *Null hypothesis :* $\mu_1 = \mu_2$

c. *Alternate hypothesis :* $\mu_1 \neq \mu_2$

d. *Test statistic :*

$$t = \frac{|\, x_1 - x_2 \,|}{\sqrt{(m+n)/n}\ \sqrt{[(n-1)s_1^2 + (m-1)s_2^2]}}$$

t follows Student's t-distribution for $(m + n - 2)$ degrees of freedom when the null hypothesis is true.

e. *Rejection criterion :* Reject the null hypothesis if the observed value of t is larger than the value of t for $(m + n - 2)$ degrees of freedom exceeded in absolute value with probability α.

Note : For the case where the distributions do not have the same standard deviation, see FISHER, R. A. and YATES, F., *Statistical Tables*, 4th edition, Hafner Publishing Company, New York, 1953, pp. 3-4.

10.6. Tests concerning the parameters of a linear law. In the case where X is free from error and the errors in the observed values of y follow a normal distribution with mean zero and standard deviation σ, the statistics

$$t = \frac{b - \beta_0}{s_b} \tag{1}$$

and

$$t = \frac{a - \alpha_0}{s_a} \tag{2}$$

follow Student's t-distribution for degrees of freedom equal to the number of degrees of freedom in the estimate of s_b and s_a, (which will be $n-2$ for both when the line does not pass through the origin, and $n-1$ for s_b when it is assumed a priori that $\alpha = 0$) when the null hypothesis $\beta = \beta_0$ and $\alpha = \alpha_0$ are true.

The tests of hypothesis concerning β can therefore be made in the same manner as tests regarding the mean of a normal distribution.

10.7. Test of the homogeneity of a set of variances. Let $s_1^2, s_2^2, \ldots, s_k^2$ be k variance estimates based on $n_1, n_2, \ldots, n_k$ observations respectively. Then the statistic

$$B = \frac{2.30259}{c}\left[N \log S^2 - \sum_{i=1}^{k} (n_i - 1) \log s_i^2\right] \tag{1}$$

where

$$N = \sum_{i=1}^{k} (n_i - 1)$$

$$S^2 = \frac{\sum_{i=1}^{k} (n_i - 1)s_i^2}{N}$$

$$c = 1 + \frac{\sum_{i=1}^{k} [1/(n_i - 1)] - 1/N}{3(k-1)}$$

has approximately the χ^2 distribution for $k-1$ degrees of freedom under the null hypothesis that the s_i^2 are all estimates of a common σ^2. The null hypothesis is rejected if the observed value of B is greater than the value of χ^2 for $(k-1)$ degrees of freedom exceeded with probability α.

10.8. Test of homogeneity of a set of averages. If $\bar{x}_1, \bar{x}_2, \ldots, \bar{x}_n$ are n averages, each based on k measurements each, the statistic

$$F = \frac{\left[k \sum_{i=1}^{n} (\bar{x} - \bar{\bar{x}})^2\right] / (n-1)}{\left[\sum_{i=1}^{n} \sum_{j=1}^{k} (x_{ij} - \bar{x}_i)^2\right] / [n(k-1)]} \tag{1}$$

where

$$\bar{\bar{x}} = \frac{1}{n} \sum_{i=1}^{n} \bar{x}_i$$

follows the F distribution for $(n-1)$, and $n(k-1)$ degrees of freedom under the null hypothesis that limiting mean of $\bar{x}_i$ is the same for all i. Reject the null hypothesis if the observed value of F is greater than the value of F for $n-1$, $n(k-1)$ degrees of freedom exceeded with probability α.

10.9. Test of whether a correlation coefficient is different from zero. If r is the observed correlation coefficient found in the linear regression case, assumed to be an estimate of ρ of a bivariate normal distribution, then in order to test the *null hypothesis* that $\rho = 0$ against the *alternative hypothesis* that $\rho \neq 0$ compute the statistic

$$t = \frac{r\sqrt{n-2}}{\sqrt{1-r^2}} \tag{1}$$

When the null hypothesis is true t follows Student's t-distribution for $(n-2)$ degrees of freedom. Reject the null hypothesis if the observed value of t is greater than the value of Student's t-distribution exceeded in absolute value with probability α.

10.10. Test of whether the correlation coefficient ρ is equal to a specified value. Let

$$\left.\begin{aligned} z &= \frac{1}{2}\log\frac{1+r}{1-r} \\ \zeta &= \frac{1}{2}\log\frac{1+\rho}{1-\rho} \end{aligned}\right\} \tag{1}$$

where r is the observed correlation found in the linear regression case and assumed to be an estimate of ρ of a bivariate normal distribution, then $(z-\zeta)$ is approximately normally distributed with mean zero and standard deviation approximately $\sqrt{1/(n-3)}$.

Thus a test for the significance of the difference of r from any arbitrary ρ is given by computing the statistic

$$t = \frac{\dfrac{1}{2}\log\left(\dfrac{1+r}{1-r}\right) - \dfrac{1}{2}\log\left(\dfrac{1+\rho_0}{1-\rho_0}\right)}{\sqrt{1/(n-3)}} \tag{2}$$

Under the null hypothesis that $\rho = \rho_0$, t will be approximately normally distributed with mean zero and standard deviation $\sigma = 1$.

Reject the null hypothesis if t is greater than the standardized normal deviate exceeded with probability $\alpha/2$ when the alternate hypothesis is that $\rho \neq \rho_0$.

11. Analysis of Variance

11.1. Analysis of Variance. Consider a two-way classification of data as shown in the table.

		Column 1	2	j	c	Row averages
Row	1	x_{11}	x_{12}	x_{1j}	x_{1c}	$\bar{x}_{R1}$
	2	x_{21}				$\bar{x}_{R2}$
				$\vdots$		
	i	x_{i1}		$\cdots$ x_{ij} $\cdots$		$\bar{x}_{Ri}$
				$\vdots$		
	r	x_{r1}		x_{rj}	x_{rc}	$\bar{x}_{Rr}$
column averages		$\bar{x}_{C1}$	$\bar{x}_{C2}$	$\bar{x}_{Cj}$	$\bar{x}_{Cc}$	$\bar{x}$ = grand average

It is a property of numbers that

$$\sum_{i=1}^{r} \sum_{j=1}^{c} (x_{ij} - \bar{x})^2$$

can be broken up into three components as follows :

$$c \text{ (Deviation of row averages about grand average)} = c \sum_{i=1}^{r} (\bar{x}_{Ri} - \bar{x})^2 \tag{1}$$

$$r \text{ (Deviation of column averages about grand average)} = r \sum_{j=1}^{c} (\bar{x}_{Cj} - \bar{x})^2 \tag{2}$$

$$\text{Residue} = \sum_{i=1}^{r} \sum_{j=1}^{c} (x_{ij} - \bar{x}_{Ri} - \bar{x}_{Cj} + \bar{x})^2 \tag{3}$$

It is assumed in what follows that the x_{ij} are normally independently distributed with variance σ^2, about means μ_{ij}

$$\mu_{ij} = \mu_0 + \delta_{Ri} + \delta_{Cj} \tag{3}$$

where μ_0, is the grand mean, δ_{Ri} and δ_{Cj} are the deviations of the row and column means, respectively, from the grand mean. Two models of analysis of variance are possible.

a. *Model I. The parameters δ_{Ri} and δ_{Cj} are fixed constants.*

For example, if the rows represent different machines and the columns different operators then the δ_{Ri} are the deviations of the individual machines performance from the grand mean, and δ_{Cj} are the operator biases.

The row averages and column averages are estimators of $\mu_0 + \delta_{Ri}$ and $\mu_0 + \delta_{Cj}$, respectively.

In order to test the hypothesis that the row means are all identical or that the column means are all identical, the following computations are made :

	Sum of squares	Degrees of freedom	Mean square	Mean square is estimate of
Deviation of row averages . . .	$R = c \sum_{i=1}^{r} (\bar{x}_{Ri} - \bar{x})^2$	$r - 1$	$\dfrac{R}{r-1}$	$\sigma^2 + c \dfrac{\sum_{i=1}^{r} \delta_{Ri}^2}{r-1}$
Deviation of column averages .	$C = r \sum_{j=1}^{c} (\bar{x}_{Cj} - \bar{x})^2$	$c - 1$	$\dfrac{C}{c-1}$	$\sigma^2 + r \dfrac{\sum_{j=1}^{c} \delta_{Cj}^2}{c-1}$
Residue	$T - R - C$	$(r-1)(c-1)$	$\dfrac{T-R-C}{(r-1)(c-1)}$	σ^2
Total	T	$rc - 1$	—	—

To test the null hypothesis that the row means are identical, that is, that

$$\sum_{i=1}^{r} \delta_{Ri}^2 = 0$$

compute the statistic

$$F = \frac{R/(r-1)}{(T-R-C)/[(r-1)(c-1)]} \tag{4}$$

Under the null hypothesis F follows the F distribution for $r - 1$, $(r-1)(c-1)$ degrees of freedom. A similar test can be made for the column means.

b. *Model II. The parameters are components of variance.*

For example, if the rows represent different samples of the same material and the columns represent different days on which the tests were conducted, then δ_{Ri} and δ_{Cj} are random variables. These random variables will have mean zero and variances σ_R^2 and σ_C^2, respectively, and in that which follows are assumed to be normally distributed. The computations are the same as shown in Model I.

	Sum of squares	Degrees of freedom	Mean square	Mean square is estimate of
Deviation of row averages	R	$r-1$	$\frac{R}{r-1}$	$\sigma^2 + c\sigma_R^2$
Deviation of column averages	C	$c-1$	$\frac{C}{c-1}$	$\sigma^2 + r\sigma_C^2$
Residue	$T-R-C$	$(r-1)(c-1)$	$\frac{T-R-C}{(r-1)(c-1)}$	σ^2
Total	T	$rc-1$	—	—

The parameters σ_R^2 and σ_C^2 are estimated by

$$s_R^2 = \frac{1}{c}\left[\frac{R}{r-1} - \frac{T-R-C}{(r-1)(c-1)}\right] \tag{5}$$

and

$$s_C^2 = \frac{1}{r}\left[\frac{C}{c-1} - \frac{T-R-C}{(r-1)(c-1)}\right] \tag{6}$$

The test of the null hypothesis that $\sigma_R^2 = 0$ is made by computing the statistic

$$F = \frac{R/(r-1)}{(T-R-C)/[(r-1)(c-1)]} \tag{7}$$

which, if the null hypothesis is true, follows the F distribution with $r-1$, $(r-1)(c-1)$ degrees of freedom.

A test of the hypothesis that $\sigma_C^2 = 0$ is conducted in a similar manner.

The analysis of variance technique of which the above are examples can be extended to n-fold classifications. The technique is also valuable in ascertaining whether the effect of a factor is maintained over different levels of other factors, i.e., for detecting the presence of interactions. See ANDERSON, R. L., and BANCROFT, T. A., *Statistical Theory in Research*, McGraw-Hill

Book Co., Inc. New York, 1952; FISHER, R. A., *Statistical Methods for Research Workers*, 11th edition, Oliver & Boyd, Ltd., London, 1950.

12. Design of Experiments

12.1. Design of experiments. The validity of inferences that can be drawn from an experiment depend upon the manner in which the experiment was conducted. In a comparison of a number of objects it has long been realized that identical conditions should be maintained during the testing of all the objects. Often this is not possible, since not all the objects to be compared can be tested in a short enough interval that conditions can be regarded as identical for all tests.

The scheduling of the tests to achieve unbiased estimates of parameters, and maximum efficiency in the testing of the hypotheses that the experimenter set out to test; the search for an optimum combination of a number of factors each of which may be at several levels; or an investigation of the influence of several factors individually and jointly on some physical quantity are treated in COCHRAN, W. G., and COX, G. M., *Experimental Designs*, John Wiley & Sons, Inc., New York, 1950; and FISHER, R. A., *The Design of Experiments*, 5th edition, Oliver & Boyd, Ltd., London, 1949.

13. Precision and Accuracy

13.1. Introduction. A measurement process is regarded as *precise* if the dispersion of values is regarded as small, i.e., σ small.

A measurement process is regarded as *accurate* if the values cluster closely about the correct value c.

The *precision of an individual measurement* is the same as the precision for the process generating the measurements.

By *accuracy of an individual measurement* or of an average of n measurements is usually meant the maximum possible error (constant and/or random) that could affect the observed value and is frequently thought of in terms of the number of significant figures to which the value can be regarded as correct.

13.2. Measure of precision. It is suggested that the measure for representing precision of an individual measurement in a set of n measurements be

$$s = \sqrt{\frac{\sum_{i=1}^{n} (x - \bar{x})^2}{n - 1}} \tag{1}$$

and for the precision of an average,

$$s_{\bar{x}} = \frac{s}{\sqrt{n}}. \tag{2}$$

13.3. Measurement of accuracy. Limits on possible constant errors due to known sources of inaccuracy can be set down from a study of the method of conducting the experiment. Such limits on errors represent a judgment on the part of the experimenter and no probability statements can be attached thereto. The setting down of the upper limit and lower limit for a quantity should involve the possible uncertainties due to any constant error plus some component due to random errors.

Suggested form for reporting results of n determinations of some quantity :

Average $\bar{x}$	=
Number of measurements n	=
s.d. of average $s/\sqrt{n}$	=
Maximum possible constant errors ; E, $-E'$	=
Lower limit to quantity : $x - E' - Ks/\sqrt{n}$	=
Lower limit to quantity : $\bar{x} + E + Ks/\sqrt{n}$	=

where K is either Student's t, $K = 2$ or 3 for approximation to 2σ or 3σ limits (for n large), or any other specified constant.

The upper and lower limit to a quantity express the experimenter's best judgment of the range within which the true value of the quantity being measured lies. A value could then be said to be accurate to as many significant figures as the upper and lower limits agree.

14. Law of Propagation of Error

14.1. Introduction. If x follows a distribution with mean μ and standard deviation σ, and $F(x)$ is some function of x, then the mean μ_F and variance $\sigma_F{}^2$ of $F(x)$, where x is the average of n values, is approximated by

$$\mu_F = F(\mu) \tag{1}$$

$$\sigma_F{}^2 = [F'(\mu)]^2\sigma^2 \tag{2}$$

provided $F(x)$, $F'(x)$ are continuous and nonzero in the neighborhood of the point $x = \mu$.

Caution must be exercised in the application of the formula. It is valid when two conditions are satisfied : (a) if the Taylor expansion

$$F(x) = F(\mu) + F'(\mu)(x - \mu) + \frac{F''(\mu)}{2!}(x - \mu)^2 + \dots \tag{3}$$

is such that $F^k(\mu)$ for $k = 2, 3, \ldots$ is small, and (b) if the expected value of $(x - \mu)^k$, for $k = 3, 4, \ldots$ is small (i.e., that the third and higher moments of the distribution of x be small). There are cases where condition (a) is not satisfied, yet the formula is valid because condition (b) holds, or where neither condition holds but the product of $F^k(\mu)$ and the kth moment is small for $k > 2$.

EXAMPLE: Let x be normally distributed with mean μ and standard deviation σ. Then the mean and variance of $y = x^2$ are according to the above formula

$$\text{mean of } y = \mu^2$$

$$\text{variance of } y = 4\mu^2\sigma^2$$

The exact answer for the mean is $\mu^2 + \sigma^2$, and for the variance, $4\mu^2\sigma^2 + 2\sigma^4$. When σ is large the " propagation of error formula " can be seriously in error in this case.

EXAMPLE: Let x have mean $\xi = e^{\mu + 1/2\sigma^2}$, and variance $\xi^2(e^{\sigma^2} - 1)$, and with third moment, or average value of $(x - \xi)^3 = \xi^3(e^{3\sigma^2} - 3e^{\sigma^2} + 2)$, fourth moment, or average value of $(x - \xi)^4 = \xi^4(e^{6\sigma^2} - 4e^{3\sigma^2} + 6e^{\sigma^2} - 3)$, etc., such that $y = \log x$ is normally distributed with mean μ and standard deviation σ. The approximation gives

for the mean of y:

$$\log \xi = \mu + \tfrac{1}{2}\sigma^2$$

for the variance of y:

$$\frac{1}{\xi^2}\, \xi^2\, (e^{\sigma^2} - 1) = e^{\sigma^2} - 1 = \sigma^2 + \frac{\sigma^4}{2!} + \frac{\sigma^6}{3!} + \cdots$$

If σ is large, the approximate formula can be seriously in error, in this case because the third and higher moments of the original distribution are not small.

If $H(x_1,x_2,x_3,\ldots)$ is a function of $x_1, x_2, x_3, \ldots$ which have mean values $\mu_1, \mu_2, \mu_3, \ldots$ and variances $\sigma_1^2, \sigma_2^2, \sigma_3^2, \ldots$ and correlations $\rho_{12}, \rho_{13}, \rho_{23}, \ldots$, then the variance, σ_H^2, of $H(x_1,x_2,x_3,\ldots)$ is given by

$$\left.\begin{aligned} \sigma_H^2 = {} & [H_{x_1}']^2\sigma_1^2 + [H_{x_2}']^2\sigma_2^2 + [H_{x_3}']^2\sigma_3^2 + \ldots \\ & - 2\rho_{12}\sigma_1\sigma_2 H_{x_1}'H_{x_2}' - 2\rho_{13}\sigma_1\sigma_3 H_{x_1}'H_{x_3}' - 2\rho_{23}\sigma_2\sigma_3 H_{x_2}'H_{x_3}' - \ldots \end{aligned}\right\} \quad (4)$$

all derivatives being evaluated at $\mu_1, \mu_2, \mu_3, \ldots$.

The restrictions on the validity of this formula are analogous to the restrictions applying to $F(x)$.

14.2. Standard deviation of a ratio of normally distributed variables. If $z = x/y$ where x is normally distributed with mean μ_x and standard deviation σ_x; y is normally distributed with mean μ_y and standard deviation σ_y; the correlation between x and y is ρ and μ_y/σ_y is sufficiently large (say greater than 5) then z will be approximately normally distributed with mean of $z = \mu_x/\mu_y$, and standard deviation of

$$z = \frac{\mu_x}{\mu_y}\sqrt{\frac{\sigma_x^2}{\mu_x^2} + \frac{\sigma_y^2}{\mu_y^2} - 2\rho\frac{\sigma_x\sigma_y}{\mu_x\mu_y}} \quad (1)$$

[FIELLER, E. C., *Biometrika*, Vol. 24 (1932), p. 428.]

14.3. Standard deviation of a product of normally distributed variables. If x and y are normally distributed with means μ_x and μ_y and standard deviations σ_x and σ_y, then $w = xy$ will have mean $\mu_x\mu_y$ and standard deviation

$$\sqrt{\mu_x^2\sigma_y^2 + \mu_y^2\sigma_x^2 + \sigma_x^2\sigma_y^2}$$

[CRAIG, C. C., *Annals of Mathematical Statistics*, Vol. 7 (1936), p. 1.]

Chapter 3

NOMOGRAMS

BY DONALD H. MENZEL

Professor of Astrophysics at Harvard University
Director of Harvard College Observatory

1. Nomographic Solutions

1.1. A nomogram (or nomograph) furnishes a graphical procedure for solving certain types of equations, primarily those containing three variables, say α, β, and γ. If any two of these quantities are known the third follows directly from the conditioning equation between the three parameters. The determinant,

$$\begin{vmatrix} x_1 & y_1 & 1 \\ x_2 & y_2 & 1 \\ x_3 & y_3 & 1 \end{vmatrix} = 0 \tag{1}$$

represents the equation :

$$\frac{y_1 - y_2}{x_1 - x_2} = \frac{y_2 - y_3}{x_2 - x_3} \tag{2}$$

The three points (x_1y_1), (x_2y_2), and (x_3y_3) are collinear because the two segments possess identical slopes and because they have a point, (x_2,y_2) in common.

Now, suppose that we can write

$$\begin{aligned} x_1 &= \phi_1(\alpha) & y_1 &= \psi_1(\alpha) \\ x_2 &= \phi_2(\beta) & y_2 &= \psi_2(\beta) \\ x_3 &= \phi_3(\gamma) & y_3 &= \psi_3(\gamma) \end{aligned} \tag{3}$$

Equations (3) define three curves, in parametric form, whose points we can label in terms of our basic parameters α, β, and γ. Thus, if we know α and β, for example, a straight line from the appropriate points on the α and β curves will intersect the γ curve at the value of γ that satisfies the equations, as indicated in Fig. 1, which represents the artificial equation

$$\begin{aligned} &2 \ln \beta \sin \alpha + 1.5\alpha \cos (\pi/2)\gamma + 2\beta^2 + 1.5\beta^2 \sin^2 (\pi/2)\gamma \\ &\qquad - 1.5 \cos (\pi/2)\gamma \ln \beta - 4 \sin \alpha - 3 \sin \alpha \sin (\pi/2)\gamma - \alpha\beta^2 = 0 \end{aligned}$$

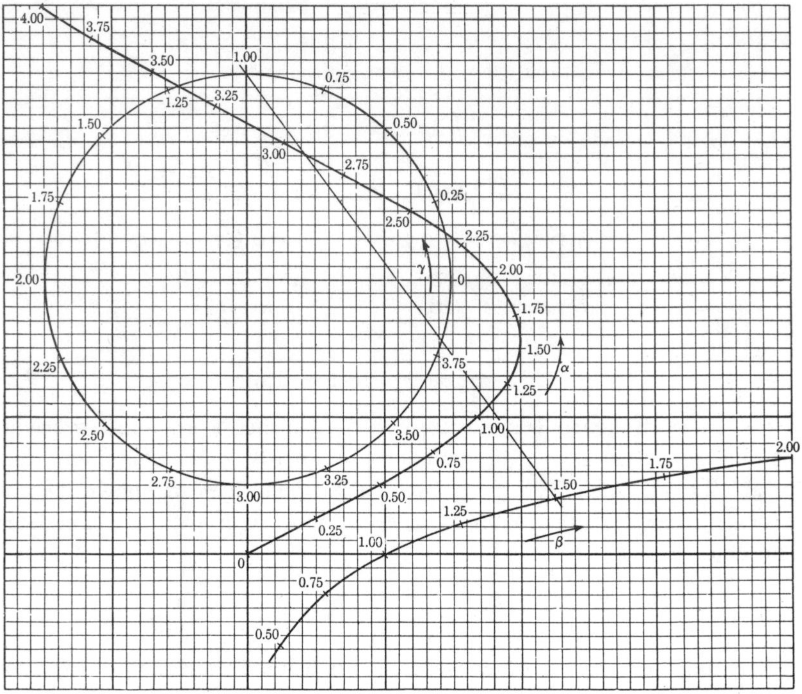

FIGURE 1

This equation fits the determinant (1), with

$$x_1 = 2 \sin \alpha, \qquad y_1 = \alpha$$

$$x_2 = \beta^2, \qquad y_2 = \ln \beta$$

$$x_3 = 1.5 \cos (\pi/2)\gamma, \quad y_3 = 2.0 + 1.5 \sin (\pi/2)\gamma$$

The line connecting given points $\beta = 1.5$ and $\gamma = 1$ intersects the α curve twice. There are thus two roots, approximately at $\alpha = 1.1$ and 2.9. By increasing the number of coordinates marking the values of α, β, γ, we can read as closely as desired. In many examples, of course, some of these curves are straight lines.

To represent an equation nomographically we must first find an equivalent determinant which we then transform by standard rules until we obtain our equation in the form of (1), viz.,

$$\begin{vmatrix} \phi_1(\alpha) & \psi_1(\alpha) & 1 \\ \phi_2(\beta) & \psi_2(\beta) & 1 \\ \phi_3(\gamma) & \psi_3(\gamma) & 1 \end{vmatrix} = 0 \tag{4}$$

We must separate the variables, as above.

Although this equation satisfies the initial condition, the diagram resulting from it is by no means the most general one possible. Let $a, b, c, d, e, f, g, h, i$ represent nine arbitrary constants. Then, using the transformation properties of a zero-valued determinant, we obtain the general form, which expansion will prove to be equivalent to (4).

$$\begin{vmatrix} \dfrac{a\phi_1 + b\psi_1 + c}{g\phi_1 + h\psi_1 + i} & \dfrac{d\phi_1 + e\psi_1 + f}{g\phi_1 + h\psi_1 + i} & 1 \\[2ex] \dfrac{a\phi_2 + b\psi_2 + c}{g\phi_2 + h\psi_2 + i} & \dfrac{d\phi_2 + e\psi_2 + f}{g\phi_2 + h\psi_2 + i} & 1 \\[2ex] \dfrac{a\phi_3 + b\psi_3 + c}{g\phi_3 + h\psi_3 + i} & \dfrac{d\phi_3 + e\psi_3 + f}{g\phi_3 + h\psi_3 + i} & 1 \end{vmatrix} = 0 \tag{5}$$

By choosing these constants $a - i$ appropriately, we can control the positions and scales of the α, β, and γ curves in order to achieve maximum accuracy for the desired range of variables.

The diagram representing Eq. (5) is a linear transformation of the diagram representing Eq. (4). In other words, (5) is a mere projection of (4) upon some arbitrary plane. Two of the constants, in effect, fix the scales of x and y. Two more determine the choice of origin, including the possibility of shifting any specific point to infinity. Four more define some specific line in the plane and the remaining constant fixes the degree of rotation of the original diagram about this arbitrary line as an axis. The nine constants thus completely determine the projection scheme. Some experimenting with constants is generally necessary to produce the best diagram for a given problem.

A few examples follow. The simple additive nomogram

$$f_1(\alpha) + f_2(\beta) + f_3(\gamma) = 0 \tag{6}$$

yields the basic determinant:

$$\begin{vmatrix} f_1 & 1 & 1 \\ -f_2 & 1 & 0 \\ -f_3 & 0 & 1 \end{vmatrix} = \begin{vmatrix} f_1 & 1 & 2 \\ -f_2 & 1 & 1 \\ -f_3 & 0 & 1 \end{vmatrix} = \begin{vmatrix} \frac{1}{2}f_1 & \frac{1}{2} & 1 \\ -f_2 & 1 & 1 \\ -f_3 & 0 & 1 \end{vmatrix} = 0 \tag{7}$$

The third determinant defines three parallel straight lines. The γ scale fits along the y axis, the β scale along the line $y = 1$, with the α scale halfway between. Now, transforming the determinant by (5) we can get a general form, wherein the three lines may no longer be parallel.

The multiplicative form

$$f_1(\alpha)f_2(\beta) = f_3(\gamma) \tag{8}$$

assumes the form of (6) if we take logarithms. However we often find the following more useful.

$$\begin{vmatrix} f_1 & 0 & 1 \\ 1 & f_2 & 0 \\ 0 & -f_3 & 1 \end{vmatrix} = \begin{vmatrix} f_1 & 0 & 1+f_1 \\ 1 & f_2 & 1 \\ 0 & -f_3 & 1 \end{vmatrix} = \begin{vmatrix} f_1/(1+f_1) & 0 & 1 \\ 1 & f_2 & 1 \\ 0 & -f_3 & 1 \end{vmatrix} = 0 \tag{9}$$

Here the γ scale coincides with the negative part of the y axis. The β scale extends in the positive direction along $y = 1$. The α scale lies along the x axis. By projecting this diagram we can make it assume a Z-shape, with the two arms of the Z not necessarily parallel.

For four variables several alternatives arise. Occasionally the nomograms for four variables are identical in form with those for three, except that one (or more) of the scales are shifted. A sort of sliderule index, to set for the fourth variable, gives the necessary flexibility.

Sometimes we may treat one of the variables as a constant, calculating a different nomogram for each value of the fourth parameter. One can then interpolate between various curves to select the appropriate point.

In still other cases, as for example, in the equations,

$$f_1(\alpha)f_2(\beta) = f_3(\gamma)f_4(\delta) = f_5(\zeta) \tag{10}$$

involving the four variables, we may introduce an arbitrary variable ζ, as above. Now we can compile two independent nomograms, one involving α, β, and ζ and the other γ, δ, and ζ. With the ζ scale identical on both, we can make a single diagram with a common ζ scale. Then, if we are given, say α, β, and γ, and wish to find δ, we first join α and β with a straight line that intersects ζ at some special point. Then connect this point with the value of δ. The intersection of this line with the γ axis determines γ.

Bibliography

1. D'OCAGNE, M., *Traité de Nomographie*, 2nd ed. Gauthier-Villars et Cie, Paris, 1921. The standard work and most complete reference.
2. DOUGLASS, R. D. and ADAMS, D. P., *Elements of Nomography*, McGraw-Hill Book Company, Inc. New York, 1947.

Chapter 4

PHYSICAL CONSTANTS

BY JESSE W. M. DUMOND

Professor of Physics, California Institute of Technology

AND E. RICHARD COHEN

Physicist, North American Aviation Inc., Downey, California

1. Constants and Conversion Factors of Atomic and Nuclear Physics *

It is an interesting paradox that whereas the physical constants, and particularly those associated with the atom, fixed as they are by nature rather than by man, are believed to be the most invariable things known to science, the best methods of arriving at reliable numerical values for these constants have undergone more change in recent years than has been seen in almost any other branch of physics. Recent rapid advances in new techniques and in our theoretical understanding have resulted in new, entirely different, and very superior approaches for ascertaining the best values. These improvements at each stage usually render quite obsolete most of the methods of the preceding stage or era, not because the earlier methods were wrong in principle, but because they admitted of so much less accuracy in practice that they are no longer worth considering in competition with the newer methods. Recently this pace has been markedly accelerated so that, in the span of only a little over four years, completely new experimental methods growing out of the development of microwave and atomic beam techniques have so greatly improved the precision of our knowledge of the atomic constants as to warrant a complete re-evaluation of the entire situation. There has resulted an improvement in accuracy of approximately tenfold as regards our knowledge of the constants and conversion factors. No guarantee can be given that another two or three years hence an equally radical revolution may not have rendered the present methods equally obsolete. It is a fortunate and reassuring fact, however, that with all these

* Values as of October, 1955.

changes in *method* the *results* obtained, *the values* of these constants and conversion factors, give every indication of tending to settle down within narrower and narrower limits of uncertainty toward stable and definite results.

Because of this ephemeral character of the methods of determining these constants, we shall say here only enough about these methods to clarify certain important points bearing on the use of the table of values. The reader who wishes to study our present methods in detail will find them described in papers by the present authors [*Rev. Mod. Phys.*, **25**, 691 (1953); **27**, 363 (1955)]. A list of references at the end of this chapter also indicates some of the key papers bearing on this subject over the last twenty-five years. The outstanding work of R. T. Birge during this period in this field has been undoubtedly the greatest single factor in bringing about its present high state of development.

The appended table of values of constants and conversion factors of atomic and nuclear physics was computed by means of formulas (given with each constant in the table), all of which can be expressed directly in terms of the values of the following five key quantities combined with certain other accurately known auxiliary constants. The five key quantities (which we call our " primitive unknowns ") are N, Avogadro's number; e, the electronic charge; α, Sommerfeld's fine structure constant; c, the velocity of light; and λ_g/λ_s, the conversion factor from the Siegbahn arbitrary scale of x-ray wavelengths (x units) to absolute (or " grating ") wavelengths in milliangstroms (10^{-11} cm). The computation of the table of values from these five primary values is a straightforward process on which we need not dwell. The computation of the values of the five primitive unknowns from various experimental data is accomplished by the following general method. A number of " observational equations " (in the present instance thirteen) is set up to express the results of experimental measurements. It is a peculiarity of the atomic constants that in but few instances are the directly measured numerical quantities which result from experimental work capable of determining a single primitive unknown (such as e or N). In most cases, it is some function of two or more of the primitive unknowns that experiments determine. The number of such determined independent functions of the unknowns should exceed the number of unknowns as much as possible to afford a considerable degree of overdetermination in order to furnish a check (the only one in fact available) on the interconsistency of the data and its freedom from systematic errors. The mathematical tool especially developed for just such a situation as this is the method of least squares.* This permits us to compute a set of " adjusted " or compromise values of the

* See Chapter 2.

unknowns that do the least violence to all the input data, according to a systematic rule for minimizing such violence. This rule allows greater latitude for deviation to the less accurate determinations. In the present instance, a judgment of the violence done to the input data by the process of least-squares adjustment can be obtained from the statement that the ratio of our " precision measure by external consistency " (a measure of the compatibility of the overdetermined input equations) to our " precision measure by internal consistency " (based on estimates of the experimental accuracy of the input data) was $r_e/r_i = 1.04$.

A rather delicate point of novel character arises, in the present least-squares adjustment. In order to minimize the numerical work, which piles up with appalling rapidity as the number of unknowns increases, an effort is made to formulate the observational equations as compactly as possible. A result of this is that, unless care is taken to avoid such a condition, the numerical values (the results of experimental measurements), to each of which is equated some function of the unknowns in each observational equation, may not always be *single* measurements but may be *a function of several* independent measurements. If, furthermore, the same independent measurement may have been used as a contributing factor in computing the numerics for several different observational equations, these numerics may thus become " observationally correlated." Great care and attention must be given either (1) to formulate the equations so these correlations are absent or at most insignificant, or (2) to take account of them by an extension of the method of least-squares which has been especially developed for this purpose.

Two scales of atomic weights exist, the physical and chemical scales. On the physical scale the O^{16} isotope of oxygen has, by definition, the atomic weight 16. On the chemical scale, atomic weight 16 is assigned to the average atomic weight of a mixture of the oxygen isotopes in their naturally occurring abundance ratios. Unfortunately, if we are interested in the highest accuracy, the chemical scale is ambiguous and indefinite now that the precision with which atomic weights can be measured (by mass spectrographic means) is amply sufficient to show that the naturally occurring abundance ratios of the oxygen isotopes is different for different natural sources of oxygen (such as water or iron ore on the one hand and limestone or air on the other). All the atomic weights used in the following table are therefore given on the *physical* scale of atomic weights.

If one wishes to be strictly correct, some care must be used in computing precision measures of quantities depending functionally upon two or more of the numerical values given in the appended tables because these values are not completely uncorrelated in the observational sense. This unavoidable complication and the simple methods of handling it (by means of correlation coefficients and cross-product contributions to the error measure) are explained in an article, [*Rev. Mod. Phys.*, **25**, 691 (1953)]. *See* note p. 154a.

2. Table of Least-Squares-Adjusted Output Values

(November, 1952)

I. AUXILIARY CONSTANTS USED:

These auxiliary constants are quantities which are uncorrelated (observationally) with the variables of the least-squares adjustment.

Rydberg wave number for infinite mass: *

$$R_\infty = 109737.309 \pm 0.012 \text{ cm}^{-1}$$

Rydberg wave numbers for the light nuclei:

$$R_H = 109677.576 \pm 0.012 \text{ cm}^{-1}$$

$$R_D = 109707.419 \pm 0.012 \text{ cm}^{-1}$$

$$R_{He^3} = 109717.345 \pm 0.012 \text{ cm}^{-1}$$

$$R_{He^4} = 109722.267 \pm 0.012 \text{ cm}^{-1}$$

Atomic mass of neutron:

$$n = 1.008982 \pm 0.000003$$

Atomic mass of hydrogen:

$$H = 1.008142 \pm 0.000003$$

Atomic mass of deuterium:

$$D = 2.014735 \pm 0.000006$$

Gas constant per mole (physical scale):

$$R_0 = 8.31696 \pm 0.00034 \times 10^7 \text{ erg mole}^{-1} \text{ deg}^{-1}$$

Standard volume of a perfect gas (physical scale):

$$V_0 = 22420.7 \pm 0.6 \text{ cm}^3 \text{ atm mole}^{-1}$$

* This differs from the value $R_\infty = 109737.311 \pm 0.012$ cm^{-1} given by E. R. Cohen (Ref. 11) and which was used in the least-squares adjustment. In Ref. 11 a tentative value $Nm = 5.48785 \times 10^{-4}$ for the atomic mass of the electron was used with the proviso, p. 359, that "... an increase of 1 part per million in the electron mass will produce an increase of 0.00005 cm^{-1} in the Rydberg." (The coefficient 0.0005 cm^{-1} in the text is a typographical error.) The present value has therefore been revised, using this coefficient to accord with our present output value of Nm and similar modifications have been made in the Rydberg values of the light nuclei.

II. Least-Squares Adjusted-Output Values:

(The quantity following each $\pm$ sign is the standard error.)

Velocity of light:

$$c = 299793.0 \pm 0.3 \text{ km sec}^{-1}$$

Avogadro's constant (physical scale):

$$N = (6.02486 \pm 0.00016) \times 10^{23} \text{ (g mole)}^{-1}$$

Loschmidt's constant (physical scale):

$$L_0 = N/V_0 = (2.68719 + 0.00010) \times 10^{19} \text{ cm}^{-3}$$

Electronic charge:

$$e = (4.80286 \pm 0.00009) \times 10^{-10} \text{ esu}$$
$$e' = e/c = (1.60206 \pm 0.00003) \times 10^{-20} \text{ emu}$$

Electron rest mass:

$$m = (9.1083 \pm 0.0003) \times 10^{-28} \text{ g}$$

Proton rest mass:

$$m_p = M_p/N = (1.67239 \pm 0.00004) \times 10^{-24} \text{ g}$$

Neutron rest mass:

$$m_n = n/\text{N} = (1.67470 \pm 0.00004) \times 10^{-24} \text{ g}$$

Planck's constant:

$$h = (6.62517 \pm 0.00023) \times 10^{-27} \text{ erg sec}$$
$$\hbar = h/2\pi = (1.05443 \pm 0.00004) \times 10^{-27} \text{ erg sec}$$

Conversion factor from Siegbahn x-units to milliangstroms:

$$\lambda_g/\lambda_s = 1.002039 \pm 0.000014$$

Faraday constant (physical scale):

$$F = Ne = (2.89366 \pm 0.00003) \times 10^{14} \text{ esu (g mole)}^{-1}$$
$$F' = Ne/c = (9652.19 \pm 0.11) \text{ emu (g mole)}^{-1}$$

Charge-to-mass ratio of the electron:

$$e/m = (5.27305 \pm 0.00007) \times 10^{17} \text{ esu gm}^{-1}$$
$$e'/m = e/(mc) = (1.75890 \pm 0.00002) \times 10^{7} \text{ emu gm}^{-1}$$

Ratio h/e:

$$h/e = (1.37942 \pm 0.00002) \times 10^{-17} \text{ erg sec (esu)}^{-1}$$

Fine structure constant:

$$\alpha = e^2/(\hbar c) = (7.29729 \pm 0.00003) \times 10^{-3}$$
$$1/\alpha = 137.0373 \pm 0.0006$$
$$\alpha/2\pi = (1.161398 \pm 0.000005) \times 10^{-3}$$
$$\alpha^2 = (5.32504 \pm 0.00005) \times 10^{-5}$$
$$1 - (1 - \alpha^2)^{1/2} = (0.266252 \pm 0.000002) \times 10^{-4}$$

Atomic mass of the electron (physical scale) :

$$Nm = (5.48763 \pm 0.00006) \times 10^{-4}$$

Ratio of mass of hydrogen to mass of proton : *

$$\mathrm{H/M}_p = \left[1 - \frac{Nm}{\mathrm{H}}(1 - \tfrac{1}{2}\alpha^2)\right]^{-1} = 1.000544613 \pm 0.000000006$$

Atomic mass of proton (physical scale) :

$$\mathrm{M}_p = \mathrm{H} - Nm = 1.007593 \pm 0.000003$$

Ratio of proton mass to electron mass :

$$\mathrm{M}_p/(Nm) = 1836.12 \pm 0.02$$

Reduced mass of electron in hydrogen atom :

$$\mu = m\mathrm{M}_p/\mathrm{H} = (9.1034 \pm 0.0003) \times 10^{-28}\ \mathrm{g}$$

Schrödinger constant for a fixed nucleus :

$$2m/\hbar^2 = (1.63836 \pm 0.00007) \times 10^{27}\ \mathrm{erg}^{-1}\ \mathrm{cm}^{-2}$$

Schrödinger constant for the hydrogen atom :

$$2\mu/\hbar^2 = (1.63748 \pm 0.00007) \times 10^{27}\ \mathrm{erg}^{-1}\ \mathrm{cm}^{-2}$$

First Bohr radius :

$$a_0 = \hbar^2/(me^2) = \alpha/(4\pi R_\infty) = (5.29172 \pm 0.00002) \times 10^{-9}\ \mathrm{cm}$$

Radius of electron orbit in normal H^1, referred to center of mass :

$$a_0' = a_0(1 - \alpha^2)^{1/2} = (5.29158 \pm 0.00002) \times 10^{-9}\ \mathrm{cm}$$

Separation of proton and electron in normal H^1 :

$$a_0'' = a_0' R_\infty / R_\mathrm{H} = (5.29446 \pm 0.00002) \times 10^{-9}\ \mathrm{cm}$$

Compton wavelength of the electron :

$$\lambda_{ce} = h/(mc) = \alpha^2/(2R_\infty) = (24.2626 \pm 0.0002) \times 10^{-11}\ \mathrm{cm}$$

$$\bar{\lambda}_{ce} = \lambda_{ce}/(2\pi) = (3.86151 \pm 0.00004) \times 10^{-11}\ \mathrm{cm}$$

Compton wavelength of the proton :

$$\lambda_{cp} = h/(m_p c) = (13.2141 \pm 0.0002) \times 10^{-14}\ \mathrm{cm}$$

$$\bar{\lambda}_{cp} = \lambda_{cp}/(2\pi) = (2.10308 \pm 0.00003) \times 10^{-14}\ \mathrm{cm}$$

Compton wavelength of the neutron :

$$\lambda_{cn} = h/(m_n c) = (13.1959 \pm 0.0002) \times 10^{-14}\ \mathrm{cm}$$

$$\bar{\lambda}_{cn} = \lambda_{cn}/(2\pi) = (2.10019 \pm 0.00003) \times 10^{-14}\ \mathrm{cm}$$

Classical electron radius :

$$r_0 = e^2/(mc^2) = \alpha^3/(4\pi R_\infty) = (2.81785 \pm 0.00004) \times 10^{-13}\ \mathrm{cm}$$

$$r_0^2 = (7.94030 \pm 0.00021) \times 10^{-26}\ \mathrm{cm}$$

* The binding energy of the electron in the hydrogen atom has been included in the quantity. The mass of the electron when found in the hydrogen atom is not m, but more correctly $m(1 - \frac{1}{2}\alpha^2 + \ldots)$.

Thompson cross section:

$$\left(\tfrac{8}{3}\right)\pi r_0^2 = (6.65205 \pm 0.00018) \times 10^{-25}\ \text{cm}^2$$

Fine-structure doublet separation in hydrogen:

$$\begin{aligned}\Delta E_{\text{H}} &= \left(\frac{1}{16}\right) R_{\text{H}}\alpha^2 \left[1 + \frac{\alpha}{\pi} + \left(\frac{5}{8} - \frac{5.946}{\pi^2}\right)\alpha^2\right] \\ &= 0.365871 \pm 0.000003\ \text{cm}^{-1} \\ &= 10968.56 \pm 0.10\ \text{Mc sec}^{-1}\end{aligned}$$

Fine-structure separation in deuterium:

$$\begin{aligned}\Delta E_{\text{D}} = \Delta E_{\text{H}}\, R_{\text{D}}/R_{\text{H}} &= 0.365970 \pm 0.000003\ \text{cm}^{-1} \\ &= 10971.54 \pm 0.10\ \text{Mc sec}^{-1}\end{aligned}$$

Zeeman displacement per gauss:

$$(e/mc)/(4\pi c) = (4.66885 \pm 0.00006) \times 10^{-5}\ \text{cm}^{-1}\ \text{gauss}^{-1}$$

Boltzmann's constant:

$$\begin{aligned}k = R_0/N &= (1.38044 \pm 0.00007) \times 10^{-16}\ \text{erg deg}^{-1} \\ k &= (8.6167 \pm 0.0004) \times 10^{-5}\ \text{ev deg}^{-1} \\ 1/k &= 11605.4 \pm 0.5\ \text{deg ev}^{-1}\end{aligned}$$

First radiation constant:

$$c_1 = 8\pi hc = (4.9918 \pm 0.0002) \times 10^{-15}\ \text{erg cm}$$

Second radiation constant:

$$c_2 = hc/k = 1.43880 \pm 0.00007\ \text{cm deg}$$

Atomic specific heat constant:

$$c_2/c = (4.79931 \pm 0.00023) \times 10^{-11}\ \text{sec deg}$$

Wien displacement law constant: *

$$\lambda_{\max} T = c_2/(4.96511423) = 0.289782 \pm 0.000013\ \text{cm deg}$$

Stefan-Boltzmann constant:

$$\sigma = (\pi^2/60)\,(k^4/\hbar^3 c^2) = (0.56687 \pm 0.00010) \times 10^{-4}\ \text{ergs cm}^{-2}\ \text{deg}^{-4}\ \text{sec}^{-1}$$

Sackur-Tetrode constant:

$$(S_0/R_0)_{ph} = \tfrac{5}{2} + \ln\left\{(2\pi R_0)^{3/2} h^{-3} N^{-4}\right\} = -\,5.57324 \pm 0.00007$$

$$(S_0)_{ph} = -\,(46.3524 \pm 0.0020) \times 10^{7}\ \text{erg mole}^{-1}\ \text{deg}^{-1}$$

Bohr magneton:

$$\mu_0 = he/(4\pi mc) = \tfrac{1}{2}e\lambda\!\!\!^{-}_{ce} = (0.92731 \pm 0.00002) \times 10^{-20}\ \text{erg gauss}^{-1}$$

* The numerical constant 4.96511423 is the root of the transcendental equation, $x = 5(1 - e^{-x})$.

Anomalous electron-moment correction :

$$\left[1 + \frac{\alpha}{(2\pi)} - 2.973\,\frac{\alpha^2}{\pi^2}\right] = \mu_e/\mu_0 = 1.001145358 \pm 0.000000005$$

(Computed using adjusted value $\alpha = (7.29729 \pm 0.00003) \times 10^{-3}$.)

Magnetic moment of the electron :

$$\mu_e = (0.92837 \pm 0.00002) \times 10^{-20} \text{ erg gauss}^{-1}$$

Nuclear magneton :

$$\mu_n = he/(4\pi m_p c) = \mu_0 Nm/\text{H}^+ = (0.505038 \pm 0.000018) \times 10^{-23} \text{ erg gauss}^{-1}$$

Proton moment :

$$\mu = 2.79275 \pm 0.00003 \text{ nuclear magnetons}$$
$$= (1.41044 \pm 0.00004) \times 10^{-23} \text{ ergs gauss}^{-1}$$

Gyromagnetic ratio of the proton in hydrogen, uncorrected for diamagnetism :

$$\gamma' = (2.67523 \pm 0.00004) \times 10^4 \text{ radians sec}^{-1} \text{ gauss}^{-1}$$

Gyromagnetic ratio of the proton (corrected) :

$$\gamma = (2.67530 \pm 0.00004) \times 10^4 \text{ radians sec}^{-1} \text{ gauss}^{-1}$$

Multiplier of (Curie constant)$^{1/2}$ to give magnetic moment per molecule :

$$(3k/N)^{1/2} = (2.62178 \pm 0.00010) \times 10^{-20} (\text{erg mole deg}^{-1})^{1/2}$$

Mass-energy conversion factors :

$$1 \text{ g} = (5.61000 \pm 0.00011) \times 10^{26} \text{ Mev}$$
$$1 \text{ electron mass} = 0.510976 \pm 0.000007 \text{ Mev}$$
$$1 \text{ atomic mass unit} = 931.141 \pm 0.010 \text{ Mev}$$
$$1 \text{ proton mass} = 938.211 \pm 0.010 \text{ Mev}$$
$$1 \text{ neutron mass} = 939.505 \pm 0.010 \text{ Mev}$$

Quantum energy conversion factors :

$$1 \text{ ev} = (1.60206 \pm 0.00003) \times 10^{-12} \text{ erg}$$
$$E/\tilde{\nu} = hc = (1.98618 \pm 0.00007) \times 10^{-16} \text{ erg cm}$$
$$E\lambda_g = (12397.67 \pm 0.22) \times 10^{-8} \text{ ev cm}$$
$$E\lambda_s = 12372.44 \pm 0.16 \text{ kilovolt } x\text{-units}$$
$$E/\nu = (6.62517 \pm 0.00023) \times 10^{-27} \text{ erg sec}$$
$$E/\nu = (4.13541 \pm 0.00007) \times 10^{-15} \text{ ev sec}$$
$$\tilde{\nu}/E = (5.03479 \pm 0.00017) \times 10^{15} \text{ cm}^{-1} \text{ erg}^{-1}$$
$$\tilde{\nu}/E = 8066.03 \pm 0.14 \text{ cm}^{-1} \text{ ev}^{-1}$$
$$\nu/E = (1.50940 \pm 0.00005) \times 10^{26} \text{ sec}^{-1} \text{ erg}^{-1}$$
$$\nu/E = (2.41814 \pm 0.00004) \times 10^{14} \text{ sec}^{-1} \text{ ev}^{-1}$$

de Broglie wavelengths, λ_D, of elementary particles : *

Electrons :

$$\lambda_{D_e} = (7.27377 \pm 0.00006)\ \text{cm}^2\ \text{sec}^{-1}/\text{v}$$
$$= (1.552257 \pm 0.000016) \times 10^{-13}\ \text{cm}\ (\text{erg})^{1/2}/(E)^{1/2}$$
$$= (1.226378 \pm 0.000010) \times 10^{-7}\ \text{cm}\ (\text{ev})^{1/2}/(E)^{1/2}$$

Protons :

$$\lambda_{D_p} = (3.96149 \pm 0.00005) \times 10^{-3}\ \text{cm}^2\ \text{sec}^{-1}/\text{v}$$
$$= (3.62253 \pm 0.00008) \times 10^{-15}\ \text{cm}\ (\text{erg})^{1/2}/(E)^{1/2}$$
$$= (2.86202 \pm 0.00004) \times 10^{-9}\ \text{cm}\ (\text{ev})^{1/2}/(E)^{1/2}$$

Neutrons :

$$\lambda_{D_n} = (3.95603 \pm 0.00005) \times 10^{-3}\ \text{cm}^2\ \text{sec}^{-1}/\text{v}$$
$$= (3.62004 \pm 0.00008) \times 10^{-15}\ \text{cm}\ (\text{erg})^{1/2}/(E)^{1/2}$$
$$= (2.86005 \pm 0.00004) \times 10^{-9}\ \text{cm}\ (\text{ev})^{1/2}/(E)^{1/2}$$

Energy of 2200 m/sec neutron :

$$E_{2200} = 0.0252973 \pm 0.0000003\ \text{ev}$$

Velocity of 1/40 ev neutron :

$$v_{0\cdot025} = 2187.036 \pm 0.012\ \text{m/sec}$$

The Rydberg and related derived constants :

$$R_\infty = 109737.309 \pm 0.012\ \text{cm}^{-1}$$
$$R_\infty c = (3.289848 \pm 0.000003) \times 10^{15}\ \text{sec}^{-1}$$
$$R_\infty hc = (2.17958 \pm 0.00007) \times 10^{-11}\ \text{ergs}$$
$$R_\infty hc^2 e^{-1} \times 10^{-8} = 13.60488 \pm 0.00022\ \text{ev}$$

Hydrogen ionization potential :

$$I_0 = R_H \left(\frac{hc^2}{e}\right)\left(1 + \frac{\alpha^2}{4} + \ldots\right) \times 10^{-8} = 13.59765 \pm 0.00022\ \text{ev}$$

* These formulas apply only to nonrelativistic velocities. If the velocity of the particle is not negligible compared to the velocity of light, c, or the energy not negligible compared to the rest mass energy, we must use $\lambda_D = \lambda_c[\epsilon(\epsilon + 2)]^{-1/2}$, where λ_c is the appropriate Compton wavelength for the particle in question and ϵ is the kinetic energy measured in units of the particle rest-mass.

Bibliography

By R. T. BIRGE

Revs. Modern. Phys., 1, 1 (1929).*
Phys. Rev., **40**, 207 (1932).
Phys. Rev., **40**, 228 (1932).
Phys. Rev., **42**, 736 (1932).
Nature, **133**, 648 (1934).
Nature, **134**, 771 (1934).
Phys. Rev., **48**, 918 (1935).
Nature, **137**, 187 (1936).
Phys. Rev., **54**, 972 (1938).
Am. Phys. Teacher, **7**, 351 (1939).
Phys. Rev., **55**, 1119 (1939).
Phys. Rev., **57**, 250 (1940).
Phys. Rev., **58**, 658 (1940).
Phys. Rev., **60**, 766 (1941).*
Reports on Progress in Physics, London, **VIII**, 90 (1942).*
Am. J. Phys., **13**, 63 (1945).*
Phys. Rev., **79**, 193 (1950).
Phys. Rev., **79**, 1005 (1950).

OTHER REFERENCES

1. F. G. DUNNINGTON, *Revs. Modern Phys.*, **11**, 65 (1939).
2. J. W. M. DUMOND, *Phys. Rev.*, **56**, 153 (1939).
3. *Phys. Rev.*, **58**, 457 (1940).
4. F. KIRCHNER, *Ergeb. exakt. Naturwiss.*, **18**, 26 (1939).
5. J. W. M. DUMOND and E. R. COHEN, *Revs. Modern Phys.*, **20**, 82-108 (1948).
6. J. W. M. DUMOND, *Phys. Rev.*, **77**, 411 (1950).
7. J. W. M. DUMOND and E. R. COHEN, *American Scientist*, **40**, 447 (1952).
8. J. A. BEARDEN and H. M. WATTS, *Phys. Rev.*, **81**, 73 (1951).
9. BEARDEN and WATTS, *Phys. Rev.*, **81**, 160 (1951).
10. E. R. COHEN, *Phys. Rev.*, **81**, 162 (1951).
11. E. R. COHEN, *Phys. Rev.*, **88**, 353 (1952).
12. J. W. M. DUMOND and E. R. COHEN, *Rev. Modern Phys.*, **25**, 691 (1953).
13. E. R. COHEN, *Revs. Modern Phys.*, **25**, 691 (1953).
14. E. S. DAYHOFF, S. TRIEBWASSER, and W. LAMB JR., *Phys. Rev.*, **89**, 106 (1953).
15. S. TRIEBWASSER, E. S. DAYHOFF, and W. LAMB JR., *Phys. Rev.*, **89**, 98 (1953).
16. E. R. COHEN, J. W. M. DUMOND, T. W. LAYTON, and J. S. ROLLETT, *Rev. Mod. Phys.*, **27**, 363 (1955).
17. E. R. COHEN and J. W. M. DUMOND, *Phys. Rev. Lett.*, **1**, 291 (1958).
18. A. PETERMANN, *Helv. Phys. Acta.*, **30**, 407 (1957).
19. C. M. SOMMERFIELD, *Phys. Rev.*, **107**, 328 (1957).
20. J. W. M. DUMOND, *Boulder Conference on Electronic Standards and Measurements*, (Paper 1 2) (August 13, 1958).

* Longer and more important review articles. The first is a classic of considerable importance since it reviews carefully all the constants of physics rather than just the atomic constants. The numerical values of the atomic constants given therein, as well as the methods of determining them, have, however, undergone considerable modification since it was written.

NOTE: In 1957 a recalculation of the anomalous magnetic moment of the electron by Petermann (Ref. 18) and Sommerfield (Ref. 19) showed that the previously accepted value was in error by approximately 14 ppm. The effect of this change has been discussed in the literature (Ref. 17, 20), but since a complete re-evaluation has not been carried out, it is felt that data as of 1955 given in the tables above should stand until such time as the new evaluation is available. The major corrections are indicated in Ref. 17.

Chapter 5

CLASSICAL MECHANICS

By Henry Zatzkis

Assistant Professor of Mathematics
Newark College of Engineering

The formulas of classical mechanics lie at the base of practically all major fields of modern physics. Although the methods of quantum mechanics are necessary for the understanding of microscopic systems, the methods of even that field are best interpreted in terms of the formulas of classical mechanics, by way of contrast. The following set of formulas, although far from complete, represents an attempt to select from the enormous number of formulas of classical mechanics those that have a particularly modern application.

Notation: A vector quantity will be indicated by bold-face italic type, e.g., $\boldsymbol{a}$. The symbol a will represent the magnitude of $\boldsymbol{a}$. The Cartesian components of $\boldsymbol{a}$ will be denoted by either a_1, a_2, a_3, or a_x, a_y, a_z. The scalar product between two vectors will be denoted by $\boldsymbol{a} \cdot \boldsymbol{b}$, and the vector product by $\mathbf{a} \times \mathbf{b}$. Cartesian coordinates will be denoted by x, y, z, or x_1, x_2, x_3, and time by t. The total time derivative will occasionally also be denoted by a dot over the quantity. The total time derivative of a vector quantity $\boldsymbol{w}(x,y,z,t)$ is given by

$$\dot{\boldsymbol{w}} = \frac{d\boldsymbol{w}}{dt} = \frac{\partial \boldsymbol{w}}{\partial t} + (\mathbf{v} \cdot \mathbf{grad})\boldsymbol{w}$$

where $\mathbf{v}$ is the velocity with which the material point, which is the carrier of the property $\boldsymbol{w}$, moves.

1. Mechanics of a Single Mass Point and a System of Mass Points

1.1. Newton's laws of motion and fundamental motions.

a. Bodies not subject to internal forces continue in a state of rest or a straight line uniform motion (law of inertia). The equations of motion are

$$\frac{d^2x}{dt^2} = \frac{d^2y}{dt^2} = \frac{d^2z}{dt^2} = 0 \tag{1}$$

Any coordinate system in which Eq. (1) holds simultaneously for all masses, not subject to forces, is called an inertial coordinate system. The collection of all reference frames which move with constant velocity with respect to a given inertial system forms the totality of all inertial frames. The coordinate system anchored in the center of mass of the fixed stars is a good approximation of an inertial system.

b. Bodies which are subject to forces undergo acceleration. The acceleration is parallel to the force and its magnitude is the ratio of the force acting on the body and the (inertial) mass of the body. The mathematical form of the second law is

$$\boldsymbol{f} = m\boldsymbol{a} \tag{2}$$

or

$$f_1 = m\ddot{x}, \quad f_2 = m\ddot{y}, \quad f_3 = m\ddot{z} \tag{3}$$

(force equals mass times acceleration.)

c. To each force exerted by a mass point A on another mass point B corresponds a force exerted by B on A. The second force is equal in magnitude to the first and opposite in direction (law of action and reaction).

$$\boldsymbol{p} = m\mathbf{v} \quad [\mathbf{v} = (\dot{x}, \dot{y}, \dot{z})]$$

is called the linear momentum. The force law can also be written

$$\boldsymbol{f} = \frac{d\boldsymbol{p}}{dt} \tag{4}$$

The acceleration $\boldsymbol{a}$ can also be decomposed into a component tangential to the path

$$\boldsymbol{b}_t = \frac{\mathbf{v}}{v}\frac{dv}{dt} \tag{5}$$

and a component normal to the path

$$\boldsymbol{b}_n = \frac{\boldsymbol{R}}{R^2}v^2 \tag{6}$$

where $\boldsymbol{R}$ is a vector that has the direction and magnitude of the radius of curvature. The expression $m\boldsymbol{b}_n$ is called the centripetal force.

If $\boldsymbol{r}$ is the position vector of the mass point m, the vector product $m(\boldsymbol{r} \times \mathbf{v})$ is called the angular momentum, and if the force $\boldsymbol{f}$ acts on the mass, the vector product $(\boldsymbol{r} \times \boldsymbol{f})$ is called the torque (or moment of force) inerted by the force $\boldsymbol{f}$.

The work dW done by the force $\boldsymbol{f}$ during a displacement $d\boldsymbol{r}$ is defined by

$$dW = \boldsymbol{f} \cdot d\boldsymbol{r} = (\boldsymbol{f} \cdot \mathbf{v})dt = m\left(\frac{d\mathbf{v}}{dt} \cdot \mathbf{v}\right) = \frac{d}{dt}\left(\frac{mv^2}{2}\right) = dT \tag{7}$$

$$T = \frac{1}{2} mv^2 = \frac{p^2}{2m} \quad \text{is called the kinetic energy.}$$

$$\frac{dW}{dt} = \boldsymbol{f} \cdot \mathbf{v} \quad \text{is called the power of the force } \boldsymbol{f}.$$

1.2. Special cases. a. *Central force.* If the force $\boldsymbol{f}(\boldsymbol{r},t)$ is parallel to $\boldsymbol{r}$, then $\boldsymbol{f} \times \boldsymbol{r} = 0$, and therefore

$$m \frac{d}{dt}(\boldsymbol{r} \times \dot{\boldsymbol{r}}) = m(\boldsymbol{r} \times \ddot{\boldsymbol{r}}) = 0 \tag{1}$$

(law of conservation of angular momentum). The vector $\boldsymbol{k} = \dfrac{\boldsymbol{r} \times \dot{\boldsymbol{r}}}{2}$ is a constant, nonlocalized vector (areal velocity vector). In planetary motion Eq. (1) expresses Kepler's second law.

b. *Mass point in a potential field.* If the force $\boldsymbol{f}$ can be written as the negative gradient of a scalar function $U(\boldsymbol{r},t)$, i.e., if

$$\boldsymbol{f} = -\,\mathbf{grad}\; U(\boldsymbol{r},t) \tag{2}$$

U is the potential energy. If U does not depend implicitly on time, i.e.,

$$U = U(\boldsymbol{r}) \tag{3}$$

the force $\boldsymbol{f}$ is called conservative, and

$$\frac{d}{dt}(T + U) = 0 \quad \text{(conservation law of energy)} \tag{4}$$

or

$$T + U = E = \text{constant} = \text{total energy.}$$

c. *Constraints.* The mass points may be constrained to move along certain surfaces or curves, called constraints. These constraints can be replaced by constraint forces $\boldsymbol{f}^*$ which keep the bodies from leaving these constraints but do themselves no work, since the bodies always move at right angles to the constraints. The law of motion for a single mass point under the combined influence of an internal force $\boldsymbol{f}$ and a constraint force $\boldsymbol{f}^*$ is

$$\boldsymbol{f} + \boldsymbol{f}^* = m\boldsymbol{a} \tag{5}$$

If the constraint is given by

$$\phi(x,y,z) = 0 \tag{6}$$

(i.e., the particle is constrained to move on a surface), the force of constraint is

$$\boldsymbol{f}^* = \lambda\, \mathbf{grad}\, \phi \tag{7}$$

The Lagrangian multiplier λ is an unknown function of space and time.

For example, consider a bead that slides frictionless on a straight wire that makes an angle with the horizontal whose tangent is A. We assume the wire in the $x-z$ plane with the positive z axis directed upward.

The equation of the constraint is

$$z - Ax - B = 0 \tag{8}$$

The equations of motion are, therefore

$$m\ddot{x} = -A\lambda \tag{9}$$

$$m\ddot{z} = -mg \tag{10}$$

We can eliminate λ from both equations and obtain

$$-\frac{m}{A}\ddot{x} = m\ddot{z} + mg \tag{11}$$

But since

$$\ddot{z} = A\ddot{x} \tag{12}$$

we have

$$-\frac{m}{A}\ddot{x} = mA\ddot{x} + mg \tag{13}$$

or

$$x = -\sin\alpha\cos\alpha g\frac{t^2}{2} \tag{14}$$

$$z = -\sin^2\alpha g\frac{t^2}{2} \tag{15}$$

and

$$\lambda = -\frac{m}{A}\ddot{x} = mg\cos^2\alpha \tag{16}$$

where $\alpha =$ arc tan A, and where we assumed the particle started from rest at the origin.

d. *Apparent or inertial forces.* All the above laws refer to inertial systems. If we refer the equations of motion to accelerated systems, additional terms appear which are called apparent or inertial forces.

The most important cases are :

1. Linearly accelerated systems :

$$\boldsymbol{r} = \boldsymbol{r}^* + \boldsymbol{b}(t) \tag{17}$$

($\boldsymbol{r}$ refers to the position vector in the inertial frame and $\boldsymbol{r}^*$ to the accelerated frame). The equation of motion is

$$m\ddot{\boldsymbol{r}}^* = \boldsymbol{f} - m\ddot{\boldsymbol{b}} \tag{18}$$

where $\boldsymbol{f}$ is the external force and $-m\ddot{\boldsymbol{b}}$ is the inertial force.

2. Rotating coordinate systems: Let $\boldsymbol{k}$ be a unit vector in the direction of the fixed axis of rotation and ω the angular velocity. The equation of motion is

$$m\ddot{\boldsymbol{r}}^* = \boldsymbol{f} - 2m\omega(\boldsymbol{k} \times \dot{\boldsymbol{r}}^*) - m\omega^2[\boldsymbol{k} \times (\boldsymbol{k} \times \boldsymbol{r}^*)] \tag{19}$$

$-2m\omega(\boldsymbol{k} \times \dot{\boldsymbol{r}}^*)$ is the Coriolis force.

$-m\omega^2[\boldsymbol{k} \times (\boldsymbol{k} \times \boldsymbol{r}^*)]$ is the centrifugal force.

1.3. Conservation laws. We shall assume a system of N mass points which exert central forces on one another and which are in addition subject to external forces $\boldsymbol{f}_n^*(n = 1,2,\ldots,N)$, where $\boldsymbol{f}_n^*$ is the force acting on the nth mass point. The following notations will be used.

$$\mathbf{v}_n = \text{velocity of the } n\text{th mass point}$$

$$m_n\mathbf{v}_n = \boldsymbol{p}_n = \text{linear momentum of the } n\text{th mass point}$$

$$\boldsymbol{P} = \sum_{n=1}^{N} \boldsymbol{p}_n = \text{total linear momentum}$$

$$\boldsymbol{I} = \sum_{n=1}^{N} m_n(\boldsymbol{r}_n \times \mathbf{v}_n) = \text{total angular momentum}$$

$$M = \sum_{n=1}^{N} m_n = \text{total mass of system}$$

$$\boldsymbol{\rho} = \frac{1}{M}\sum_{n=1}^{N} m_n\boldsymbol{r}_n = \text{position of center of mass}$$

$$\boldsymbol{F}^* = \sum_{n=1}^{N} \boldsymbol{f}_n^* = \text{total external force}$$

All quantities are assumed to be taken with respect to a Cartesian inertial reference frame.

The following three conservation laws hold

$$\frac{d\mathbf{P}}{dt} = M\frac{d\boldsymbol{\rho}}{dt} = \mathbf{F}^* \tag{1}$$

$$\frac{d\mathbf{I}}{dt} = M(\boldsymbol{\rho} \times \mathbf{F}^*) \tag{2}$$

$$\frac{dE}{dt} = \sum_{n=1}^{N} \mathbf{f}_n{}^* \cdot \mathbf{v}_n \tag{3}$$

If the internal forces vanish, the linear momentum, angular momentum, and energy of the system remain constant, hence the name "conservation laws."

1.4. Lagrange equations of the second kind for arbitrary curvilinear coordinates. The equations of motions containing the forces of constraint (see § 1.2*c*) contain supernumerary coordinates, since the constraints actually reduce the number of degrees of freedom. They also use Cartesian coordinates. Often these Lagrange equations of the first kind (as they are called) are difficult to use. The generalized coordinates $q_1, \ldots, q_f$, where f is the true number of degrees of freedom, are often more convenient. All the q_k must be independent of each other, and

$$\left.\begin{aligned} x_i &= x_i(q_1,\ldots,q_f) \\ y_i &= y_i(q_1,\ldots,q_f) \\ z_i &= z_i(q_1,\ldots,q_f) \end{aligned}\right\} \quad (i = 1, 2, \ldots, N) \tag{1}$$

and

$$\dot{x}_i = \sum_{s=1}^{f} \frac{\partial x_i}{\partial q_s}\dot{q}_s \tag{2}$$

etc. The total kinetic energy

$$T = \frac{1}{2}\sum_{i=1}^{N} m_i(\dot{x}_i{}^2 + \dot{y}_i{}^2 + \dot{z}_i{}^2)$$

expressed in terms of the generalized velocities $\dot{q}$, will be a homogeneous quadratic function of the $\dot{q}$, where the coefficients will in general be functions of the q. We assume the system conservative and denote the potential energy by U and define the Lagrange function or the Lagrangian L to be

$$L \equiv T - U \tag{3}$$

The Lagrange equations of the second kind read

$$\frac{d}{dt}\left(\frac{\partial L}{\partial \dot{q}_k}\right) - \frac{\partial L}{\partial q_k} = 0 \tag{4}$$

The quantities $p_k = \partial L/\partial \dot{q}_k$ are called the generalized momenta.

Note: Equation (4) assumes that all forces are conservative. If there exist in addition nonconservative forces F_k (e.g., magnetic forces or friction forces), the Lagrange equations are to be generalized to

$$\frac{d}{dt}\left(\frac{\partial L}{\partial \dot{q}_k}\right) - \frac{\partial L}{\partial q_k} = F_k \tag{5}$$

Sometimes we can find a function M of q and $\dot{q}$ so that even the nonconservative forces F_k appear in the form

$$\frac{d}{dt}\frac{\partial M}{\partial \dot{q}_k} - \frac{\partial M}{\partial q_k} - F_k \tag{6}$$

In that case we introduce a new Lagrangian $\bar{L} \equiv L - M$, and the equation of motion again assumes the form (4):

$$\frac{d}{dt}\frac{\partial \bar{L}}{\partial \dot{q}_k} - \frac{\partial \bar{L}}{\partial q_k} = 0 \tag{7}$$

Consider the motion of an electric charge ϵ in an electromagnetic field. In that case

$$M = \frac{\epsilon}{c}(A \cdot \mathbf{v}) \tag{8}$$

where A is the so-called magnetic vector potential and $\mathbf{v}$ is the velocity of the charge ϵ.

1.5. The canonical equations of motion. If in the Lagrangian L the generalized velocities $\dot{q}$ are expressed in terms of the generalized momenta p, we obtain a function L^* which is numerically equal to L:

$$L^*(q_k,p_k,t) = L(q_k,\dot{q}_k,t) \tag{1}$$

We introduce a new function $H(q_k,p_k,t)$ defined by

$$H(q_k,p_k,t) \equiv -L^*(q_k,p_k,t) + \sum_{s=1}^{f} p_s\dot{q}_s(q_k,p_k,t) \tag{2}$$

The equations of motion (§ 1.4-5) become then

$$\dot{p}_k = -\frac{\partial H}{\partial q_k} \tag{3}$$

$$\dot{q}_k = \frac{\partial H}{\partial p_k} \tag{4}$$

where H the Hamiltonian,

$$H = T + U \tag{5}$$

hence equal to the total energy in a conservative system. The symbolical $2f$-dimensional space of q and p comprises the " phase space." Through each point in phase space passes exactly one mechanical trajectory. The f-dimensional subspace the coordinates q is called the configuration space.

1.6. Poisson brackets. If F is any dynamical variable of the mechanical system, e.g., the angular momentum, assumed to be expressed in terms of q, p, and t, its time rate of change is given by

$$\frac{dF}{dt} = \sum_{k=1}^{f} \left(\frac{\partial F}{\partial q_k} \dot{q}_k + \frac{\partial F}{\partial p_k} \dot{p}_k \right) + \frac{\partial F}{\partial t} = \sum_{k=1}^{f} \left(\frac{\partial F}{\partial q_k} \frac{\partial H}{\partial p_k} - \frac{\partial F}{\partial p_k} \frac{\partial H}{\partial q_k} \right) + \frac{\partial F}{\partial t} \tag{1}$$

An expression such as

$$\sum_{k=1}^{f} \frac{\partial F}{\partial q_k} \frac{\partial H}{\partial p_k} - \frac{\partial F}{\partial p_k} \frac{\partial H}{\partial q_k}$$

is called a Poisson bracket. The Poisson bracket of any two dynamical variables F and G is defined by

$$[F,G] \equiv \sum_{k=1}^{f} \left(\frac{\partial F}{\partial q_k} \frac{\partial G}{\partial p_k} - \frac{\partial F}{\partial p_k} \frac{\partial G}{\partial q_k} \right) \tag{2}$$

Equation (1) can be written

$$\dot{F} = [F,H] + \frac{\partial F}{\partial t} \tag{3}$$

In particular,

$$\dot{H} = [H,H] + \frac{\partial H}{\partial t} = \frac{\partial H}{\partial t} \tag{4}$$

or

$$\frac{dH}{dt} = \frac{\partial H}{\partial t} \tag{5}$$

In a conservative system, in which the energy does not depend explicitly on time, H is constant. Any constant of the motion (or " integral of the motion ") has a vanishing Poisson bracket with H.

The Poisson brackets satisfy the identities

$$[F,G] + [G,F] \equiv 0 \tag{6}$$

and

$$\Big[F,[G,M]\Big] + \Big[G,[M,F]\Big] + \Big[M,[F,G]\Big] \equiv 0 \tag{7}$$

The latter is called Jacobi's identity. The Poisson brackets of the canonical coordinates q and p themselves have the values

$$\left.\begin{aligned} [q_k,q_l] &= 0 \\ [p_k,p_l] &= 0 \\ [q_k,p_k] &= \delta_{kl} \end{aligned}\right\} \tag{8}$$

These Poisson brackets are of fundamental importance in the Heisenberg formulation of quantum mechanics, where they express the fundamental uncertainty principle.

1.7. Variational principles. a. *Hamilton's principle.* The Lagrange equations (§ 1.4-4) are the Euler-Lagrange equations of the variational principle

$$\delta S = \delta \int_{t_1}^{t_2} L\, dt = 0 \tag{1}$$

The variation is taken in configuration space only, subject to the restriction that the time (and therefore also the end points of the path) is not varied. This formulation of the Lagrange equations shows their invariance in any coordinate system i.e., if $L(q,\dot{q},t) = L^*(q^*,\dot{q}^*,t)$, then in the new coordinate system the same equations hold again :

$$\frac{d}{dt}\left(\frac{\partial L^*}{\partial \dot{q}^*}\right) - \frac{\partial L^*}{\partial q^*} = 0 \tag{2}$$

$$S = \int_{t_1}^{t_2} L(q,\dot{q},t)dt$$

is called the action function, or Hamilton's principal function, and (1) defines Hamilton's "principle of least action," the analogue of Fermat's principle in optics, (see § 1.11).

b. *Variational principle in phase space.* The canonical equations of motion (§ 1.5-3.4) are the Euler-Lagrange equations of the variational principle in phase space (i.e. q and p are considered the coordinates of a point) :

$$\delta \int_{t_1}^{t_2} \left[\sum_{k=1}^{f} p_k\dot{q}_k - H(q_k,p_k,t)\right] dt = 0 \quad [\dot{q}_k = \dot{q}_k(q_k,p_k,t)] \tag{3}$$

The time and the endpoints are again held fixed.

1.8. Canonical transformations. The form of the equations is independent of any specific coordinate system. The basic principle serves as a unifying guide for expressing more general theories, e.g., in the general theory of relativity or quantum mechanics.

Any transformation law from the canonical coordinates q and p to new canonical coordinates $\bar{q}$ and $\bar{p}$, by which the canonical equations remain invariant, is called a canonical transformation. Mathematically the canonical transformations are a special case of the contact transformations. The necessary and sufficient condition that the transformation be canonical is that the two linear differential forms

$$\sum_{k=1}^{f} p_k dq_k - H(q_k,p_k,t)dt \quad \text{and} \quad \sum_{k=1}^{f} \bar{p}_k d\bar{q}_k - \bar{H}(\bar{q}_k,\bar{p}_k,t)dt$$

differ by a total differential, denoted by dV:

$$(\Sigma p_k dq_k - H dt) - (\Sigma \bar{p}_k d\bar{q}_k - \bar{H} dt) = dV \tag{1}$$

$$\bar{q}_k = \bar{q}_k(q_1 q_2 \cdots p_1 p_2 \cdots t) \tag{2}$$

$$\bar{p}_k = \bar{p}_k(q_1 q_2 \cdots p_1 p_2 \cdots t) \tag{3}$$

where V is called the generating function. The canonical transformations are particular mappings of phase space upon itself. The following choices are possible for the function V.

$$\left.\begin{aligned} &V = V(q,\bar{q},t) \\ &p_k = \frac{\partial V}{\partial q_k}; \quad \bar{p}_k = -\frac{\partial V}{\partial \bar{q}_k}; \quad H = \bar{H} - \frac{\partial V}{\partial t} \end{aligned}\right\} \tag{4}$$

$$\left.\begin{aligned} &V = V(q,\bar{p},t) \\ &p_k = \frac{\partial V}{\partial q_k}; \quad \bar{q}_k = \frac{\partial V}{\partial \bar{p}_k}; \quad H = \bar{H} - \frac{\partial V}{\partial t} \end{aligned}\right\} \tag{5}$$

$$\left.\begin{aligned} &V = V(\bar{q},p,t) \\ &\bar{p}_k = -\frac{\partial V}{\partial \bar{q}_k}; \quad q_k = -\frac{\partial V}{\partial p_k}; \quad H = \bar{H} - \frac{\partial V}{\partial t} \end{aligned}\right\} \tag{6}$$

$$\left.\begin{aligned} &V = V(p,\bar{p},t) \\ &q_k = -\frac{\partial V}{\partial p_k}; \quad \bar{q}_k = \frac{\partial V}{\partial \bar{p}_k}; \quad H = H - \frac{\partial V}{\partial t} \end{aligned}\right\} \tag{7}$$

For example,

$$V = q_1\bar{p}_1 + q_2\bar{p}_2 + \dots \tag{8}$$

leads to the identity transformation

$$q_1 = \bar{q}_1 \qquad p_1 = \bar{p}_1 \qquad \text{etc.}$$

$$V = p_x r \cos\phi + p_y r \sin\phi + p_z z \tag{9}$$

gives us the transformation from Cartesian to cylindrical coordinates:

$$\begin{aligned} x &= r\cos\phi, & p_r &= p_x \cos\phi + p_y \sin\phi \\ y &= r\sin\phi, & p_z &= -p_x r \sin\phi + p_y r \cos\phi \\ z &= z, & p_z &= p_z \end{aligned}$$

The expression $p_x^2 + p_y^2$ goes over into $p_r^2 + p_\phi^2/r^2$.

If we use $$V = p_x r \cos\phi \sin\theta + p_y r \sin\phi \sin\theta + p_z r \cos\theta \tag{10}$$

we obtain the transition to polar coordinates:

$$\begin{aligned} x &= r\cos\phi\sin\theta, & p_r &= p_x\cos\phi\sin\theta + p_y\sin\phi\sin\theta + p_z\cos\theta \\ y &= r\sin\phi\sin\theta, & p_z &= -p_x r\sin\phi\sin\theta + p_y r\cos\phi\sin\theta \\ z &= r\cos\theta, & p_\theta &= p_x r\cos\phi\cos\theta + p_y r\sin\phi\cos\theta - p_z r\sin\theta \end{aligned}$$

and $$p_x^2 + p_y^2 + p_z^2 = p_r^2 + \frac{p_\theta^2}{r^2} + \frac{p_\phi^2}{r^2\sin^2\theta}\,.$$

Rotation of a Cartesian coordinate system (x_1,x_2,x_3) to $(\bar{x}_1,\bar{x}_2,\bar{x}_3)$:

$$V = \Sigma\alpha_{ik}x_i\bar{p}_k \qquad (i,k = 1,2,3) \tag{11}$$

where $$\Sigma\alpha_{ij}\alpha_{kj} = \delta_{ik}$$

$$\bar{x}_i = \sum_{r=1}^{3} \alpha_{ri}x_r, \quad \text{and} \quad p_i = \sum_{r=1}^{3} \alpha_{ri}x_r$$

A frequently used transformation is

$$\left.\begin{aligned} V &= \frac{m}{2}\omega q^2 \cot\bar{q} \\ q &= \sqrt{\frac{2\bar{p}}{m\omega}}\sin\bar{q}; \quad p = \sqrt{2m\omega\bar{p}}\cos\bar{q} \end{aligned}\right\} \tag{12}$$

and $p^2/2m + m\omega^2q^2/2$ goes over into $\omega\bar{p}$. This last transformation shows how we can easily find the equations of motion of the linear harmonic oscillator. Since $T = m\dot{q}^2/2$ and $U = kq^2/2 = m\omega^2 q^2/2$, we obtain for the Hamiltonian the expression

$$H = \frac{p^2}{2m} + \frac{m\omega^2}{2} q^2$$

We shall call the new canonical variables ϕ and α and thus have the relations :

$$q = \sqrt{\frac{2\alpha}{m\omega}} \sin\phi \qquad p = \sqrt{2m\omega\alpha} \cos\phi$$

Hence the new Hamiltonian is $\bar{H} = \omega\alpha$, and the equations of motion are

$$\alpha = \text{constant} \qquad \phi = \omega t + \beta$$

and therefore

$$q = \sqrt{\frac{2\alpha}{m\omega}} \sin(\omega t + \beta)$$

The canonical transformations were characterized by the fact that they leave a certain integral in phase space, the action integral, invariant. Still other quantities that are invariant in phase space under canonical transformations, include the so-called integral invariants. They are

$$F_1 = \iint \sum_{k=1}^{f} dp_k dq_k \tag{13}$$

taken over an arbitrary two-dimensional submanifold in phase space. Similarly

$$F_2 = \iiiint \Sigma\, dp_i dp_k dq_i dq_k \tag{14}$$

where the summation is taken over every combination of two indices and the integral is again taken over any arbitrary four-dimensional submanifold in phase space. Similarly

$$F_3 = \iiint \iiint \Sigma\, dp_i dp_k dp_l dq_i dq_k dq_l \tag{15}$$

where the summation is taken over every combination of three indices and the integral is taken over any arbitrary six-dimensional submanifold. The last integral in this sequence is

$$F_f = \int \dots \int dp_1 dp_2 \dots dp_f dq_1 dq_2 \dots dq_f \tag{16}$$

The last mentioned integral invariant, known as Liouville's theorem in statistical mechanics, says that the volume in phase space is an invariant. The integral invariants were of great significance in the earlier Bohr-Sommerfeld formulation of quantum theory. They were those dynamical quantities of classical mechanics which had to be quantized.

1.9. Infinitesimal contact transformations. * A transformation

$$\left.\begin{aligned}\bar{q}_k &= q_k + \epsilon\phi_k(q_1q_2\cdots p_1p_2\cdots)\\ \bar{p}_k &= p_k + \epsilon\psi_k(q_1q_2\cdots p_1p_2\cdots)\end{aligned}\right\}\quad (k = 1,2,\ldots,f) \tag{1}$$

is called infinitesimal if ϵ is a small quantity such that its higher powers can be neglected as compared to the first power. Thus if we have a dynamical function $F(\bar{q}_k,\bar{p}_k)$, we can write

$$F(q_k + \epsilon\phi_k, p_k + \epsilon\psi_k) = F(q_k,p_k) + \epsilon\sum_{r=1}^{f}\frac{\partial F}{\partial q_r}\phi_r + \epsilon\sum_{r=1}^{f}\frac{\partial F}{\partial p_r}\psi_r \tag{2}$$

This infinitesimal transformation will be canonical, if

$$\Sigma p_r\delta q_r - \Sigma\bar{p}_r\delta\bar{q}_r = \delta F$$

or

$$\Sigma p_r\delta q_r - \Sigma(p_r + \epsilon\psi_r)(\delta q_r + \epsilon\delta\phi_r) = \delta F$$

or omitting terms containing ϵ^2, if

$$-\epsilon\Sigma p_r\delta\phi_r + \psi_r\delta q_r = \delta F$$

or

$$\epsilon\sum_r\left[\sum_l p_r\left(\frac{\partial\phi_r}{\partial p_l}\delta p_l + \frac{\partial\phi_r}{\partial q_l}\delta q_l\right) + \psi_r\delta q_l\right] = -\delta F$$

If we put $F - \epsilon W$, we obtain

$$\left.\begin{aligned}\psi_r + \sum_l p_l\frac{\partial\phi_l}{\partial q_r} &= -\frac{\partial W}{\partial q_r}\\ \sum_l p_l\frac{\partial\phi_l}{\partial p_r} &= -\frac{\partial W}{\partial p_r}\end{aligned}\right\} \tag{3}$$

or

$$\psi_r + \frac{\partial}{\partial q_r}\sum p_l\phi_l = -\frac{\partial W}{\partial q_r}$$

$$\frac{\partial}{\partial p_r}\sum p_l\phi_l - \phi_r = -\frac{\partial W}{\partial p_r} \tag{4}$$

If we put

$$W + \Sigma p_l\phi_l = V \tag{5}$$

then

$$\psi_r = -\frac{\partial V}{\partial q_r},\quad \phi_r = \frac{\partial V}{\partial p_r} \tag{6}$$

* See, e. g., HAMEL, G., *Theoretische Mechanik*, Julius Springer, Berlin, 1949, p. 299.

Therefore the infinitesimal canonical transformation is

$$\bar{q}_k - q_k \equiv \Delta q_k = \epsilon \frac{\partial V}{\partial p_k} \tag{7}$$

$$\bar{p}_k - p_k \equiv \Delta p_k = -\epsilon \frac{\partial V}{\partial q_k} \tag{8}$$

where V is the generating function. Therefore we can regard the canonical equations

$$dq_k = \frac{\partial H}{\partial p_k} dt \quad \text{and} \quad dp_k = -\frac{\partial H}{\partial p_k} dt$$

as infinitesimal transformations with the Hamiltonian H as the generating function. Any finite canonical transformation can be built up out of successive infinitesimal transformations.

Any arbitrary dynamical function $F = F(q,p)$ undergoes during an infinitesimal canonical transformation, a change ΔF.

$$\begin{aligned} \Delta F &= \sum \frac{\partial F}{\partial q_r} \Delta q_r + \sum \frac{\partial F}{\partial p_r} \Delta p_r \\ &= \epsilon \sum \left(\frac{\partial F}{\partial q_r} \frac{\partial V}{\partial p_r} - \frac{\partial F}{\partial p_r} \frac{\partial V}{\partial q_r} \right) = \epsilon[F,V] \end{aligned} \tag{9}$$

This equation brings out again the meaning of the Poisson bracket. If V is the Hamiltonian and if F is a constant of the motion,

$$\Delta F = 0 \tag{10}$$

1.10. Cyclic variables. *The Hamilton-Jacobi partial differential equation.* If the Hamiltonian H does not contain a certain coordinate, e.g., q_1,

$$H = H(p_1,q_2,p_2,\ldots,q_f,p_f,t) \tag{1}$$

then

$$\dot{p}_1 = 0 \tag{2}$$

$$p_1 = \text{constant} \tag{3}$$

Therefore p_1 is a constant of the motion. Helmholtz has introduced the name of cyclic coordinate for such a " hidden " coordinate. We shall try to find such a canonical transformation wherein all the new coordinates are cyclic. We denote the original canonical variables by q_k and p_k and the corresponding new cyclic canonical coordinates and momenta, respectively, by ϕ_k and α_k. Denote the generating function by $S = S(q_1,\ldots,\alpha_1,\ldots,t)$.

Then

$$p_k = \frac{\partial S}{\partial q_k} \quad \text{and} \quad \phi_k = \frac{\partial S}{\partial \alpha_k} \tag{4}$$

(compare § 1.8, Case c).

By appropriately disposing of the time dependence of the generating function, we can always make the new Hamiltonian $\bar{H}$ vanish.* Therefore

$$\bar{H} = H + \frac{\partial S}{\partial t} = 0 \tag{5}$$

or

$$H\left(q_k, \frac{\partial S}{\partial q_k}, t\right) + \frac{\partial S}{\partial t} = 0 \tag{6}$$

This is the Hamilton-Jacobi partial differential equation. Since the ϕ_k are cyclic variables, the equations of motion can be written down at once :

$$\alpha_k = \text{constant} \tag{7}$$

and

$$\phi_k = \omega_k t + \beta_k \tag{8}$$

where

$$\omega_k = \frac{\partial S}{\partial \alpha_k} \tag{9}$$

Since $p_k = \partial S/\partial q_k$, we have

$$dS = \sum_{k=1}^{f} \frac{\partial S}{\partial q_k} dq_k + \frac{\partial S}{\partial t} dt = \sum p_k \dot{q}_k dt + \frac{\partial S}{\partial t} dt = 2T dt + \frac{\partial S}{\partial t} dt$$

We assume furthermore a conservative system. Therefore $H = E$ or $\partial S/\partial t = -E$; since $E = T + U$,

$$\left.\begin{aligned} dS &= 2T\,dt - E\,dt = (T - U)dt = L\,dt \\ \text{and} \qquad S &= \int_{t_1}^{t_2} L\,dt \end{aligned}\right\} \tag{10}$$

This shows that for a conservative system the generating function S is identical with the action function introduced in § 1.7, $S = S(q_1, \ldots, q_f, t)$ which we can write

$$S = S^*(q_1, \ldots, q_f, \alpha_1, \ldots, \alpha_f) - Et \tag{11}$$

$$p_k = \frac{\partial S}{\partial q_k} = \frac{\partial S^*}{\partial q_k} \tag{12}$$

Here S^* is sometimes called the " reduced " action function because it

* See, e.g., BERGMANN, P. G., *Basic Theories of Physics: Mechanics and Electrodynamics*, Prentice-Hall, Inc., New York, 1949, p. 38.

depends only on the coordinates but not on the time. The Hamilton-Jacobi partial differential equation can also be written

$$H(q_1,q_2,\ldots,\frac{\partial S^*}{\partial q_1},\frac{\partial S^*}{\partial q_2},\ldots) = W(\alpha_1,\alpha_2,\ldots,\alpha_f) = E \tag{13}$$

$$\omega_k = \frac{\partial W}{\partial \alpha_k} \tag{14}$$

An important special case arises if the Hamiltonian is " separable," i.e., if

$$H = H_1(q_1,p_1) + H_2(q_2,p_2) + \ldots + H_f(q_f,p_f) \tag{15}$$

Set

$$H_k(q_k,p_k) = W_k$$

where

$$W_1 + W_2 + \ldots + W_f = W \tag{16}$$

If, furthermore,

$$S^*\,(q_1,\ldots,q_f) = S_1{}^*(q_1) + \ldots + S_f{}^*(q_f) \tag{17}$$

then

$$p_k = \frac{\partial S^*}{\partial q_k} = \frac{dS_k{}^*}{dq_k} \tag{18}$$

and instead of the Hamilton-Jacobi partial differential equation we have f ordinary differential equation to solve, each of the type

$$\left(\frac{dS_k{}^*}{dq_k}\right)^2 + g_k(q_k) = \alpha_k \tag{19}$$

The total energy E of the system is then a function of the $\alpha_1, \ldots, \alpha_f$:

$$E = E(\alpha_1,\ldots,\alpha_f) \tag{20}$$

As an example, consider the motion of a mass point in a central force field $U(r)$. Using the canonical transformation for spherical coordinates, we obtain

$$H = T + U = \frac{1}{2m}\left(p_r{}^2 + \frac{p_\theta{}^2}{r^2} + \frac{p\ {}^2}{r^2 \sin^2\theta}\right) + U(r) \tag{21}$$

We put

$$S^* = S_r{}^*(r) + S_\theta{}^*(\theta) + S\ {}^*(\phi) \tag{22}$$

and the Hamilton-Jacobi equation

$$\left(\frac{\partial S^*}{\partial r}\right)^2 + \frac{1}{r^2}\left(\frac{\partial S^*}{\partial \theta}\right)^2 + \frac{1}{r^2\sin^2\theta}\left(\frac{\partial S^*}{\partial \phi}\right)^2 + 2m[U(r) - W] = 0 \tag{23}$$

falls apart into three ordinary differential equations :

$$\frac{dS_\phi^*}{d\phi} = \alpha_\phi \tag{24}$$

$$\left(\frac{dS_\theta^*}{d\theta}\right)^2 + \frac{\alpha_\phi^2}{\sin^2\theta} = \alpha_\theta^2 \tag{25}$$

$$\left(\frac{dS_r^*}{dr}\right)^2 + \frac{\alpha_\theta^2}{r^2} + 2m[U(r) - W] = 0 \tag{26}$$

or

$$\frac{dS_r^*}{dr} = p_r = \sqrt{2m[W - U(r)] - \frac{\alpha_\theta^2}{r^2}} \tag{27}$$

$$\frac{dS_\theta^*}{d\theta} = p_\theta = \sqrt{\alpha_\theta^2 - \frac{\alpha_\phi^2}{\sin^2\theta}} \tag{28}$$

$$\frac{dS_\phi^*}{d\phi} = p_\phi = \alpha_\phi \tag{29}$$

The three constants of integration are $\alpha_z = p_z = mr^2 \sin^2 \theta\dot{\phi}$, the angular momentum component along the polar axis; α_θ, the total angular momentum ($\alpha_\phi = \alpha_\theta \cos \gamma$) where γ is the angle between the orbital plane and the equatorial plane); and $W = E$, the total energy.

1.11. Transition to wave mechanics. The optical-mechanical analogy. The Hamilton-Jacobi partial differential equation is of central importance in the transition from classical to wave mechanics. Before discussing the analogy with wave optics we shall discuss the relation to the limiting case of wave optics, viz., geometrical optics. Let μ denote the index of refraction, which may be a function of the coordinates. The paths of the light rays are governed by Fermat's principle :

$$\delta \int_{p_1}^{p_2} \mu\, ds = 0 \quad \text{or} \quad \delta \int_{p_1}^{p_2} \frac{ds}{u} = 0 \quad (u = \text{phase velocity}) \tag{1}$$

We shall now consider all trajectories of a particle belonging to the same constant total energy E. According to Hamilton's principle

$$0 = \delta \int_{t_1}^{t_2} (2T - E)dt - \delta \int_{t_1}^{t_2} 2T\, dt - \delta \int_{t_1}^{t_2} mv^2 dt - \delta \int_{p_1}^{p_2} mv\, ds$$

or

$$\delta \int_{p_1}^{p_2} v\, ds = 0 \tag{2}$$

This equation is of the same structure as (1), but the material velocity (or "group velocity") v is inversely proportional to the phase velocity u.

$$v = \sqrt{\frac{2}{m}(E - U)} \tag{3}$$

The index of refraction is defined by

$$\mu = \frac{u_0}{u} = \frac{v}{v_0} = \sqrt{\frac{E - U}{E}} \tag{4}$$

("Vacuum" is here defined as the region where the potential energy $U = 0$). Thus we can associate with a moving mass point a "wave" whose phase velocity is inversely proportional to the velocity of the particle and whose index of refraction is $\sqrt{(E - U)/U}$. It remains now to recover the "wave equation." We shall use Cartesian coordinates. We assume again a single mass point with the total energy E moving in the potential field U. The action function $S = S(x,y,z,t)$ represents at every instant t a surface in space on which the value of S is constant. Let the velocity with which this surface of a fixed value for S propagates itself in space be denoted by $\boldsymbol{u}$. An observer travelling with the same velocity $\boldsymbol{u}$ will not see any change in value of S. Therefore

$$0 = \frac{dS}{dt} = \frac{\partial S}{\partial t} + \boldsymbol{u} \cdot \nabla S = \frac{\partial S}{\partial t} + \boldsymbol{u} \cdot \boldsymbol{p}$$

or

$$-\frac{\partial S}{\partial t} = \boldsymbol{u} \cdot \boldsymbol{p} = m\boldsymbol{u} \cdot \mathbf{v} \tag{5}$$

But

$$-\frac{\partial S}{\partial t} = E \tag{6}$$

and

$$E = m\mathbf{v} \cdot \boldsymbol{u} \tag{7}$$

But $\boldsymbol{u}$ is parallel to $\mathbf{v}$ since $\boldsymbol{u}$ is perpendicular to the surfaces $S =$ constant and $\boldsymbol{p} = \mathbf{grad}\, S$, which means that $\boldsymbol{p}$ is also perpendicular to the surfaces $S =$ constant.

Therefore

$$E = mvu \tag{8}$$

or

$$u = \frac{E}{mv} = \frac{E}{\sqrt{2m(E - U)}} \tag{9}$$

If we solve Eq. (9) for U and replace E again by $-\partial S/\partial t$, the Hamilton-Jacobi equation assumes the form

$$\frac{1}{2m}(|\,\mathbf{grad}\, S\,|)^2 + U + \frac{\partial S}{\partial t} = 0 \tag{10}$$

or

$$(|\,\mathbf{grad}\; S\,|)^2 = \frac{1}{u^2}\left(\frac{\partial S}{\partial t}\right)^2 \tag{11}$$

This is the " wave equation " of classical mechanics. In optics the law according to which a scalar quantity $g(x,y,z,t)$ (e.g., the component of the electric field strength in an electromagnetic wave) is propagated is governed by the wave equation :

$$\nabla^2 g = \frac{\mu^2}{c^2}\frac{\partial^2 g}{\partial t^2} \tag{12}$$

where μ = index of refraction, c = velocity of light in empty space. We try a solution of the form

$$g(x,y,z,t) = f(x,y,z)e^{i\phi} \tag{13}$$

The phase ϕ is given by

$$\phi = 2\pi\left(\frac{l}{\lambda} - \nu t\right) = 2\pi\nu\left(\frac{l}{\mu} - t\right) = 2\pi\nu\left(\frac{\mu l}{c} - t\right) \tag{14}$$

where ν = frequency, λ = wavelength, l = geometrical path length of optical path.

We shall assume that the amplitude function $f(x,y,z)$ changes so slowly that we can assume it constant. Substituting Eq. (13) into Eq. (12) we obtain for ϕ,

$$i \operatorname{div} \mathbf{grad}\,\phi - (|\,\mathbf{grad}\,\phi\,|)^2 = i\frac{\mu^2}{c^2}\frac{\partial^2\phi}{\partial t^2} - \frac{\mu^2}{c^2}\left(\frac{\partial\phi}{\partial t}\right)^2 \tag{15}$$

But

$$\frac{\partial\phi}{\partial t} = -2\pi\nu, \qquad \frac{\partial^2\phi}{\partial t^2} = 0$$

We shall assume also that the index of refraction changes very slowly ($\mathbf{grad}\,\mu = 0$) and that the rays are nearly parallel (div $\boldsymbol{S}_0 = 0$, where $\boldsymbol{S}_0$ is a unit tangent vector along the optical path.) Then Eq. (15) reduces to

$$(\mathbf{grad}\,\phi)^2 = \frac{1}{u^2}\left(\frac{\partial\phi}{\partial t}\right)^2 \tag{16}$$

But this is an equation of the same type as Eq. (11). Schrödinger therefore assumed that S is proportional to ϕ. Since S has the dimension of an action and ϕ is a pure number, Schrödinger put

$$\phi = \frac{2\pi}{h}S = \frac{2\pi}{h}(S^* - Et) \tag{17}$$

since, according to Planck's law,

$$\frac{E}{h} = \nu \tag{18}$$

Schrödinger assumed that the behavior of a particle can also be described by a wave equation :

$$\nabla^2\Psi = \frac{1}{u^2}\cdot\frac{\partial^2\Psi}{\partial t^2} \tag{19}$$

He put

$$\Psi = \psi(x,y,z)e^{-2\pi iEt/h} \tag{20}$$

where

$$\psi(x,y,z) = e^{2\pi iS^*/h}$$

and therefore, if this value (20) is substituted into Eq. (19), we obtain

$$\nabla^2\psi = -\frac{4\pi^2E^2}{h^2u^2}\psi \tag{21}$$

but

$$u^2 = \frac{E^2}{2m(E-U)}$$

and thus we obtain the celebrated Schrödinger equation :

$$\nabla^2\psi + \frac{8\pi^2m}{h^2}(E-U)\psi = 0 \tag{22}$$

1.12. The Lagrangian and Hamiltonian formalism for continuous systems and fields. So far only systems with a finite (or perhaps denumerably many) degrees of freedoms have been considered. We shall deal now with a continuous system, such as a fluid or an elastic solid. The ensuing relations could be derived as limiting cases of the corresponding laws for discrete systems. The classical treatment of continuous systems is a very important preliminary step to the quantization of wave fields, the so-called " second quantization."

The dynamical variables of particle mechanics, which are *functions* of a finite number of degrees of freedom, become now *functionals*, i.e., functions of infinitely many variables, namely the values of the field variables at each space-time point. We shall denote the field variables by $y_A(A = 1,\ldots,N)$. The Cartesian coordinates will be denoted by x_1, x_2, and x_3. The y_A play now the role of the generalized coordinates.

$$y_A = y_A(x_1,x_2,x_3,t) \tag{1}$$

Partial derivatives with respect to spatial coordinate, will be denoted by commas :

$$y_{A,r} \equiv \frac{\partial y_A}{\partial r} \tag{2}$$

and partial derivatives with respect to time by dots :

$$\dot{y}_A \equiv \frac{\partial y_A}{\partial t} \tag{3}$$

The $\dot{y}_A$ play the role of the generalized velocities $\dot{q}$. In the case of the electromagnetic field we have four field variables ($N = 4$), viz., the three components of the vector potential A and the scalar potential ϕ.

We assume again that the " equations of motion " are the Euler-Lagrange equations of a variational principle :

$$\delta L = 0 \tag{4}$$

$$L = \int_{t_1}^{t_2} dt \int_V \mathcal{L}\, dV \tag{5}$$

where dV is an ordinary three-dimensional volume element. The variations are taken only in the interior of the " world-domain " t, V, and the independent variables x, t are not varied. We shall assume that $\mathcal{L}$, (the " Lagrangian density "), is a function of the y_A, $y_{A,r}$, and $\dot{y}_A$ only and not of x and t, or any higher derivatives of the $\dot{y}_A$:

$$\mathcal{L} = \mathcal{L}(y_A, y_{A,r}, \dot{y}_A) \tag{6}$$

The resulting Euler-Lagrange equations are

$$\frac{\partial \mathcal{L}}{\partial y_A} - \sum_{r=1}^{3} \frac{\partial}{\partial x_r}\left(\frac{\partial \mathcal{L}}{\partial y_{A,r}}\right) - \frac{\partial}{\partial t}\frac{\partial \mathcal{L}}{\partial \dot{y}_A} = 0 \tag{7}$$

These equations are called the field equations. One calls

$$\frac{\partial L}{\partial y_A} - \sum_r \frac{\partial}{\partial x_r}\frac{\partial \mathcal{L}}{\partial y_{A,r}} \equiv \frac{\delta \mathcal{L}}{\delta y_A} \equiv \frac{\partial L}{\vartheta y_A}$$

the variational derivative of $\mathcal{L}$ with respect to y_A or the functional derivatives of L with respect to y_A. The energy-stress tensor $t_\sigma{}^\varrho$ is defined by

$$t_\sigma{}^\varrho = \mathcal{L}\delta_\sigma{}^\varrho - y_{A,\sigma}\frac{\partial \mathcal{L}}{\partial y_{A,\varrho}} \tag{8}$$

where $\rho, \sigma = 1, 2, 3, 4$. The index 4 refers to the time coordinate t. On account of the field equations (7) the tensor $t_\sigma{}^\varrho$ obeys the conservation laws :

$$\sum_{\varrho=1}^{4} t_\sigma{}^\varrho{}_{,\varrho} = 0 \tag{9}$$

These are four ordinary divergency relations, expressing the conservation of linear momentum ($\sigma = 1, 2, 3$) and energy ($\sigma = 4$).

In analogy to particle mechanics the momentum densities Π_A are defined to be

$$\Pi_A \equiv \frac{\partial \mathcal{L}}{\partial \dot{y}_A} \tag{10}$$

The next step is to define the Hamiltonian density $\mathcal{H}$. The Hamiltonian H itself is then

$$H = \iiint \mathcal{H}\, dV \tag{11}$$

$\mathcal{H}$ is defined the same way as in particle mechanics :

$$\mathcal{H}(\Pi_A, y_A, y_{A,r}) \equiv \sum_{A=1}^{N} \Pi_A \dot{y}_A - \mathcal{L}[y_A, y_{A,r}, \dot{y}_A(y_A, y_{A,r}, \Pi_A)] \tag{12}$$

In this expression we assume that the velocities $\dot{y}_A$ have been expressed in terms of the momenta. The canonical equations are

$$\dot{y}_A = \frac{\partial \mathcal{H}}{\partial \Pi_A} \equiv \frac{\delta \mathcal{H}}{\delta \Pi_A} \tag{13}$$

$$\dot{\Pi}_A = -\frac{\partial \mathcal{H}}{\partial y_A} + \frac{\partial}{\partial x_r}\left(\frac{\partial \mathcal{H}}{\partial y_{A,r}}\right) \equiv -\frac{\delta \mathcal{H}}{\delta y_A} \quad (r = 1,2,3) \tag{14}$$

If F and G are any two dynamical functionals, i.e.,

$$F = \iiint \mathcal{F}(y_A, y_{A,r}, \Pi_A) dV \tag{15}$$

and

$$G = \iiint \mathcal{G}(y_A, y_{A,r}, \Pi_A) dV \tag{16}$$

then the Poisson bracket of F and G is defined by

$$[F,G] = \iiint \sum_A \left(\frac{\delta \mathcal{F}}{\delta y_A}\frac{\delta \mathcal{G}}{\delta \Pi_A} - \frac{\delta \mathcal{F}}{\delta \Pi_A}\frac{\delta \mathcal{G}}{\delta y_A}\right) dV \tag{17}$$

In particular, the time rate of change of F is given by

$$\frac{dF}{dt} = [F,H] = \iiint \sum_A \left(\frac{\delta \mathcal{F}}{\delta y_A}\frac{\delta \mathcal{H}}{\delta \Pi_A} - \frac{\delta \mathcal{F}}{\delta \Pi_A}\frac{\delta \mathcal{H}}{\delta y_A}\right) dV \tag{18}$$

For example, in the electromagnetic field in free space, the field variables y_A are A_1, A_2, A_3, ϕ (i.e., the vector and the scalar potential). Maxwell's equations

$$\left.\begin{array}{c} \mathbf{curl}\, \boldsymbol{E} + \dfrac{1}{c}\dfrac{\partial \boldsymbol{H}}{\partial t} = 0; \quad \mathbf{curl}\, \boldsymbol{H} - \dfrac{1}{c}\dfrac{\partial \boldsymbol{E}}{\partial t} = 0 \\ \text{div}\, \boldsymbol{E} = 0; \qquad \text{div}\, \boldsymbol{H} = 0 \end{array}\right\} \tag{19}$$

are the Euler-Lagrange equations belonging to the Lagrangian density

$$\mathcal{L} = \frac{1}{8\pi}\left(\left|\frac{1}{c}\frac{\partial \boldsymbol{A}}{\partial t} + \mathbf{grad}\,\phi\right|\right)^2 - \frac{1}{8\pi}(|\,\mathbf{curl}\, \boldsymbol{A}\,|)^2 \tag{20}$$

The conjugate momentum density to A_1 is Π_1, where

$$\Pi_1 = \frac{1}{4\pi c}\left(\frac{1}{c}\frac{\partial A_1}{\partial t} + \frac{\partial \phi}{\partial x}\right) \tag{21}$$

The conjugate momentum to ϕ vanishes identically since $\partial\phi/\partial t$ does not occur in the Hamiltonian. The Hamiltonian density is

$$H = \mathbf{\Pi} \cdot \frac{\partial \boldsymbol{A}}{\partial t} - \mathcal{L} = 2\pi c^2 \mid \mathbf{\Pi} \mid^2 + \frac{1}{8\pi}(\mid \mathbf{curl}\, \boldsymbol{A} \mid)^2 - c\mathbf{\Pi} \cdot \nabla\phi \tag{22}$$

The canonical equations are

$$\frac{\partial \boldsymbol{A}}{\partial t} = 4\pi c^2 \mathbf{\Pi} - c\, \mathbf{grad}\, \phi \tag{23}$$

and

$$\frac{\partial \mathbf{\Pi}}{\partial t} = -\frac{1}{4\pi} \mathbf{curl}\, \mathbf{curl}\, \boldsymbol{A} \tag{24}$$

These equations give us all of Maxwell equations except the third

$$\operatorname{div} \boldsymbol{E} = 0$$

We must require that $\operatorname{div} \boldsymbol{E} = 0$ at some instant of time. But then the canonical equations give us automatically the result that this restriction is maintained at all times.

Bibliography

1. Corben, H. C. and Stehle, P., *Classical Mechanics*, John Wiley & Sons, Inc., New York, 1950. An excellent account of the mechanics of continuous systems and fields.
2. Goldstein, H., *Classical Mechanics*, Addison-Wesley Press, Inc., Cambridge, Mass., 1950.
3. Lanczos, C., *The Variational Principles of Mechanics*, University of Toronto Press, Toronto, 1949.
4. Synge, J. L. and Griffith, B. A., *Principles of Mechanics*, 2d ed., McGraw-Hill Book Company, Inc., New York, 1949.
5. Whittaker, E. T., *A Treatise on the Analytical Dynamics of Rigid Bodies*, 4th ed., Cambridge University Press, London, 1937. The standard work.

Chapter 6

SPECIAL THEORY OF RELATIVITY

By Henry Zatzkis

Assistant Professor of Mathematics
Newark College of Engineering

1. The Kinematics of the Space-Time Continuum

1.1. The Minkowski "world." Ordinary three-dimensional space plus the time form the four-dimensional "world." A "world point" is an ordinary point at a certain time. Its four coordinates are the Cartesian coordinates, x, y, z and the time t, which will also be denoted by x^1, x^2, x^3 and x^4. An "event" is a physical occurrence at a certain world point. In what follows, the Einstein summation convention will be used: Whenever an index appears twice, as an upper (contravariant) and a lower (covariant) index, it is to be summed over. If the index is a lower-case Greek letter, the summation extends from 1 to 4; if it is an italic lower-case letter, the summation extends from 1 to 3. (Generally, the indices 1, 2, and 3 will refer to the spatial dimensions and the index 4 to the timelike dimension of the continuum.) Examples are

$$a^\tau b_{\tau\lambda} = a^1 b_{1\lambda} + a^2 b_{2\lambda} + a^3 b_{3\lambda} + a^4 b_{4\lambda}$$

$$a^{\tau r} b_r = a^{\tau 1} b_1 + a^{\tau 2} b_2 + a^{\tau 3} b_3$$

The two most fundamental invariants of the special theory of relativity are the magnitude c of the speed of light in vacuo and the four-dimensional "distance" τ_{12} of any two world points (x_1, y_1, z_1, t_1) and (x_2, y_2, z_2, t_2) defined by

$$\begin{aligned} {\tau_{12}}^2 &= (t_2 - t_1)^2 - \frac{1}{c^2}[(x_2 - x_1)^2 + (y_2 - y_1)^2 + (z_2 - z_1)^2] \\ &= (\Delta t)^2 - \frac{1}{c^2}[(\Delta x)^2 + (\Delta y)^2 + (\Delta z)^2] \end{aligned} \tag{1}$$

All the kinematic properties of the theory are a consequence of these invariants.

1.2. The Lorentz transformation. We introduce a covariant " metric tensor " $\eta_{\mu\nu}$ and its corresponding contravariant tensor $\eta^{\mu\nu}$, defined by

$$\eta^{\mu\lambda}\eta_{\lambda\nu} = \delta^{\mu}_{\nu} \tag{1}$$

where

$$\delta_{\nu}^{\mu} = \begin{cases} 1 & \text{if} \quad \mu = \nu \\ 0 & \text{if} \quad \mu \neq \nu \end{cases} \tag{2}$$

and

$$\left.\begin{aligned} \eta_{\mu\nu} &= \begin{bmatrix} -\frac{1}{c^2}, & 0, & 0, & 0 \\ 0, & -\frac{1}{c^2}, & 0, & 0 \\ 0, & 0, & -\frac{1}{c^2}, & 0 \\ 0, & 0, & 0, & +1 \end{bmatrix} \\ &\text{and} \\ \eta^{\mu\nu} &= \begin{bmatrix} -c^2, & 0, & 0, & 0 \\ 0, & -c^2, & 0, & 0 \\ 0, & 0, & -c^2, & 0 \\ 0, & 0, & 0, & +1 \end{bmatrix} \end{aligned}\right\} \tag{3}$$

In terms of the metric tensor, the " world distance " or " world interval " τ_{12} becomes

$$\tau_{12}^{\ 2} = \eta_{\mu\nu}\Delta x^{\mu}\Delta x^{\nu} \tag{4}$$

When we go from one reference system Σ with the coordinates x^{λ} to a new system $\overset{*}{\Sigma}$ with the coordinates $\overset{*}{x}{}^{\lambda}$ the only admissible systems are those resulting from a nonsingular, linear transformation that leaves $\tau_{12}^{\ 2}$ invariant. Since only coordinate differences and not the coordinates themselves are involved, the transformations may be homogeneous or inhomogeneous in the coordinates. Usually, only homogeneous transformations are considered: the " Lorentz transformation." Think of them as rotations in the four-dimensional continuum (the inhomogeneous transformations would correspond to rotations plus translations. The transformation equations are

$$\left.\begin{aligned} \overset{*}{x}{}^{\lambda} &= \gamma^{\lambda\cdot}_{\cdot\tau} x^{\tau} \\ x^{\lambda} &= \gamma^{\cdot\lambda}_{\tau\cdot} \overset{*}{x}{}^{\tau} \end{aligned}\right\} \tag{5}$$

where the $\gamma^{\cdot\lambda}_{\tau\cdot}$ denote the coefficients of the inverse transformation and are solutions of the equations

$$\gamma^{\lambda\cdot}_{\cdot\tau}\gamma^{\cdot\tau}_{\nu\cdot} = \delta^{\lambda}_{\nu} \tag{6}$$

The $\gamma^{\lambda\cdot}_{\cdot\tau}$ themselves are the solutions of the equations

$$\eta_{\mu\nu}\gamma^{\mu\cdot}_{\cdot\lambda}\gamma^{\nu\cdot}_{\cdot\tau} = \eta_{\lambda\tau} \tag{7}$$

A spatial rotation in three-dimensional space is a special case, characterized by $\gamma^{r\cdot}_{\cdot 4} = 0$ and $\gamma^{4\cdot}_{\cdot 4} = 1$. Another special case, the most important one in the subsequent discussion, is that one giving the relations between a coordinate system Σ at rest and another system $\overset{*}{\Sigma}$ moving in the direction of the positive x axis with the constant speed v and whose origin $\overset{*}{O}$ coincides at the time $t = 0$ with the origin O of Σ. The transformation coefficients $\gamma^{\lambda\cdot}_{\cdot\tau}$ are in this case.

$$\gamma^{\lambda\cdot}_{\cdot\tau} = \begin{bmatrix} \dfrac{1}{\sqrt{1 - v^2/c^2}}, & 0, & 0, & \dfrac{-v}{\sqrt{1 - v^2/c^2}} \\ 0, & 1, & 0, & 0 \\ 0, & 0, & 1, & 0 \\ \dfrac{-v/c^2}{\sqrt{1 - v^2/c^2}}, & 0, & 0, & \dfrac{1}{\sqrt{1 - v^2/c^2}} \end{bmatrix} \tag{8}$$

and their inverse matrix:

$$\gamma^{\cdot\lambda}_{\tau\cdot} = \begin{bmatrix} \dfrac{1}{\sqrt{1 - v^2/c^2}}, & 0, & 0, & \dfrac{v}{\sqrt{1 - v^2/c^2}} \\ 0, & 1, & 0, & 0 \\ 0, & 0, & 1, & 0 \\ \dfrac{v/c^2}{\sqrt{1 - v^2/c^2}}, & 0, & 0, & \dfrac{1}{\sqrt{1 - v^2/c^2}} \end{bmatrix} \tag{9}$$

The relations between the two coordinate systems are

$$\left.\begin{aligned} \overset{*}{x}{}^1 &= \overset{*}{x} = \frac{x - vt}{\sqrt{1 - v^2/c^2}} \\ \overset{*}{x}{}^2 &= \overset{*}{y} = y \\ \overset{*}{x}{}^3 &= \overset{*}{z} = z \\ \overset{*}{x}{}^4 &= \overset{*}{t} = \frac{t - vx/c^2}{\sqrt{1 - v^2/c^2}} \end{aligned}\right\} \tag{10}$$

$$\left.\begin{aligned} x^1 &= x = \frac{\overset{*}{x} + v\overset{*}{t}}{\sqrt{1 - v^2/c^2}} \\ x^2 &= y = \overset{*}{y} \\ x^3 &= z = \overset{*}{z} \\ x^4 &= t = \frac{\overset{*}{t} + v\overset{*}{x}/c^2}{\sqrt{1 - v^2/c^2}} \end{aligned}\right\} \tag{11}$$

The ratio v/c will be denoted by β.

The Lorentz transformations form a group, i.e., if the system $\overset{*}{\Sigma}$ is related to the system Σ by a Lorentz transformation, and if the system $\overset{**}{\Sigma}$ is related to $\overset{*}{\Sigma}$ by another Lorentz transformation, then there exists a third Lorentz transformation which relates $\overset{**}{\Sigma}$ directly to Σ.

Lorentz's transformation laws [*Proceedings of the Academy of Sciences*, Amsterdam, **6**, 809 (1904)] are however, not completely equivalent to Einstein's. His formulas (4) and (5) read:

$$\overset{*}{x} = \frac{x}{\sqrt{1-\beta^2}}; \quad t = \overset{*}{t}\sqrt{1-\beta^2} - \frac{xv/c^2}{\sqrt{1-\beta^2}}$$

Therefore $x^2 - c^2t^2$ is not an invariant, since

$$\overset{*}{x}{}^2 - c^2\overset{*}{t}{}^2 = x^2 - c^2t^2 + 2\beta xct + c^2\beta^2t^2.$$

1.3. Kinematic consequences of the Lorentz transformation.

a. *Relativity of simultaneity. Order of events.* If an event occurs in Σ at the points x_1 and x_2 at the times t_1 and t_2, then

$$\overset{*}{t}_1 = \frac{t_1 - vx_1/c^2}{\sqrt{1-\beta^2}}, \quad \overset{*}{t}_2 = \frac{t_2 - vx_2/c^2}{\sqrt{1-\beta^2}} \tag{1}$$

and therefore

$$\overset{*}{t}_2 - \overset{*}{t}_1 = \frac{t_2 - t_1 - (v/c^2)(x_2 - x_1)}{\sqrt{1-\beta^2}} \tag{2}$$

If $t_2 = t_1$, we obtain

$$\overset{*}{t}_2 - \overset{*}{t}_1 = \frac{(-v/c^2)(x_2 - x_1)}{\sqrt{1-\beta^2}} \neq 0 \tag{3}$$

Thus two events that are simultaneous in Σ are no longer simultaneous in $\overset{*}{\Sigma}$. If the two events in Σ can be connected by a signal whose speed w is less than the speed of light, i.e., if

$$\frac{x_2 - x_1}{t_2 - t_1} = w > c \tag{4}$$

then $$\overset{*}{t}_2 - \overset{*}{t}_1 = (t_2 - t_1)\frac{1 - vw/c^2}{\sqrt{1 - v^2/c^2}} \tag{5}$$

In other words, $\overset{*}{t}_2 - \overset{*}{t}_1$ will always be positive if $t_2 - t_1$ is positive. The temporal order of events is not destroyed by a Lorentz transformation.

b. *The time dilatation.* If a clock, e.g., an atomic oscillator, vibrates with the period T at a fixed point x, a moving observer will observe the period

$$\overset{*}{T} = \frac{T}{\sqrt{1 - v^2/c^2}} \tag{6}$$

c. *The Lorentz contraction.* The spatial distance between two world points $\overset{*}{P}_1 = (\overset{*}{x}_1, \overset{*}{y}_1, \overset{*}{z}_1, \overset{*}{t})$ and $\overset{*}{P}_2 = (\overset{*}{x}_2, \overset{*}{y}_2, \overset{*}{z}_2, \overset{*}{t})$ is

$$\left.\begin{aligned} \overset{*}{L} &= \sqrt{(\overset{*}{x}_2 - \overset{*}{x}_1)^2 + (\overset{*}{y}_2 - \overset{*}{y}_1)^2 + (\overset{*}{z}_2 - \overset{*}{z}_1)^2} \\ &= \sqrt{(x_2 - x_1)^2(1 - \beta)^2 + (y_2 - y_1)^2 + (z_2 - z_1)^2} \end{aligned}\right\} \tag{7}$$

The linear dimensions in the direction of motion appear contracted in the ratio $1 : \sqrt{1 - \beta^2}$ to a moving observer. Distances orthogonal to the direction of motion remain unchanged. This contraction explains the negative outcome of the Michelson-Morley experiment, historically the starting point for the creation of the theory of relativity. Suppose that an observer on a material system, moving through space with a velocity v, makes observations on the velocity of light, and determines the time taken by light to pass from a point A to a point B at a distance d from A and to be reflected back to A. Let the velocity v of his system be in the direction from A to B. Let the light start from A at time t_1, arrive at B at time t_2, and get back to A at t_3. Then if the light travels through space with velocity c, we have

$$d + v(t_2 - t_1) = c(t_2 - t_1) \tag{8}$$

$$d - v(t_3 - t_2) = c(t_3 - t_2) \tag{9}$$

or $$t_2 - t_1 = \frac{d}{c - v} \tag{10}$$

$$t_3 - t_2 = \frac{d}{c + v} \tag{11}$$

Hence $$\Delta t = t_3 - t_1 = \frac{2d}{c(1 - v^2/c^2)} = \frac{2d}{c(1 - \beta^2)} \tag{12}$$

Suppose now that AB is perpendicular to the velocity v. In this case the distance traversed by the light signal, as it travels from A to B and back to A again is

$$2\sqrt{d^2 + \frac{v^2}{4}(t_3' - t_1')^2} = c(t_3' - t_1') = c\Delta t' \tag{13}$$

Therefore

$$\Delta t' = \frac{2d}{c\sqrt{1-\beta^2}} \tag{14}$$

The two time intervals Δt and $\Delta t'$ are not the same. However, because of the Lorentz contraction, the required time Δt is shortened, and becomes

$$\Delta t = \frac{2d\sqrt{1-\beta^2}}{c(1-\beta^2)} = \Delta t' \tag{15}$$

It was experimentally observed that $\Delta t = \Delta t'$.

d. *Velocity addition theorem.* If the velocity of a material particle has the Cartesian components u_1, u_2, and u_3 in Σ, its components in $\overset{*}{\Sigma}$ are given by

$$\overset{*}{u}_1 = \frac{u_1 + v}{1 + vu_1/c^2} \tag{16}$$

$$\overset{*}{u}_2 = \frac{u_2\sqrt{1-\beta^2}}{1 + vu_1/c^2} \tag{17}$$

$$\overset{*}{u}_3 = \frac{u_3\sqrt{1-\beta^2}}{1 + vu_1/c^2} \tag{18}$$

Applications. (1) Assume that a light ray is traveling in the y direction, $u_1 = u_3 = 0$; $u_2 = c$. Then $\overset{*}{u}_1 = -v$; $\overset{*}{u}_2 = c\sqrt{1-\beta^2}$; $\overset{*}{u}_3 = 0$. The angle α which the ray includes with the $\overset{*}{y}$ axis is given by

$$\tan\alpha = \frac{\overset{*}{u}_1}{\overset{*}{u}_2} = -\frac{v}{c\sqrt{1-\beta^2}} = -\beta\left(1 + \frac{\beta^2}{2} + \dots\right) \tag{19}$$

For small values of β, the absolute value of α becomes β (angle of aberration).

(2) Let N be the refractive index of a substance having the velocity v in the x direction. The speed of light in this substance is $u = c/N$. Consequently the speed $\overset{*}{u}$ of a light ray traveling in the $\overset{*}{x}$ direction is

$$\overset{*}{u} = \overset{*}{u}_1 = \frac{u + v}{1 + vu/c^2} = \frac{c}{N} + v\left(1 - \frac{1}{N^2}\right) + \dots \tag{20}$$

the factor $(1 - 1/N^2)$ is called the Fresnel-drag coefficient.

e. *Doppler effect and general formula for the aberration.* If a plane light wave in vacuum travels along a direction which forms an angle $\overset{*}{\theta}$ with the $\overset{*}{x}$ axis and has a frequency $\overset{*}{\nu}$ in the system $\overset{*}{\Sigma}$, an observer in the system Σ will observe the frequency ν and the angle θ, where

$$\nu = \frac{(1 + \beta \cos \overset{*}{\theta})}{\sqrt{1 - \beta^2}} \overset{*}{\nu} \tag{21}$$

$$\cos \theta = \frac{\cos \overset{*}{\theta} + \beta}{1 + \beta \cos \overset{*}{\theta}} \tag{22}$$

$$\sin \theta = \frac{\sqrt{1 - \beta^2} \sin \overset{*}{\theta}}{1 + \beta \cos \overset{*}{\theta}} \tag{23}$$

f. *Reflection at moving mirror.* Let a mirror have the speed u and a plane wave be reflected by it (Fig. 1). If ψ_1 is the angle of incidence, and ψ_2 the angle of reflection,

$$\frac{\tan\left(\frac{\psi_2}{2}\right)}{\tan\left(\frac{\psi_1}{2}\right)} = \frac{c - u}{c + u} \tag{24}$$

If ν_1 is the frequency of the incident wave and ν_2 that of the reflected wave,

$$\frac{\nu_2}{\nu_1} = \frac{\sin \psi_1}{\sin \psi_2} \tag{25}$$

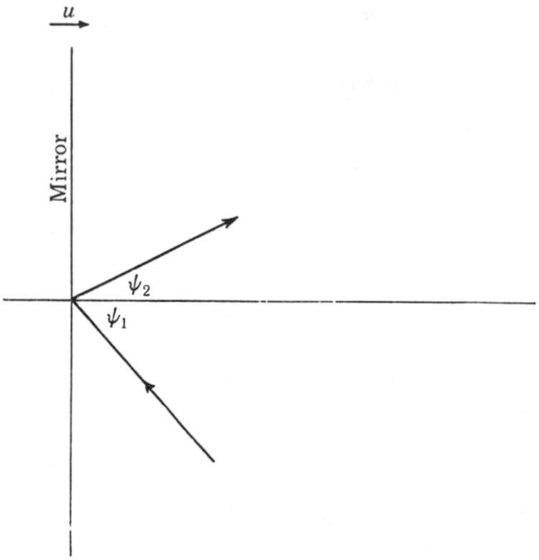

FIGURE 1

If the mirror moves opposite to the incident wave, we note that $\psi_2 < \psi_1$ and $\nu_2 < \nu_1$. These formulas played an important part in Planck's derivation of black-body radiation and in the thermodynamics of cavity radiation.

g. *Graphical representation of the Lorentz transformation.* We project the four-dimensional world upon the x,t plane (Fig. 2) and choose as abscissa

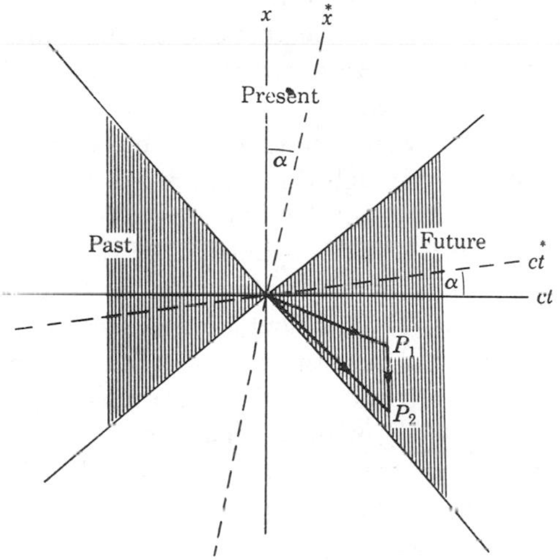

FIGURE 2

not the time t but $\xi_4 = ct$ and as ordinate $x = \xi_1$. The relations between $x' = \xi_1'$ and $\overset{*}{ct} = \overset{*}{\xi}_4$, and ξ_1 and ξ_4 are given by

$$\overset{*}{\xi}_1 = \xi_1 \cos\alpha - i\xi_4 \sin\alpha \tag{26}$$

$$\overset{*}{\xi}_4 = -i\xi_1 \sin\alpha + \xi_4 \cos\alpha \tag{27}$$

or if we put $\xi_1 = \eta_1$ and $i\xi_4 = \eta_4$,

$$\overset{*}{\eta}_1 = \eta_1 \cos\alpha - \eta_4 \sin\alpha \tag{28}$$

$$\overset{*}{\eta}_4 = \eta_1 \sin\alpha + \eta_4 \cos\alpha \tag{29}$$

where $i \tan\alpha = v/c$.

Introducing the real angle γ, defined by $\gamma = i\alpha$, we can rewrite these formulas in the form

$$\overset{*}{\xi}_1 = \xi_1 \cosh \gamma - \xi_4 \sinh \gamma \tag{30}$$

$$\overset{*}{\xi}_4 = -\xi_1 \sinh \gamma + \xi_4 \cosh \gamma \tag{31}$$

The expression $c \tanh \gamma = v$ is the speed with which the reference frame $\overset{*}{\Sigma}$ moves with respect to the reference frame Σ.

A uniform motion of a particle, passing through the origin of either coordinate system at the time zero is represented by a straight line through the origin. Its slope is given by $\tan \phi = w/c$ where w is the speed with respect to the system Σ. Since the speed cannot exceed the speed of light, the two 45° lines passing through the origin represent the motion of a light signal. In the Minkowski world the light signals reaching the origin at the time zero form the surface of a three-dimensional cone with its vertex at the origin. Likewise the light signals leaving O at the time zero form another cone with the vertex at O. This double cone is called the " light cone." Any physical event either in the past or in the future that can be linked to an event happening at the origin at the time zero must take place at a world point located in the interior or on the boundary of the light cone. The left-hand cone represents the past and the right-hand cone the future. All other events are located in the " present," since they can neither have caused nor be the consequence of the event at the origin at time zero. In the sense of the Minkowski metric the " distance " between two points $P_1(\xi_4,\xi_1)$ and $P_2(\xi_4',\xi_1')$ is $d(P_1P_2) = +\sqrt{(\xi_4' - \xi_4)^2 - (\xi_1' - \xi_1)^2}$, where it is assumed that both points lie within the same part of the light cone, i.e., both points lie in the past or in the future. It is an immediate consequence of this definition that

$$d(OP_1) + d(OP_2) \leqq d(P_1P_2) \tag{32}$$

In other words, the usual triangle-inequality is just reversed. The straight-line connection between two points is not the shortest, but the longest connection.

2. Dynamics

2.1. Conservation laws. Classical mechanics knows two independent conservation laws, the conservation of a vector p^s, the linear momentum, and the conservation of a scalar, the energy E. Relativistic mechanics knows only one conservation law, the conservation of a world vector P^ϱ.

More precisely, it states that if there are n particles present with rest masses $\underset{1}{m}, \underset{2}{m}, \ldots, \underset{n}{m}$, then in the absence of external forces four laws of inertia hold :

$$\sum_{k=1}^{n} \underset{k}{P}^{\varrho} = \text{constant} \qquad (\rho = 1,2,3,4) \tag{1}$$

where
$$\underset{k}{P}^{\varrho} = \left(\frac{\underset{k}{m}\underset{k}{u}^{s}}{\sqrt{1 - \underset{k}{u}^{2}/c^{2}}}, \frac{\underset{k}{E}}{c^{2}} \right) \tag{2}$$

The corresponding covariant vector is

$$\underset{k}{P}_{\varrho} = \left(\frac{-\underset{k}{m}\underset{k}{u}^{s}}{c^{2}\sqrt{1 - \underset{k}{u}^{2}/c^{2}}}, \frac{\underset{k}{E}}{c^{2}} \right) \qquad \underset{k}{E} = \frac{\underset{k}{m}c^{2}}{\sqrt{1 - \underset{k}{u}^{2}/c^{2}}} \tag{4}$$

The relativistic kinetic energy T is defined by

$$T = \frac{mc^{2}}{\sqrt{1 - u^{2}/c^{2}}} - mc^{2} \tag{5}$$

which for small velocities becomes the classical value $\frac{1}{2}mu^{2}$. Since the vector P^{ϱ} is conserved, its magnitude, and hence also the square of its magnitude are also conserved. Thus

$$P^{\varrho}P_{\varrho} = \frac{1}{c^{2}} \left[\frac{E^{2}}{c^{2}} - \sum_{s=1}^{3} (p^{s})^{2} \right] = \text{constant} \tag{6}$$

When the expression for E and p^{s} are substituted into Eq. (6), one finds that

$$P^{\varrho}P_{\varrho} = m^{2} \tag{7}$$

If the rest-mass is zero, e.g., in the case of a photon, we obtain the result that

$$\sum_{s=1}^{3} (p^{s})^{2} = \frac{E^{2}}{c^{2}} \tag{8}$$

In words : the magnitude of the linear momentum is $p = E/c$. On using the Planck relation that $E = h\nu$ (ν is the frequency of the photon), we obtain

$$p = \frac{h\nu}{c} \tag{9}$$

To illustrate the application to the Compton effect, let us assume an electron, of rest-mass m, is initially at rest and struck by a photon of frequency ν.

The scattered photon has the frequency ν' and forms an angle θ with the direction of the incident photon. The electron acquires the velocity v in a direction making an angle ϕ with the direction of the incident photon. The conservation law for the energy states that

$$h\nu = h\nu' + \frac{mc^2}{\sqrt{1 - v^2/c^2}} \tag{10}$$

The conservation law for the momentum states that

$$\frac{h\nu}{c} = \frac{h\nu'}{c} \cos \theta + \frac{mv}{\sqrt{1 - v^2/c^2}} \cos \phi \tag{11}$$

and

$$0 = \frac{h\nu'}{c} \sin \theta - \frac{mv}{\sqrt{1 - v^2/c^2}} \sin \phi \tag{12}$$

By eliminating ϕ, we obtain

$$m(\nu - \nu') - \frac{h}{c^2} (1 - \cos \theta)\nu\nu' = 0 \tag{13}$$

On writing $1 - \cos \theta = 2 \sin^2 \theta/2$; $\nu = c/\lambda$; $\nu' = c/\lambda'$ we finally obtain

$$\lambda' - \lambda = \frac{2h}{mc} \sin^2 \frac{\theta}{2} \tag{14}$$

where h/mc is usually called the Compton wavelength.

2.2. Dynamics of a free mass point. The dynamics of a free mass point of rest mass m follows from a variational principle :

$$I = \int_{t_1}^{t_2} L(x^s, \dot{x}^s) dt, \quad \left(\dot{x}^s = \frac{dx^s}{dt}\right) \tag{1}$$

$$\delta I = 0$$

The variation is to be performed only in the interior, the end points are kept fixed. The Lagrangian function L is given by

$$L = - mc^2 \sqrt{1 - v^2/c^2} \tag{2}$$

$$v^2 = \left(\frac{dx^1}{dt}\right)^2 + \left(\frac{dx^2}{dt}\right)^2 + \left(\frac{dx^3}{dt}\right)^2$$

The corresponding Euler-Lagrange equations are

$$\frac{\partial L}{\partial x^s} - \frac{d}{dt} \left(\frac{\partial L}{\partial \dot{x}^s}\right) = 0 \tag{3}$$

The momenta are, as usual, defined by

$$p_s \equiv \frac{\partial L}{\partial \dot{x}^s} = \frac{m\dot{x}^s}{\sqrt{1 - v^2/c^2}} \tag{4}$$

Equations (2.2) and (2.3) can be brought into the canonical form. We introduce the Hamiltonian funtion H, defined by

$$\left.\begin{aligned} H = -L + p_s\dot{x}^s = \frac{mc^2}{\sqrt{1 - u^2/c^2}} = mc^2\sqrt{1 + \frac{p^2}{m^2c^2}} \\ (p^2 = p_1{}^2 + p_2{}^2 + p_3{}^2) \end{aligned}\right\} \tag{5}$$

The canonical equations are

$$\dot{x}^s = \frac{\partial H}{\partial p_s} = \frac{p_s/m}{\sqrt{1 + p^2/m^2}} \tag{6}$$

$$\dot{p}_s = -\frac{\partial H}{\partial x^s} = 0 \tag{7}$$

2.3. Relativistic force. The force f_s acting on an accelerated mass point can be defined, just as in classical mechanics, as the time rate change of linear momentum :

$$\left.\begin{aligned} f_s = \frac{d}{dt}\left(\frac{m\dot{x}^s}{\sqrt{1 - v^2/c^2}}\right) = \left(1 - \frac{v^2}{c^2}\right)^{-3/2} m\left[\delta_{st}\left(1 - \frac{v^2}{c^2}\right) + \frac{\dot{x}^s\dot{x}^t}{c^2}\right]\ddot{x}^t \\ (v^2 = \dot{x}^2 + \dot{y}^2 + \dot{z}^2) \end{aligned}\right\} \tag{1}$$

This force is in general not parallel to the acceleration. It is parallel only when the acceleration is either parallel or perpendicular to the velocity. When it is parallel, Eq. (1) takes the form

$$f_s = \left(1 - \frac{v^2}{c^2}\right)^{-3/2} m\ddot{x}^s \tag{2}$$

If the acceleration is orthogonal to the velocity, Eq. (1) takes the form

$$f_s = \left(1 - \frac{v^2}{c^2}\right)^{-1/2} m\ddot{x}^s \tag{3}$$

The coefficients $(1 - v^2/c^2)^{-3/2}m$ and $(1 - v^2/c^2)^{-1/2}m$ are called the " longitudinal mass " and the " transversal mass " in the older literature.

2.4. Relativistic electrodynamics. We introduce the covariant tensor $\phi = (A_s, -c\phi)$,

$$A_s = \text{vector potential}, \quad \phi = \text{scalar potential}$$

the world current-vector $s^{\omega} = (I_s, \rho)$,

$$I_s = \text{current density}, \quad \rho = \text{charge density}$$

and the world force density $F^{\omega} = (k_s, \ \lambda/c^2)$,

$$k_s = \text{ordinary force density}, \quad \lambda = \text{power density}$$

Comma means differentiation, e.g., $\phi_{\varrho,\sigma} = \partial\phi_{\varrho}/\partial x^{\sigma}$ and

$$\phi_{\varrho\sigma} \equiv \frac{\partial\phi_{\varrho}}{\partial x^{\sigma}} - \frac{\partial\phi_{\sigma}}{\partial x^{\varrho}} = \phi_{\varrho,\sigma} - \phi_{\sigma,\varrho} = -\phi_{\sigma\varrho}$$

$$\begin{aligned}
\boldsymbol{E} &= (E_1, E_2, E_3) = \text{electric field strength} \\
\boldsymbol{D} &= (D_1, D_2, D_3) = \text{electric displacement vector} \\
\boldsymbol{H} &= (H_1, H_2, H_3) = \text{magnetic field strength} \\
\boldsymbol{B} &= (B_1, B_2, B_3) = \text{magnetic displacement vector}
\end{aligned}$$

We finally introduce two antisymmetric tensors $\Phi_{\varrho\sigma}$ and $\Psi^{\varrho\sigma}$:

$$\begin{aligned}
(\Phi_{12}, \Phi_{23}, \Phi_{31}) &= (-B_3, -B_1, -B_2) \\
(\Phi_{14}, \Phi_{24}, \Phi_{34}) &= (-cE_1, -cE_2, -cE_3) \\
(\Psi^{12}, \Psi^{23}, \Psi^{31}) &= (-c^4H_3, -c^4H_1, -c^4H_2) \\
(\Psi^{41}, \Psi^{42}, \Psi^{43}) &= (-c^3D_1, -c^3D_2, -c^3D_3)
\end{aligned}$$

The fundamental equations of electrodynamics are

$$s^{\lambda}{}_{,\lambda} = 0 \quad \text{or} \quad \operatorname{div} \boldsymbol{I} + \frac{\partial\rho}{\partial t} = 0 \tag{1}$$

$$\phi^{\lambda}{}_{,\lambda} = 0 \quad \text{or} \quad \operatorname{div} \boldsymbol{A} + \frac{1}{c}\frac{\partial\phi}{\partial t} = 0 \tag{2}$$

$$\phi_{\lambda.\varrho} - \phi_{\varrho.\lambda} = +\Phi_{\lambda\varrho} \quad \text{or} \quad \boldsymbol{B} = \textbf{curl}\, \boldsymbol{A}; \qquad \boldsymbol{E} = -\textbf{grad}\,\phi - \frac{1}{c}\frac{\partial \boldsymbol{A}}{\partial t} \tag{3}$$

$$\Phi_{\lambda\varrho,\sigma} + \Phi_{\varrho\sigma,\lambda} + \Phi_{\sigma\lambda,\varrho} = 0 \quad \text{or} \quad \operatorname{div} \boldsymbol{B} = 0; \quad \textbf{curl}\, \boldsymbol{E} = -\frac{1}{c}\frac{\partial \boldsymbol{B}}{\partial t} \tag{4}$$

$$\Psi^{\varrho\lambda}{}_{,\lambda} = -4\pi c^3 s^{\varrho} \quad \text{or} \quad \textbf{curl}\, \boldsymbol{H} = \frac{4\pi \boldsymbol{I}}{c} + \frac{1}{c}\frac{\partial \boldsymbol{D}}{\partial t}; \quad \operatorname{div} \boldsymbol{D} = 4\pi\rho \tag{5}$$

$$F^{\varrho} = \frac{1}{c^3}\,\Phi^{\varrho\lambda}s_{\lambda} \quad \text{or} \quad \begin{cases} k = \rho E + \dfrac{1}{c}[I \times B] & (6) \\ \lambda = I \cdot E & (7) \end{cases}$$

In vacuum $E = D$ and $H = B$ and therefore $\Phi^{\varrho\lambda} = \Psi^{\varrho\lambda}$. In this case we can eliminate $\Phi_{\lambda\varrho}$ by combining Eq. (3) and Eq. (5) and utilizing (2) we obtain

$$\Box \phi^{\varrho} = 4\pi c s^{\varrho} \tag{8}$$

where $\Box$ is the wave operator or the " D'Alembertian "

$$\Box = \frac{\partial^2}{\partial x^2} + \frac{\partial^2}{\partial y^2} + \frac{\partial^2}{\partial z^2} - \frac{1}{c^2}\frac{\partial^2}{\partial t^2}$$

If we put

$$\phi^{\varrho} = Z^{\varrho\lambda}{}_{,\lambda} \quad (Z^{\varrho\lambda} = -Z^{\lambda\varrho})$$

$$s^{\varrho} = Q^{\varrho\lambda}{}_{,\lambda} \quad (Q^{\varrho\lambda} = -Q^{\lambda\varrho})$$

we obtain

$$4\pi c Q^{\varrho\lambda} = \Box Z^{\varrho\lambda} \tag{9}$$

where $Z^{\varrho\lambda}$ corresponds to the Hertz vector in electrodynamics.

2.5. Gauge invariance. The Maxwell equations (1), (4) and (5) of the preceding paragraph remain unaltered if the world vector ϕ_{ϱ} is changed into a new world vector $\bar{\phi}_{\varrho}$ by adding to it an arbitrary four-dimensional gradient field $\psi_{,\varrho}$:

$$\bar{\phi}_{\varrho} = \phi_{\varrho} + \psi_{,\varrho} \tag{1}$$

Such a transformation is called gauge transformation. The condition (2.4-2) is not gauge-invariant. But since ψ is arbitrary and does not affect the Maxwell equations, we are permitted to make this choice. The advantage of having the " Lorentz gauge condition " (2.4-2) satisfied is that we can solve for the highest time derivatives of the potentials and thus obtain a system of differential equations of the Cauchy-Kowalewski type. In such a system the initial conditions on a function guarantee its unique continuation into the future (and its past).

If one works with the field strengths E and H themselves (which are the only physically observable quantities), rather than with the potentials, then Maxwell's equations are automatically of the Cauchy-Kowalewski type with respect to the time derivatives, regardless of any gauge condition. We introduce potentials primarily for mathematical convenience.

2.6. Transformation laws for the field strengths

$$\left.\begin{aligned}
\overset{*}{E}_1 &= E_1 & \overset{*}{B}_1 &= B_1 \\
\overset{*}{E}_2 &= \frac{1}{\sqrt{1-\beta^2}}(E_2 - \beta B_3) & \overset{*}{B}_2 &= \frac{1}{\sqrt{1-\beta^2}}(B_2 + \beta E_3) \\
\overset{*}{E}_3 &= \frac{1}{\sqrt{1-\beta^2}}(E_3 + \beta B_2) & \overset{*}{B}_3 &= \frac{1}{\sqrt{1-\beta^2}}(B_3 - \beta E_2) \\
\overset{*}{D}_1 &= D_1 & \overset{*}{H}_1 &= H_1 \\
\overset{*}{D}_2 &= \frac{1}{\sqrt{1-\beta^2}}(D_2 - \beta H_3) & \overset{*}{H}_2 &= \frac{1}{\sqrt{1-\beta^2}}(H_2 + \beta D_3) \\
\overset{*}{D}_3 &= \frac{1}{\sqrt{1-\beta^2}}(D_3 + \beta H_2) & \overset{*}{H}_3 &= \frac{1}{\sqrt{1-\beta^2}}(H_3 - \beta D_2)
\end{aligned}\right\} \quad (1)$$

Here $\beta = v/c$, where v is the velocity of $\overset{*}{\Sigma}$ with respect to Σ and $v = (v,0,0)$. The total charge is an invariant. If the world current vector s^ϱ is purely convective, and if the charge has the velocity u in the system Σ and $\overset{*}{u}$ in Σ, then

$$\left.\begin{aligned}
\overset{*}{\rho}\overset{*}{u}_1 &= \rho\,\frac{u_1 - v}{\sqrt{1-\beta^2}} \\
\overset{*}{\rho}\overset{*}{u}_2 &= \rho u_2 \\
\overset{*}{\rho}\overset{*}{u}_3 &= \rho u_3 \\
\overset{*}{\rho} &= \rho\,\frac{1 - vu_1/c^2}{\sqrt{1-\beta^2}}
\end{aligned}\right\} \quad (2)$$

If the charge is at rest in Σ, i.e., if $u_1 = u_2 = u_3 = 0$, then

$$\overset{*}{\rho} = \frac{\rho}{\sqrt{1-\beta^2}}$$

2.7. Electrodynamics in moving, isotropic ponderable media (Minkowski's equation). The relations between $\Phi^{\varrho\lambda}$ and $\Psi^{\varrho\lambda}$ must be of such a nature that, in a reference system at rest with respect to the medium, the relations must go over into

$$\boldsymbol{D} = \epsilon \boldsymbol{E} \quad (1) \qquad \boldsymbol{B} = \mu \boldsymbol{H} \quad (2) \qquad \boldsymbol{I}_c = \sigma \boldsymbol{E} \quad (3)$$

ϵ = dielectric constant, μ = magnetic permeability, σ = electric conductivity, $\boldsymbol{I}_c$, the " metallic " current density, is obtained from the total current density s^ϱ through subtraction of the convective current density $U^\varrho \cdot (s^\lambda U_\lambda)$, where

$$U^\varrho = \left(\frac{u^s}{c\sqrt{1-u^2/c^2}}, \frac{1}{c\sqrt{1-u^2/c^2}}\right)$$

We shall denote the corresponding world vector of $\boldsymbol{I}_c$ by l^ϱ. Therefore

$$l^\varrho = s^\varrho - U^\varrho \cdot (s^\lambda U_\lambda) \tag{3}$$

becomes

$$\boldsymbol{l} = \sigma \boldsymbol{E} \tag{4}$$

We introduce two new world vectors

$$e^\varrho = \Phi^{\varrho\lambda} U_\lambda \quad \text{and} \quad d^\varrho = \Psi^{\varrho\lambda} U_\lambda$$

Then

$$d^\varrho = \epsilon \varepsilon^\varrho \tag{5}$$

is an invariant relation which reduces to $\boldsymbol{D} = \epsilon \boldsymbol{E}$ in a system at rest with respect to the medium. We also introduce two antisymmetric tensors

$$b^{\varrho\lambda\tau} = \Phi^{\varrho\lambda} U^\tau + \Phi^{\lambda\tau} U^\varrho + \Phi^{\tau\varrho} U^\lambda$$

and

$$h^{\varrho\lambda\tau} = \Psi^{\varrho\lambda} U^\tau + \Psi^{\lambda\tau} U^\varrho + \Psi^{\tau\varrho} U^\lambda$$

Then

$$b^{\varrho\tau\lambda} = \mu h^{\varrho\tau\lambda} \tag{6}$$

is another invariant relation, which reduces to $\boldsymbol{B} = \mu \boldsymbol{H}$ in a system with respect to the medium. We introduce certain abbreviations :

$$\left.\begin{aligned} \boldsymbol{E}' &\equiv \boldsymbol{E} + \frac{1}{c}(\boldsymbol{u} \times \boldsymbol{B}) \\ \boldsymbol{D}' &\equiv \boldsymbol{D} + \frac{1}{c}(\boldsymbol{u} \times \boldsymbol{H}) \end{aligned}\right\} \text{(electromotive force)}$$

$$\left.\begin{aligned} \boldsymbol{H}' &\equiv \boldsymbol{H} - \frac{1}{c}(\boldsymbol{u} \times \boldsymbol{D}) \\ \boldsymbol{B}' &\equiv \boldsymbol{B} - \frac{1}{c}(\boldsymbol{u} \times \boldsymbol{E}) \end{aligned}\right\} \text{(magnetomotive force)}$$

Then the Minkowski equations read

$$D' = \epsilon E' \tag{7}$$

$$B' = \mu H' \tag{8}$$

The generalization of Ohm's law is

$$l^\varrho = \sigma \varepsilon^\varrho \tag{9}$$

In vectorial form this means

$$\boldsymbol{I} + \frac{\boldsymbol{u}[(\boldsymbol{u}\cdot\boldsymbol{I}) - \rho c^2]}{c^2 - u^2} = \frac{\sigma \boldsymbol{E}'}{\sqrt{1 - u^2/c^2}} \tag{10}$$

If the current is purely convective, then $\boldsymbol{I} = \rho\boldsymbol{u}$ and $s^\varrho = U^\varrho \cdot (s^\lambda U_\lambda)$, and the Lorentz-force expression becomes

$$k^\varrho = \frac{\rho_0}{c^4}\,\Phi^{\varrho\lambda} U_\lambda \tag{11}$$

The foregoing equations are to be regarded as representative of various competing tensors and not as completely verified formulas.

2.8. Field of a uniformly moving point charge in empty space. Force between point charges moving with the same constant velocity. We shall assume a point charge ε moving with the constant speed v in the positive x direction. In a reference system $\overset{*}{\Sigma}$, chosen so that the charge is at rest with respect to it, and permanently at its origin, we have

$$\overset{*}{D}_s = \overset{*}{E}_s = \varepsilon\frac{\overset{*}{x}{}^s}{\overset{*}{r}{}^3}; \qquad \overset{*}{B}_s = \overset{*}{H}_s = 0 \tag{1}$$

Using the transformation equations of F, we obtain

$$\left.\begin{aligned} E_1 &= \overset{*}{E}_1, & H_1 &= 0 \\ E_2 &= \frac{\overset{*}{E}_2}{\sqrt{1-\beta^2}}, & H_2 &= -\frac{\beta}{\sqrt{1-\beta^2}}\overset{*}{E}_3 \\ E_3 &= \frac{\overset{*}{E}_3}{\sqrt{1-\beta^2}}, & H_3 &= \frac{\beta}{\sqrt{1-\beta^2}}\overset{*}{E} \end{aligned}\right\} \tag{2}$$

Since $\overset{*}{r}{}^2 = \overset{*}{x}{}^2 + \overset{*}{y}{}^2 + \overset{*}{z}{}^2$ and $\overset{*}{x} = \dfrac{x - vt}{\sqrt{1-\beta^2}}$; $\quad \overset{*}{y} = y$; $\overset{*}{z} = z$

we obtain

$$\left.\begin{aligned} E_1 &= \frac{\varepsilon}{\sqrt{1-\beta^2}}\cdot\frac{x - vt}{R^3}, & H_1 &= 0 \\ E_2 &= \frac{\varepsilon}{\sqrt{1-\beta^2}}\cdot\frac{y}{R^3}, & H_2 &= -\frac{\varepsilon\beta}{\sqrt{1-\beta^2}}\cdot\frac{z}{R^3} \\ E_3 &= \frac{\varepsilon}{\sqrt{1-\beta^2}}\cdot\frac{z}{R^3}, & H_3 &= \frac{\varepsilon\beta}{\sqrt{1-\beta^2}}\cdot\frac{y}{R^3} \\ R^2 &= \frac{(x - vt)^2}{1-\beta^2} + y^2 + z^2 \end{aligned}\right\} \tag{3}$$

These expressions show that the electromagnetic field is carried rigidly along by the charge (" convective field "). For simplicity, consider the field at the time zero, when the charge passes through the origin. Then

$$R^2 = \frac{x^2}{1-\beta^2} + y^2 + z^2$$

and

$$\left.\begin{aligned} E_1 &= \frac{\varepsilon}{\sqrt{1-\beta^2}} \cdot \frac{x}{R^3}, & H_1 &= 0 \\ E_2 &= \frac{\varepsilon}{\sqrt{1-\beta^2}} \cdot \frac{y}{R^3}, & H_2 &= -\frac{\varepsilon\beta}{\sqrt{1-\beta^2}} \cdot \frac{z}{R^3} \\ E_3 &= \frac{\varepsilon}{\sqrt{1-\beta^2}} \cdot \frac{z}{R^3}, & H_3 &= \frac{\varepsilon\beta}{\sqrt{1-\beta^2}} \cdot \frac{y}{R^3} \end{aligned}\right\} \quad (4)$$

The magnetic field lies entirely in the " equatorial plane," i.e., the plane orthogonal to the direction of motion. Let us compute the electric field strength at a distance s from the origin. If the point of observation lies on the x axis, $R = s/\sqrt{1-\beta^2}$, we obtain

$$E_1 = \frac{\varepsilon(1-\beta^2)}{s^2}; \quad E_2 = E_3 = 0 \quad (5)$$

If the point lies on the y axis at a distance S from the origin, we obtain $R = S$ and

$$E_2 = \frac{\varepsilon}{s^2\sqrt{1-\beta^2}}; \quad E_1 = E_3 = 0 \quad (6)$$

As the speed increases to the speed of light, in the first case $E_1 = E_2 = E_3 = 0$ and in the second case $E_2 = \infty$; $E_1 = E_3 = 0$. At high speeds the electric field tends more and more to crowd in the equatorial plane. The energy density becomes zero outside and infinite inside the equatorial plane in such a manner that the total energy of the field becomes infinite, and therefore also the electromagnetic mass of the charge becomes infinite. The Lorentz force $\boldsymbol{F}$ exerted on a second charge ε' which moves with the same velocity $\boldsymbol{u}$, is

$$\boldsymbol{F} = \varepsilon' \left\{ E + \frac{1}{c}[\boldsymbol{u} \times \boldsymbol{H}] \right\} \quad (7)$$

and can be written $\boldsymbol{F} = \textbf{grad}\, \psi$,

where $\psi = (1-\beta^2)\dfrac{\varepsilon}{R}$ (8)

Here ψ is called the " convective potential."

The surfaces of constant convective potential are

$$R^2 = x^2 + (1 - \beta^2)(y^2 + z^2) = \text{constant} \tag{9}$$

They are ellipsoids of revolution with the charge at their center. The direction of motion is the axis of revolution and the principal axis is smaller than the two minor axes. These ellipsoids are called the "Heaviside ellipsoids."

If the charge ε' lies on the x axis at a distance s from e, the magnitude of the force exerted on it, is

$$F = \frac{\varepsilon\varepsilon'}{s^2}(1 - \beta^2) \tag{10}$$

If the charge is located on the y axis at a distance s from the charge e, the force is

$$F = \frac{\varepsilon\varepsilon'}{s^2}\sqrt{1 - \beta^2} \tag{11}$$

In either case the force between the two charges tends to zero as their velocity approaches the velocity of light.

2.9. The stress energy tensor and its relation to the conservation laws. We introduce the tensor $M^{\varrho\sigma}$, defined by

$$M^{\varrho\sigma} = \frac{1}{4\pi c^\sigma}\left[\frac{\eta^{\varrho\sigma}}{4}\Phi_{\lambda\mu}\Phi^{\lambda\mu} - \frac{1}{2}(\Phi^{\varrho\lambda}\Psi^{\cdot\sigma\mu} + \Phi^{\sigma\lambda}\Psi^{\cdot\varrho\mu})\eta_{\lambda\mu}\right] \tag{1}$$

$$\Phi_{\lambda\mu}\Psi^{\cdot\lambda\mu} = 2c^4(\boldsymbol{B}\cdot\boldsymbol{H} - \boldsymbol{E}\cdot\boldsymbol{D}) \tag{2}$$

$$M^{rs} = \frac{1}{4\pi}\left[\frac{1}{2}\delta_{rs}(\boldsymbol{B}\cdot\boldsymbol{H} + \boldsymbol{D}\cdot\boldsymbol{E}) - B_r H_s - D_r E_s\right] \tag{3}$$

$$M^{14} = \frac{1}{8\pi c}(\boldsymbol{D}\times\boldsymbol{B} + \boldsymbol{E}\times\boldsymbol{H})_x \tag{4}$$

$$M^{44} = \frac{1}{8\pi c^2}(\boldsymbol{E}\cdot\boldsymbol{D} + \boldsymbol{H}\cdot\boldsymbol{B}) \tag{5}$$

The M^{rs} are the components of the Maxwell stress tensor; c^2M^{4s} are the components of the Poynting vector; c^2M^{44} is the energy density of the electromagnetic field. The energy stress tensor obeys the four conservation laws

$$M^{\varrho\lambda}{}_{,\lambda} = 0 \tag{6}$$

If in addition to the electrodynamic stresses $M^{\varrho\lambda}$ any other stresses $P^{\varrho\lambda}$ of nonelectrical origin are present, the conservation laws take on the form

$$T^{\varrho\lambda}{}_{,\lambda} = 0 \tag{7}$$

where $\quad T^{\varrho\lambda} \equiv P^{\varrho\lambda} + M^{\varrho\lambda} =$ "total stress energy tensor."

As a special example assume matter of constant rest density σ moving with uniform velocity v along the x axis. Assume furthermore that the mechanical stresses are given by the tensor t^{rs}. Then

$$\left.\begin{aligned}
&P^{11} = \frac{t^{11} + \sigma v^2}{1 - v^2/c^2}; \quad P^{22} = t^{22}; \quad P^{33} = t^{33}; \quad P^{44} = \frac{\sigma + v^2 t^{11}/c^4}{1 - v^2/c^2}\\
&P^{12} = \frac{t^{12}}{\sqrt{1 - v^2/c^2}}; \quad P^{13} = \frac{t^{13}}{\sqrt{1 - v^2/c^2}}; \quad P^{14} = -\frac{\sigma v + v t^{11}/c^2}{1 - v^2/c^2}\\
&P^{23} = t^{23}; \quad P^{24} = -\frac{v}{c^2}\frac{t^{12}}{1 - v^2/c^2}; \quad P^{34} = -\frac{v}{c^2}\frac{t^{13}}{1 - v^2/c^2}
\end{aligned}\right\} \tag{8}$$

In the case of a perfect fluid of pressure p, we have

$$\left.\begin{aligned}
&t^{rs} = p\delta^{rs}; \quad p^{12} = p^{13} = p^{24} = p^{34} = p^{23} = 0\\
&p^{11} = \frac{p + \sigma v^2}{1 - v^2/c^2}; \quad p^{22} = p^{33} = p; \quad p^{44} = \frac{\sigma + v^2 p/c^4}{1 - v^2/c^2}\\
&p^{14} = -\frac{\rho v + v p/c^2}{1 - v^2/c^2}
\end{aligned}\right\} \tag{9}$$

3. Miscellaneous Applications

3.1. The ponderomotive equation. The Lorentz-force on a particle of rest-mass m and charge ε can be derived as the Euler-Lagrange equation of the variational principle

$$I = \int_{P_1}^{P_2} L\,dt = \int_{P_1}^{P_2} \left\{ -mc^2\sqrt{1 - \frac{u^2}{c^2}} - \varepsilon\phi + \frac{\varepsilon}{c}\,u^s A_s \right\} dt \quad (\delta I = 0) \tag{1}$$

In this variation the end points are kept fixed. The Euler-Lagrange equations belonging to Eq. (1) are

$$\frac{d}{dt}\left(\frac{mu^s}{\sqrt{1 - u^2/c^2}}\right) = \varepsilon E_s + \frac{\varepsilon}{c}\,u^r H_{sr} \qquad H_{12} = -H_{21} = H_3 \quad \text{etc.} \tag{2}$$

The " relativistic momenta " p_s are defined by

$$p_s = \frac{\partial L}{\partial u^s} = \frac{mu^s}{\sqrt{1 - u^2/c^2}} + \frac{\varepsilon}{c}\,A_s \tag{3}$$

The velocities can be expressed in terms of the momenta :

$$u^s = \frac{1}{m}\left(p_s - \frac{\varepsilon}{c}\,A_s\right)\left[1 + \frac{(p_k - \varepsilon A_k/c)^2}{m^2c^2}\right]^{-1/2} \tag{4}$$

Therefore the Lagrangian L can be written in terms of the momenta and is given by

$$L = \left[1 + \frac{(p_k - \varepsilon A_k/c)^2}{m^2c^2}\right]^{-1/2} \left[-mc^2 + \frac{\varepsilon A_s}{c} \cdot \frac{p_s - (\varepsilon A_s/c)}{m}\right] - \varepsilon\phi \tag{5}$$

The Hamiltonian function H, defined by $H = p_s u^s - L$ is given by

$$H = mc^2 \left[1 + \frac{(p_k - \varepsilon A_k/c)^2}{m^2c^2}\right]^{1/2} + \varepsilon\phi \tag{6}$$

The canonical equations of motion are

$$\dot{u}^s = \frac{\partial H}{\partial p_s} = \frac{1}{m}\left[1 + \frac{(p_k - \varepsilon A_k/c)^2}{m^2c^2}\right]^{-1/2} \left(p_s - \frac{\varepsilon}{c} A_s\right) \tag{7}$$

$$\dot{p}_s = -\frac{\partial H}{\partial x^s} = \frac{\varepsilon}{mc}\left[1 + \frac{(p_k - \varepsilon A_k/c)^2}{m^2c^2}\right]^{-1/2} \left(p_r - \frac{\varepsilon}{c} A_r\right) A_{r,s} - \varepsilon\phi_{,s} \tag{8}$$

Equations (8) are equivalent to Eq. (2).

3.2. Application to electron optics. We consider the motion of a charged particle in a combined stationary electric and magnetic field.

$$\boldsymbol{E} = -\,\mathbf{grad}\,\phi; \quad \boldsymbol{H} = \mathbf{curl}\,\boldsymbol{A} \tag{1}$$

The Lagrangian is

$$L = mc^2 \left(1 - \sqrt{1 - \frac{u^2}{c^2}}\right) - \varepsilon\phi + \frac{\varepsilon}{c}(\boldsymbol{u} \cdot \boldsymbol{A}) \tag{2}$$

The variational principle $\delta \int_{P_1}^{P_2} \leq dt = 0$ yields

$$\frac{d}{dt}\left(\frac{m\boldsymbol{u}}{\sqrt{1 - u^2/c^2}}\right) = -\,\varepsilon\,\mathbf{grad}\,\phi + \frac{\varepsilon}{c}(\boldsymbol{u} \times \mathbf{curl}\,\boldsymbol{A}) \tag{3}$$

The total rate of change of the potential energy ϕ of the particle along its path is given by

$$\frac{d\phi}{dt} = \frac{\partial\phi}{\partial t} + \boldsymbol{u} \cdot \mathbf{grad}\,\phi \tag{4}$$

Since we assume a stationary electromagnetic field,

$$\frac{\partial\phi}{\partial t} = 0 \tag{5}$$

and thus

$$\boldsymbol{u} \cdot \mathbf{grad}\,\phi = \frac{d\phi}{dt} \tag{6}$$

Multiplying both sides of Eq. (3) scalarly by $\boldsymbol{u}$, we obtain

$$mc^2 \frac{d}{dt}\left(\frac{1}{\sqrt{1-u^2/c^2}}\right) = -\varepsilon \frac{d\phi}{dt} \tag{7}$$

or

$$\frac{d}{dt}\left(\frac{mc^2}{\sqrt{1-u^2/c^2}} + \varepsilon\phi\right) = 0 \tag{8}$$

Or introducing the relativistic kinetic energy

$$I = \frac{mc^2}{\sqrt{1-u^2/c^2}} - mc^2$$

Eq. (8) can also be written

$$\frac{d}{dt}\left(\frac{mc^2}{\sqrt{1-u^2/c^2}} - mc^2 + \varepsilon\phi\right) = 0 \tag{9}$$

And thus one obtains one integral of the motion, the energy E,

$$\frac{mc^2}{\sqrt{1-u^2/c^2}} - mc^2 + \varepsilon\phi = E \tag{10}$$

or

$$\frac{mc^2}{\sqrt{1-u^2/c^2}} + \varepsilon\phi = E + mc^2 \tag{11}$$

This equation enables us to rewrite the Lagrangian in the form

$$L = (E - \varepsilon\phi)\left(1 + \sqrt{1 - \frac{u^2}{c^2}}\right) + \frac{\varepsilon}{c}(\boldsymbol{u} \cdot \boldsymbol{A}) - E \tag{12}$$

Since $\boldsymbol{u} = ds/dt$, we obtain from Eq. (10)

$$\frac{ds}{dt} = c\,\frac{\sqrt{(E + 2mc^2 - \varepsilon\phi)(E - \varepsilon\phi)}}{E + mc^2 - \varepsilon\phi} \tag{13}$$

and

$$L\,dt = \left[\frac{1}{c}\sqrt{(\mathrm{E} - \varepsilon\phi)(E + 2mc^2 - \varepsilon\phi)} + \frac{\varepsilon}{c}(\boldsymbol{s} \cdot \boldsymbol{A})\right] ds - E\,dt \tag{14}$$

where $\boldsymbol{s}$ is a unit vector in the direction of the tangent of the trajectory of the particle : $\boldsymbol{u}\,dt = \boldsymbol{s}\,ds$. The variational principle can be reformulated

$$\delta \int_{P_1}^{P_2} \left[\frac{1}{c}\sqrt{(E - \varepsilon\phi)(E + 2mc^2 - \varepsilon\phi)} + \frac{\varepsilon}{c}(\boldsymbol{s} \cdot \boldsymbol{A})\right] ds = 0 \tag{15}$$

since E is a constant and t is not varied, and therefore

$$\delta \int_{P_1}^{P_2} E\,dt = 0.$$

If we call the dimensionless quantity μ

$$\mu = \frac{1}{mc^2}\left[\sqrt{(E-\varepsilon\phi)(E+2mc^2-\varepsilon\phi)} + \varepsilon(\boldsymbol{s}\cdot A)\right] \tag{16}$$

the variational principle is

$$\delta \int_{P_1}^{P_2} \mu \, ds = 0 \tag{17}$$

This is identical with Fermat's principle in geometrical optics. μ is the "index of refraction." It depends through ϕ and A on the position (x,y,z) and through $\boldsymbol{s}$ also on the direction. The index of refraction is thus seen to be both inhomogeneous and unisotropic. Equation (16) is the fundamental equation of electron optics.

3.3. Sommerfeld's theory of the hydrogen fine structure. We shall assume that an electron of rest-mass m and charge ε moves in the field of a proton which is assumed at rest. Let r be the distance between electron and proton. The energy E is given by

$$mc^2\left(\frac{1}{\sqrt{1-u^2/c^2}}-1\right)-\frac{\varepsilon^2}{r} = E \tag{1}$$

The angular momentum

$$I_\theta = \frac{mr^2\dot{\theta}}{\sqrt{1-u^2/c^2}}$$

is also a constant of the motion. According to Sommerfeld's quantization rule

$$\oint I_\theta d\theta = n_\theta h \tag{2}$$

The integration must be carried out over a complete cycle of the motion, $n_\theta = 1, 2, 3, \ldots$ ($n_\theta = 0$ would imply that the electron penetrates through the nucleus.) Therefore

$$\frac{mr^2\dot{\theta}}{\sqrt{1-u^2/c^2}}\cdot 2\pi = n_\theta \cdot h \tag{3}$$

or

$$mr^2\dot{\theta} = n_\theta \cdot \hbar\sqrt{1-u^2/c^2} \qquad \left(\hbar = \frac{h}{2\pi}\right) \tag{4}$$

The radial momentum p_r is given by

$$p_r = \frac{m\dot{r}}{\sqrt{1-u^2/c^2}} \tag{5}$$

It must again obey the same quantization rule

$$\oint p_r dr = n_r h \qquad (n_r = 0,1,2,...) \tag{6}$$

If Eq. (4) is substituted into Eq. (6) and the integration is carried out, we obtain

$$E = mc^2\left[1 - \frac{1}{\sqrt{1 + \alpha^2/(n_r + \sqrt{n_\theta^2 - \alpha^2})}}\right] \tag{7}$$

where $\alpha = \varepsilon^2/\hbar c$ is the fine-structure constant. If E is expanded in a power series of α, we obtain

$$E = \frac{m\varepsilon^4}{2\hbar^2(n_r + n_\theta)^2}\left[1 + \frac{\alpha^2}{(n_r + n_\theta)^2}\left(\frac{n_r}{n_\theta} + \frac{1}{4}\right) + ...\right] \tag{8}$$

Usually $n_r + n_\theta = n$ is called the principal or orbital quantum number, and $n_\theta = k$ the azimuthal quantum number. Hence

$$E = \frac{m\varepsilon^4}{2\hbar^2 n^2}\left[1 + \frac{\alpha^2}{n^2}\left(\frac{n}{k} - \frac{3}{4}\right)\right] \tag{9}$$

Thus the energy depends not only on n, as in the classical theory, but also on k. If the speed of light were infinite, then $\alpha = 0$, and E would depend on n only.

4. Spinor Calculus

4.1. Algebraic properties. The tensor formalism does not include all the objects that form relations that are invariant under Lorentz transformations. Besides the world tensors there exist " spinors " whose transformation group is also Lorentz-covariant and whose representation admits an even larger possibility than that of the tensors. In that sense the spinors are more fundamental than the tensors. In order to facilitate the reading of the literature * we shall change slightly our previous notation. The coordinates (x,y,z,ct) will be denoted by (x^1,x^2,x^3,x^4) and the metric tensor will be assumed

$$\eta^{11} = \eta^{22} = \eta^{33} = 1; \quad \eta^{44} = -1$$

$$\eta_{11} = \eta_{22} = \eta_{33} = 1; \quad \eta_{44} = -1$$

* See especially the paper, " Application of Spinor Analysis to the Maxwell and Dirac Equations," by LAPORTE, O. and UHLENBECK, G. E., *Phys. Rev.*, **37**, 1380 (1931). The following discussion follows this paper closely and the reader will find further discussion and applications in it.

We shall call any two-dimensional vector $g = (g_1, g_2)$ a spinor of rank one. Its components obey the following linear transformation law

$$\left.\begin{aligned} \overset{*}{g}_1 &= \alpha_{11} g_1 + \alpha_{12} g_2 \\ \overset{*}{g}_2 &= \alpha_{21} g_1 + \alpha_{22} g_2 \end{aligned}\right\} \tag{1}$$

and where the determinant of the transformation coefficients has the value unity

$$\begin{vmatrix} \alpha_{11} & \alpha_{12} \\ \alpha_{21} & \alpha_{22} \end{vmatrix} = 1 \tag{2}$$

the spinor g will also be denoted by $g_k (k = 1, 2)$. Spinor indices will always be denoted by lower-case italic letters and always assume the values 1 and 2 only. The coefficients $\alpha_{rs}(r = 1,2; s = 1,2)$ can be real or complex numbers.

We shall also define a spinor $f = (f_{\dot{1}}, f_{\dot{2}})$ with dotted indices, which obeys the transformation law

$$\left.\begin{aligned} \overset{*}{f}_{\dot{1}} &= \bar{\alpha}_{11} f_{\dot{1}} + \bar{\alpha}_{12} f_{\dot{2}} \\ \overset{*}{f}_{\dot{2}} &= \bar{\alpha}_{21} f_{\dot{1}} + \bar{\alpha}_{22} f_{\dot{2}} \end{aligned}\right\} \tag{3}$$

This spinor will also be denoted by $f_{\dot{k}}(\dot{k} = 1,2)$. A bar will always denote the conjugate complex.

Spinors of higher ranks are quantities that transform like products of spinors of first rank, e.g., the spinor $a_{kl}(k,l = 1,2)$ shall transform like $a_k b_l$, or the "mixed" spinor $a_{\dot{k}l}$ shall transform like $a_{\dot{k}} b_l$. We raise or lower spin indices in the usual fashion by introducing a metric spinor ϵ_{kl} of second rank, symmetric in its indices, such that

$$\left.\begin{aligned} a^k &= \epsilon^{kl} a_l \\ a_k &= \epsilon_{kl} a^l \end{aligned}\right\} \tag{4}$$

$$\left.\begin{aligned} \epsilon_{kl} = \epsilon_{lk} = \epsilon_{\dot{k}\dot{l}} = \epsilon_{\dot{l}\dot{k}} &= \begin{pmatrix} 0 & -1 \\ 1 & 0 \end{pmatrix} \\ \epsilon^{kl} = \epsilon^{lk} = \epsilon^{\dot{k}\dot{l}} = \epsilon^{\dot{l}\dot{k}} &= \begin{pmatrix} 0 & 1 \\ -1 & 0 \end{pmatrix} \end{aligned}\right\} \tag{5}$$

$$\epsilon^{kl} \epsilon_{lm} = \delta^k{}_m \tag{6}$$

Summations will be performed only over undotted or dotted indices, but not over mixed indices.

Thus e.g.,
$$a^l b_l = a^1 b_1 + a^2 b_2$$
and
$$a^{\dot{l}} b_{\dot{l}} = a^{\dot{1}} b_{\dot{1}} + a^{\dot{2}} b_{\dot{2}}$$
but no summation for
$$a^l b_{\dot{l}}$$

From the above definitions follow the fundamental relations

$$\left.\begin{aligned} a^1 &= a_2; \qquad b^{\dot{1}} = b_{\dot{2}} \\ a^2 &= -a_1; \quad b^{\dot{2}} = -b_{\dot{1}} \end{aligned}\right\} \tag{7}$$

This ensures that $a_k b^k$ and $a_{\dot{k}} b^{\dot{k}}$ are invariants. From the above definitions,

$$a_l b^l = -a^l b_l \tag{8}$$

Also, the absolute value of any spinor of odd rank is zero.

$$a_l a^l = 0 \quad \text{or} \quad a_{lmn} a^{lmn} = 0 \tag{9}$$

$$a^l b_l c_m + a_l b_m c^l + a_m b^l c_l = 0 \tag{10}$$

All these rules hold of course also for dotted indices. The positions of dotted and undotted indices may be interchanged without changing the spinor, e.g.,

$$a_{\dot{r}st} = a_{s\dot{r}t} = a_{st\dot{r}} \tag{11}$$

However, the interchange of two dotted or two undotted indices will in general change the spinor unless it possesses certain additional symmetry properties. To obtain the complex conjugate of any spinor equation replace all dotted indices by dotted ones and vice versa.

4.2. Connection between spinors and world tensors. To each contravariant world tensor A^ϱ a mixed spinor $a_{\dot{r}s}$ can be uniquely associated, and vice versa. This permits us to rewrite each tensor equation as a spinor equation and vice versa. The relations are the following :

$$\left.\begin{aligned} A^1 &= A_1 = \frac{1}{2}(a_{\dot{2}1} + a_{\dot{1}2}) \\ A^2 &= A_2 = \frac{1}{2i}(a_{\dot{2}1} - a_{\dot{1}2}) \\ A^3 &= A_3 = \frac{1}{2}(a_{\dot{1}1} - a_{\dot{2}2}) \\ A^4 &= -A_4 = \frac{1}{2}(a_{\dot{1}1} + a_{\dot{2}2}) \end{aligned}\right\} \tag{1}$$

Conversely, we can also express the spinor components in terms of the vector components

$$\left.\begin{aligned} a_{\dot{2}1} &= -a^{\dot{1}2} = A^1 + iA^2 = A_1 + iA_2 \\ a_{\dot{1}2} &= -a^{\dot{2}1} = A^1 - iA^2 = A_1 - iA_2 \\ a_{\dot{1}1} &= +a^{\dot{2}2} = A^3 + A^4 = A_3 - A_4 \\ -a_{\dot{2}2} &= -a^{\dot{1}1} = A^3 - A^4 = A_3 + A_4 \end{aligned}\right\} \quad (2)$$

$$A_\varrho A^\varrho = -\frac{1}{2} a_{\dot{r}t} a^{\dot{r}t} \quad (3)$$

The above scheme also permits to introduce spinor differential operators

$$\left.\begin{aligned} \partial^{\dot{1}}{}_{1} &= \partial_{\dot{2}1} = \frac{\partial}{\partial x^1} + i\frac{\partial}{\partial x^2} \\ -\partial^{\dot{2}}{}_{2} &= \partial_{\dot{1}2} = \frac{\partial}{\partial x^1} - i\frac{\partial}{\partial x^2} \\ -\partial^{\dot{2}}{}_{1} &= \partial_{\dot{1}1} = \frac{\partial}{\partial x^3} - \frac{\partial}{\partial x^4} \\ -\partial^{\dot{1}}{}_{2} &= -\partial_{\dot{2}2} = \frac{\partial}{\partial x^3} + \frac{\partial}{\partial x^4} \end{aligned}\right\} \quad (4)$$

By means of the relations (4) all the usual vector differential operators that occur in vector analysis can be translated into equivalent spinor relations, e.g., let $\phi_{\dot{r}s}$ be the spinor that corresponds to the world vector ϕ^ϱ; then $\Phi^\varrho{}_{,\varrho} \equiv \dfrac{\partial\phi^1}{\partial x} + \dfrac{\partial\phi^2}{\partial y} + \dfrac{\partial\phi^3}{\partial z} + \dfrac{1}{c}\dfrac{\partial\phi^4}{\partial t}$ corresponds to $-\frac{1}{2}\,\partial_{\dot{r}s}\phi^{\dot{r}s}$, and if the ordinary scalar S corresponds to the spinor scalar s, then

$$\Box S = \eta^{\varrho\sigma}\frac{\partial^2 S}{\partial x^\varrho \partial x^\sigma} = \nabla^2 S - \frac{1}{c^2}\frac{\partial^2 S}{\partial t^2}$$

corresponds to

$$-\frac{1}{2}\,\partial_{\dot{r}t}\partial^{\dot{r}t}s$$

4.3. Transformation laws for mixed spinors of second rank. Relation between spinor and Lorentz transformations. Since mixed spinors of rank two are the fundamental " building stones " for world tensors, we shall write out their transformation laws explicity. The symbol $R(\alpha)$ shall denote the real part of α and $I(\alpha)$ shall denote the imaginary part of α. Thus if $\alpha = \beta + i\gamma$, where β and γ are real numbers,

$$R(\alpha) = \beta \quad \text{and} \quad I(\alpha) = \gamma$$

The absolute value of α will be denoted by $|\alpha| = +\sqrt{\beta^2 + \gamma^2}$.

Let $g_{\dot{r}s}$ be a mixed spinor of rank two, and $\overset{*}{g}_{\dot{r}s}$ the corresponding spinor in the new coordinate system. Then

$$\left.\begin{aligned}
\overset{*}{g}_{\dot{1}1} &= |\alpha_{11}|^2 g_{\dot{1}1} + \bar{\alpha}_{11}\alpha_{12} g_{\dot{1}2} + \bar{\alpha}_{12}\alpha_{11} g_{\dot{2}1} + |\alpha_{12}|^2 g_{\dot{2}2} \\
\overset{*}{g}_{\dot{1}1} &= \bar{\alpha}_{11}\alpha_{21} g_{\dot{1}1} + \bar{\alpha}_{11}\alpha_{22} g_{\dot{1}2} + \bar{\alpha}_{12}\alpha_{21} g_{21} + \bar{\alpha}_{12}\alpha_{22} g_{\dot{2}2} \\
\overset{*}{g}_{\dot{2}1} &= \bar{\alpha}_{21}\alpha_{11} g_{\dot{1}1} + \bar{\alpha}_{21}\alpha_{12} g_{\dot{1}2} + \bar{\alpha}_{22}\alpha_{11} g_{\dot{2}1} + \bar{\alpha}_{22}\alpha_{12} g_{22} \\
\overset{*}{g}_{22} &= |\alpha_{21}|^2 g_{\dot{1}1} + \bar{\alpha}_{21}\alpha_{22} g_{\dot{1}2} + \bar{\alpha}_{22}\alpha_{21} g_{\dot{2}1} + |\alpha_{22}|^2 g_{\dot{2}2}
\end{aligned}\right\} \quad (1)$$

The relations between the α_{rs} and the coefficients $\gamma^{\lambda\cdot}_{\cdot\tau}$ of the Lorentz transformation follow:

$$\left.\begin{aligned}
\gamma^{1\cdot}_{\cdot 1} &= R(\bar{\alpha}_{11}\alpha_{22}) + R(\bar{\alpha}_{12}\alpha_{21}) \\
\gamma^{2\cdot}_{\cdot 1} &= I(\bar{\alpha}_{11}\alpha_{22}) + I(\alpha_{12}\bar{\alpha}_{21}) \\
\gamma^{3\cdot}_{\cdot 1} &= R(\alpha_{11}\bar{\alpha}_{21}) - R(\bar{\alpha}_{12}\alpha_{22}) \\
\gamma^{4\cdot}_{\cdot 1} &= -R(\alpha_{11}\bar{\alpha}_{21}) - R(\bar{\alpha}_{12}\alpha_{22}) \\
\gamma^{1\cdot}_{\cdot 2} &= I(\bar{\alpha}_{22}\alpha_{11}) + I(\bar{\alpha}_{21}\alpha_{12}) \\
\gamma^{2\cdot}_{\cdot 2} &= R(\bar{\alpha}_{11}\alpha_{22}) - R(\bar{\alpha}_{12}\alpha_{21}) \\
\gamma^{3\cdot}_{\cdot 2} &= I(\bar{\alpha}_{21}\alpha_{11}) + I(\bar{\alpha}_{12}\alpha_{22}) \\
\gamma^{4\cdot}_{\cdot 2} &= I(\bar{\alpha}_{11}\alpha_{21}) + I(\bar{\alpha}_{12}\alpha_{22}) \\
\gamma^{1\cdot}_{\cdot 3} &= R(\bar{\alpha}_{11}\alpha_{12}) - R(\bar{\alpha}_{21}\alpha_{22}) \\
\gamma^{2\cdot}_{\cdot 3} &= I(\bar{\alpha}_{11}\alpha_{12}) + I(\bar{\alpha}_{22}\alpha_{21}) \\
\gamma^{3\cdot}_{\cdot 3} &= \frac{1}{2}(|\alpha_{11}|^2 + |\alpha_{22}|^2 - |\alpha_{12}|^2 - |\alpha_{21}|^2) \\
\gamma^{4\cdot}_{\cdot 3} &= \frac{1}{2}(|\alpha_{21}|^2 + |\alpha_{22}|^2 - |\alpha_{11}|^2 - |\alpha_{12}|^2)
\end{aligned}\right\} \quad (2)$$

$$\left.\begin{aligned}
\gamma^{1\cdot}_{\cdot 4} &= -R(\alpha_{11}\bar{\alpha}_{12}) - R(\bar{\alpha}_{21}\alpha_{22})\\
\gamma^{2\cdot}_{\cdot 4} &= -I(\alpha_{11}\bar{\alpha}_{12}) + I(\alpha_{21}\bar{\alpha}_{22})\\
\gamma^{3\cdot}_{\cdot 4} &= \frac{1}{2}(|\alpha_{12}|^2 + |\alpha_{22}|^2 - |\alpha_{11}|^2 - |\alpha_{21}|^2)\\
\gamma^{4\cdot}_{\cdot 4} &= \frac{1}{2}(|\alpha_{11}|^2 + |\alpha_{12}|^2 + |\alpha_{21}|^2 + |\alpha_{22}|^2)
\end{aligned}\right\}$$

The $\gamma^{\lambda\cdot}_{\cdot\tau}$ are quadratic in the α_{rs}. To each Lorentz transformation belong two spinor transformations which differ only in sign. But to each spinor transformation belongs only one Lorentz transformation. In this sense the spinors are more fundamental than the tensors.

Two special examples will illustrate the meaning of Eq. (2).

$$\alpha_{11} = e^{-i\theta/2}; \quad \alpha_{22} = e^{i\theta/2}; \quad \alpha_{12} = \alpha_{21} = 0 \tag{3}$$

This yields

$$\gamma^{1\cdot}_{\cdot 1} = \gamma^{2\cdot}_{\cdot 2} = \cos\theta$$

$$\gamma^{1\cdot}_{\cdot 2} = -\gamma^{2\cdot}_{\cdot 1} = \sin\theta$$

$$\gamma^{3\cdot}_{\cdot 3} = \gamma^{4\cdot}_{\cdot 4} = 1$$

All other γ are zero.

This corresponds to a purely spatial rotation about the z axis through an angle θ.

$$\alpha_{11} = e^{\theta/2}; \quad \alpha_{22} = e^{-\theta/2}; \quad \alpha_{12} = \alpha_{21} = 0 \tag{4}$$

This yields

$$\gamma^{1\cdot}_{\cdot 1} = \gamma^{2\cdot}_{\cdot 2} = 1$$

$$\gamma^{3\cdot}_{\cdot 4} = -\gamma^{4\cdot}_{\cdot 3} = \sinh\theta$$

$$\gamma^{3\cdot}_{\cdot 3} = \gamma^{4\cdot}_{\cdot 4} = \cosh\theta$$

All other γ are zero. This corresponds to a coordinate system $\overset{*}{\Sigma}$ moving in the positive z direction with the velocity v, where $\tan i\theta = iv/c$.

4.4. Dual tensors. We assume the existence of an antisymmetric tensor $F^{\alpha\beta}$ whose components are all assumed real. The dual of $F^{\alpha\beta}$, denoted by $\overset{\Delta}{F}{}^{\alpha\beta}$, is defined by

$$\overset{\Delta}{F}{}^{\alpha\beta} = \eta^{\alpha\varrho}\eta^{\beta\sigma}\overset{\Delta}{F}_{\varrho\sigma} \tag{1}$$

where $\overset{\Delta}{F}_{\varrho\sigma}$ is defined by

$$\overset{\Delta}{F}_{\varrho\sigma} = \frac{i}{2}\,\delta_{\varrho\sigma\lambda\tau}F^{\lambda\tau} \tag{2}$$

Here $\delta_{\varrho\sigma\lambda\tau}$ is either +1 or −1, depending on whether the indices $\rho\sigma\lambda\tau$ are an even or odd permutation of the array (1234). If any two or more indices are the same, then $\delta_{\varrho\sigma\lambda\tau} = 0$. From these definitions, we get

$$\left.\begin{aligned} \overset{\Delta}{F}{}^{12} &= iF^{34}; & \overset{\Delta}{F}{}^{14} &= -iF^{23} \\ \overset{\Delta}{F}{}^{23} &= iF^{14}; & \overset{\Delta}{F}{}^{24} &= -iF^{31} \\ \overset{\Delta}{F}{}^{31} &= iF^{24}; & \overset{\Delta}{F}{}^{34} &= -iF^{12} \end{aligned}\right\} \tag{3}$$

The same relations hold true also for a covariant antisymmetric tensor $F_{\varrho\sigma}$ and its dual tensor $\overset{\Delta}{F}_{\varrho\sigma}$. A tensor $F^{\varrho\sigma}$ is self-dual if

$$F^{\varrho\sigma} = \overset{\Delta}{F}{}^{\varrho\sigma}$$

From a given antisymmetric tensor $F^{\varrho\sigma}$ one can always construct a self-dual tensor $G^{\varrho\sigma}$ by forming

$$G^{\varrho\sigma} \equiv F^{\varrho\sigma} + \overset{\Delta}{F}{}^{\varrho\sigma}$$

Whereas an antisymmetric tensor possesses six linearly independent components, a self-dual tensor possesses only three linearly independent components. They can most conveniently be expressed in terms of a three-dimensional, complex-valued vector $k = (k_1,k_2,k_3)$. The components of $G^{\varrho\sigma}$ are then

$$\left.\begin{aligned} (G^{14},G^{24},G^{34}) &= (-ik_1,-ik_2,-ik_3) \\ (G^{23},G^{31},G^{12}) &= (k_1,k_2,k_3) \end{aligned}\right\} \tag{4}$$

We may write the vector k as a complex sum of two real three-dimensional vectors a and b, i.e.,

$$\boldsymbol{k} = \boldsymbol{a} + i\boldsymbol{b} \tag{5}$$

Then

$$\left.\begin{aligned} (G^{14},G^{24},G^{34}) &= (b_1 - ia_1, b_2 - ia_2, b_3 - ia_3) \\ (G^{23},G^{31},G^{12}) &= (a_1 + ib_1, a_2 + ib_2, a_3 + ib_3) \end{aligned}\right\} \tag{6}$$

It is possible to associate with a self-dual tensor $G^{\varrho\sigma}$ a symmetric spinor g_{rs} of rank two which also possesses only three linearly independent components, viz.,

$$\left.\begin{aligned} g_{11} &= 2(\bar{k}_2 - i\bar{k}_1) = 2[a_2 - b_1 - i(b_2 + a_1)] \\ g_{22} &= 2(\bar{k}_2 + i\bar{k}_1) = 2[a_2 + b_1 - i(b_2 - a_1)] \\ g_{12} &= g_{21} = 2i\bar{k}_3 = 2[b_3 + ia_3] \\ g_{\dot{1}\dot{1}} &= \bar{g}_{11} = 2(k_2 + ik_1) = 2[(a_2 - b_1) + i(b_2 + a_1)] \\ g_{\dot{2}\dot{2}} &= \bar{g}_{22} = 2(k_2 - ik_1) = 2[(a_2 + b_1) + i(b_2 - a_1)] \\ g_{\dot{1}\dot{2}} &= g_{\dot{2}\dot{1}} = \bar{g}_{12} = \bar{g}_{21} = 2[b_3 - ia_3] \end{aligned}\right\} \tag{7}$$

4.5. Electrodynamics of empty space in spinor form. We construct a self-dual tensor $G^{\varrho\sigma}$, where

$$G^{\varrho\sigma} = F^{\varrho\sigma} + \overset{\Delta}{F}{}^{\varrho\sigma} \tag{1a}$$

The antisymmetric tensors $F^{\varrho\sigma}$ and $\overset{\Delta}{F}{}^{\varrho\sigma}$ are defined by

$$F^{\varrho\sigma} = \begin{bmatrix} 0 & H_3 & -H_2 & -E_1 \\ & 0 & H_1 & -E_2 \\ & & 0 & -E_3 \\ & & & 0 \end{bmatrix} \text{ and } \overset{\Delta}{F}{}^{\varrho\sigma} = -i \begin{bmatrix} 0 & E_3 & -E_2 & H_1 \\ & 0 & E_1 & H_2 \\ & & 0 & H_3 \\ & & & 0 \end{bmatrix} \tag{1b}$$

In our present notation, however, we set $c = 1$. Therefore the vector $\boldsymbol{k}$ is in this case

$$\boldsymbol{k} = \boldsymbol{H} - i\boldsymbol{E} \tag{2}$$

The corresponding spinor $g_{\dot{r}\dot{s}}$ has the components

$$\left.\begin{aligned} g_{\dot{1}\dot{1}} &= 2[H_2 + E_1 + i(H_1 - E_2)] \\ g_{\dot{2}\dot{2}} &= 2[H_2 - E_1 - i(H_1 + E_2)] \\ g_{\dot{1}\dot{2}} &= -2[E_3 + iH_3] \end{aligned}\right\} \tag{3}$$

We also introduce a current-density spinor $s_{\dot{r}s}$ corresponding to the contravariant current density world-vector

$$s^\lambda = \frac{\rho}{c} v_1, \quad \frac{\rho}{c} v_2, \quad \frac{\rho}{c} v_3, \quad \rho, \quad \text{viz.,}$$

$$\left.\begin{aligned} s_{\dot{1}1} &= \frac{\rho}{c} v_3 + \rho & \qquad s_{\dot{2}1} &= \frac{\rho}{c} v_1 + \frac{i\rho}{c} v_2 \\ s_{\dot{1}2} &= \frac{\rho}{c} v_1 - \frac{i\rho}{c} v_2 & \qquad s_{\dot{2}2} &= -\frac{\rho}{c} v_3 + \rho \end{aligned}\right\} \tag{4}$$

Maxwell's eight equations read then in spinor language

$$\sigma^{r}{}_{i}g_{\dot{r}\dot{m}} = 2s_{l\dot{m}} \tag{5}$$

2. Fundamental Relativistic Invariants

1. Speed of light in empty space c
2. Length of a world line element ds:

$$ds^2 = dx^2 + dy^2 + dz^2 - c^2dt^2$$

3. Phase of an electromagnetic wave
4. Rest mass of a particle m
5. Electric charge ε
6. Action integral $\int_{t_1}^{t_2} L\,dt$, if it is understood that t means the proper time of the observer and that the Lagrangian L is meant to be

$$L = m_0c^2(1 - \sqrt{1 - \beta^2}) - V$$

(V is the potential energy).
7. Planck's constant h
8. Entropy $S = S_0$ (S = entropy of moving system; S_0 = entropy of system at rest)
9. Boltzmann's constant k
10. Rest temperature $T_0 = T/\sqrt{1 - \beta^2}$
11. Heat $Q_0 = Q/\sqrt{1 - \beta^2}$

Bibliography

1. Bergmann, P. G., *Introduction to the Theory of Relativity*, Prentice-Hall, Inc., New York, 1942. Probably the best modern representation of the subject.
2. Eddington, Sir A. S., *The Mathematical Theory of Relativity*, Cambridge University Press, London, 1923.
3. Laue, M. V., *Das Relativitätsprinzip*, Vol. 1 (special theory), 4th ed., 1921; Vol. 2 (general theory), 2d ed., 1923; F. Vieweg und Sohn, Braunschweig.
4. Lorentz, A. H., Einstein, A. and Minkowski, H., *Das Relativitätsprinzip*, 4th ed., B. G. Teubner, Leipzig, 1922. (An English translation, *The Principle of Relativity*, is available from Dover Publications, Inc., New York
5. Pauli, W., Jr., *Relativitätstheorie*, 4th ed., B. G. Teubner, Leipzig, 1922. (Article in *Enzyklopädie der mathematischen Wissenschaften*, Vol. 5, Part 2, Sec. 4.) An exhaustive treatment of the earlier parts of the theory.
6. Weyl, H., *Raum, Zeit, Materie*, 5th ed., Julius Springer, Berlin, 1923. Discusses the fundamental philosophical principles of space and time. (A translation was published by Dover Publications, New York, under the title *Space, Time, Matter*.)

Chapter 7

THE GENERAL THEORY OF RELATIVITY

BY HENRY ZATZKIS

Assistant Professor of Mathematics
Newark College of Engineering

1. Mathematical Basis of General Relativity

1.1. Mathematical introduction. The world continuum of space-time is no longer considered " flat " or " Euclidean," as was done in the special theory of relativity, but " curved " or " Riemannian." This means there exists a " metric tensor " $g_{\varrho\sigma} = g_{\sigma\varrho}$, such that the distance $d\tau$ between two infinitesimally close world points (x^1,x^2,x^3,x^4) and $(x^1 + dx^1, x^2 + dx^2, x^3 + dx^3, x^4 + dx^4)$ is expressed by

$$\left.\begin{aligned} d\tau^2 &= g_{\varrho\sigma}\, dx^\varrho\, dx^\sigma, \\ g_{\varrho\sigma} &= g_{\sigma\varrho}(x^1,x^2,x^3,x^4), \quad x^4 = t \end{aligned}\right\} \tag{1}$$

whereas in flat space the distance is

$$dt^2 = \eta_{\varrho\sigma}\, dx^\varrho\, dx^\sigma \tag{2}$$

The essential point is that the $g_{\varrho\sigma}$ are functions of the coordinates and that no special coordinate system exists in which $g_{\varrho\sigma}$ would be constant in a finite domain.

It was Einstein's fundamental idea to establish a relation between the metric tensor and the matter distribution in the world continuum. This relation is the content of Einstein's field equations. The $g_{\varrho\sigma}$ play the role of gravitational potentials and are the direct generalization of the Newtonian gravitational potential. More specifically, if the deviation from flatness is small, all $g_{\varrho\sigma}$ vanish, except g_{44}, which then becomes identical with the Newtonian potential of a mass distribution.

Another important geometrical object in a Riemanian manifold, besides the metric tensor, is the Riemanian curvature tensor

$$R_{\lambda\varrho\tau.}{}^{\sigma} = \Gamma^{\sigma}_{\tau\lambda,\varrho} - \Gamma^{\sigma}_{\tau\varrho,\lambda} - \Gamma^{\sigma}_{\omega\lambda}\Gamma^{\omega}_{\tau\varrho} + \Gamma^{\sigma}_{\omega\varrho}\Gamma^{\omega}_{\tau\lambda} \tag{3}$$

A comma denotes as usual the derivative with respect to the corresponding coordinate.

The Γ symbols are called the Christoffel three-index symbols of the second kind, and are defined by

$$\Gamma^{\lambda}_{\varrho\sigma} = \tfrac{1}{2} g^{\lambda\omega}(g_{\varrho\omega,\sigma} + g_{\sigma\omega,\varrho} - g_{\varrho\sigma,\omega}) \tag{4}$$

The Γ's vanish identically if the $g_{\varrho\sigma}$ are constants, i.e., if the space is flat. However, even in flat space the $g_{\varrho\sigma}$ may not be constants as a result of a special choice of coordinates, e.g., if we choose polar coordinates instead of Cartesian coordinates in ordinary three-dimensional Euclidean space. It is therefore useful to have the following criterion for the curvature of a given space: The necessary and sufficient condition for a space to be flat is that all the components of the Riemannian curvature tensor vanish.

From the curvature tensor we can obtain tensors of lower rank by the process of "contraction." Of special interest is the tensor of rank two, denoted by $R_{\lambda\sigma}$ and the completely contracted curvature tensor, which gives us the curvature scalar R.

$$R_{\lambda\sigma} = R_{\sigma\lambda} = R^{\omega}_{\omega\lambda\sigma} \tag{5}$$

$$R = g^{\lambda\sigma} R_{\lambda\sigma} \tag{6}$$

Written in terms of the Christoffel symbols, $R_{\lambda\sigma}$ takes the form

$$R_{\lambda\sigma} = \Gamma^{\omega}_{\sigma\omega,\lambda} - \Gamma^{\omega}_{\sigma\lambda,\omega} - \Gamma^{\omega}_{\tau\omega}\Gamma^{\tau}_{\sigma\lambda} + \Gamma^{\omega}_{\tau\lambda}\Gamma^{\tau}_{\sigma\lambda} \tag{7}$$

For later purposes we need the expression denoted by $G^{\lambda\sigma}$

$$G^{\lambda\sigma} = R^{\lambda\sigma} - \tfrac{1}{2} g^{\lambda\sigma} R \tag{8}$$

By contracting the above expression we obtain

$$g_{\lambda\sigma} G^{\lambda\sigma} = R - (\tfrac{1}{2}) \cdot 4R = -R \tag{9}$$

Therefore

$$G^{\lambda\sigma} = 0 \tag{10}$$

is equivalent to

$$R^{\lambda\sigma} = 0 \tag{11}$$

Finally we shall denote the determinant of the $g_{\varrho\sigma}$ matrix by g.

1.2. The field equations. If matter can be represented by a continuous distribution $\rho = \rho(x,y,z)$ which depends on space but not on time, the gravitational potential G, generated by the mass distribution, is given by Poisson's equation

$$\nabla^2 G = -4\pi\rho \tag{1}$$

In regions where the space is empty, this reduces to Laplace's equation

$$\nabla^2 G = 0 \tag{2}$$

If the matter distribution depended on time also, it would be natural to assume that Eq. (1) should be generalized to the inhomogeneous wave equation

$$\Box G = -4\pi\rho(x,y,z,t) \tag{3}$$

This equation, however, is not Lorentz-covariant, since the right side is not Lorentz-covariant. In order to be so, the right side should read ρ_0 instead ρ, i.e., should include only the rest density. On the other hand, it is an empirical fact that the total mass is responsible for gravitational forces and not only the rest mass. Thus ρ must include the contributions from the kinetic energy as well. But the latter terms are no longer scalars.

Another possibility would have been to introduce four gravitational potentials and choose as the right-hand side the energy-momentum vector. But this is again not feasible. In the presence of matter, the mass density ρ is not a scalar, but one component of a tensor $P^{\lambda\sigma}$ (see the discussion of the stress energy tensor in the special theory of relativity).

In his search for the correct formulation Einstein postulated the field equations

$$G^{\lambda\sigma} = -8\pi K P^{\lambda\sigma} \tag{4}$$

The $G^{\lambda\sigma}$ are the expressions defined by Eq. (8). K is the gravitational constant and $P^{\lambda\sigma}$ is the stress-energy tensor and represents a given distribution of matter. If the gravitating matter is concentrated in small regions, " mass points," and zero elsewhere, the field equations reduce to

$$G^{\lambda\sigma} = 0 \tag{5}$$

which is equivalent to

$$R^{\lambda\sigma} = 0 \tag{6}$$

Equation (4) corresponds to Poisson's equation, and Eq. (5) to Laplace's equation. And just as Laplace's equation has solutions which are regular everywhere, except at the origin, also Eq. (4) possesses solutions $g_{\varrho\sigma}$ which are regular everywhere except for a singularity at the origin.

1.3. The variational principle. Laplace's equation

$$\nabla^2 G = 0 \tag{1}$$

is the Euler-Lagrange equation of the variational principle

$$\delta \int_V (|\, \mathbf{grad}\, G\, |)^2\, dV = 0 \tag{2}$$

As usual, G is varied only in the interior of the three-dimensional volume V and not on its boundary. Similarly, Einstein's field equations for empty space can be derived as the Euler-Lagrange equations of the variational principle

$$\delta \int_D R\sqrt{-g}\, dx^1\, dx^2\, dx^3\, dx^4 = 0 \tag{3}$$

Again the field variables $g_{\varrho\sigma}$ and their derivatives are to be varied only in the interior of the four-dimensional domain D.

If matter is present, the variational principle is to be generalized to

$$\delta \int_D (R + 8\pi KM)\sqrt{-g}\, dx^1\, dx^2\, dx^3\, dx^4 = 0 \tag{4}$$

where M is a function of the undifferentiated $g_{\varrho\sigma}$ and the distribution of matter. If $P^{\varrho\sigma}$ denotes the gravitational part of the stress-energy tensor, then the gravitational part of M, denoted by M_{grav} must obey the relation

$$\frac{\partial(\sqrt{-g}M_{\text{grav}})}{\partial \mathscr{G}^{\varrho\sigma}} = P_{\varrho\sigma} \qquad (\mathscr{G}^{\varrho\sigma} = \sqrt{-g}g^{\varrho\sigma}) \tag{5}$$

The electrodynamic part of M is given by

$$M_{\text{el}} = -\frac{1}{8\pi c^{\sigma}}\, \Phi^{\varrho\sigma}\Phi_{\varrho\sigma} \tag{6}$$

Here $\Phi^{\varrho\sigma}$ is the antisymmetric tensor defined in the chapter on the special theory of relativity. In this case the field variables that are to be varied are the $g_{\varrho\sigma}$ and the electromagnetic potentials. Maxwell's equations are then part of the resulting Euler-Lagrange equations.

1.4. The ponderomotive law. Einstein, Infeld, and Hoffmann * have shown that the motion of bodies (considered as singularities in the field) follows from the field equations alone. This is in strong contrast to the Lorentz theory of electrodynamics where the ponderomotive law is an additional postulate, logically independent of the field equations of Maxwell. In the limiting case where the rest mass of the particle is so small that its contribution to the field is negligible, its motion under the influence of a gravitational field alone becomes indistinguishable from a force-free or

* See EINSTEIN, A. and INFELD, L., *Canadian Journal of Mathematics*, 1, No. 3, 209 (1949).

inertial motion. In curved space, it describes a geodesic line, which corresponds to a straight line in flat space. Therefore its path is

$$\frac{d^2x^\varrho}{d\tau^2} + \Gamma^\varrho_{\lambda\sigma} \frac{dx^\lambda}{d\tau} \frac{dx^\sigma}{d\tau} = 0 \tag{1}$$

the equation of a geodesic in a Riemannian manifold.

If the space is flat, all the Γ vanish and the classical law of inertia results :

$$\frac{d^2x^\varrho}{d\tau^2} = 0 \tag{2}$$

1.5. The Schwarzschild solution. Only a few rigorous solutions of Einstein's field equations are known. One of them is the Schwarzschild solution, which describes the static field of a mass point at rest. Let m be its mass; then the equation of a line element in polar coordinates is

$$d\tau^2 = c^2\left(1 - \frac{2Km}{c^2r}\right)dt^2 - \left(1 - \frac{2Km}{c^2r}\right)^{-1} dr^2 - r^2\, d\theta^2 - r^2 \sin^2\theta\, d\phi^2 \tag{1}$$

We shall introduce dimensionless variables by letting

$$\frac{d\tau}{\alpha} = ds, \quad \frac{ct}{\alpha} = T, \quad \frac{r}{\alpha} = R, \quad 2\alpha = \frac{2Km}{c^2} = \text{“ gravitational radius ”}$$

and obtain

$$ds^2 = \left(1 - \frac{2}{R}\right)dT^2 - \left(1 - \frac{2}{R}\right)^{-1} dR^2 - R^2\, d\theta^2 - R^2 \sin^2\theta\, d\phi^2 \tag{2}$$

H. Weyl and T. Levi-Civita * succeeded in finding the static solutions that have only rotational but not spherical symmetry.

1.6. The three “ Einstein effects ” †

a. *The motion of the perihelion.* Suppose that an infinitesimal particle is moving in the field of a Schwarzschild singularity at the origin. The equation of motion is

$$\frac{d^2x^\varrho}{ds^2} = -\Gamma^\varrho_{\lambda\sigma} \frac{dx^\lambda}{ds} \frac{dx^\sigma}{ds} \tag{1}$$

The Γ can be computed from Eq. (45-2). One obtains a plane motion

* Weyl, H., *Ann. Physik*, **54**, 117 (1917); **59**, 185 (1919). Bach and Weyl, H., *Math. Z.*, **13**, 142 (1921). Levi-Civita, T., *Atti accad. naz. Lincei* (several notes, 1918-1919).

† See Wilson, H. A., *Modern Physics*, 3d ed., Blackie & Son, Ltd., London, 1948, p. 392 (the treatment we follow here).

which can be chosen in the plane $\theta = \pi/2$. Two further integrals of the motion are

$$\frac{h}{\alpha c} = H = R^2 \frac{d\phi}{ds} \qquad \text{(conservation of angular momentum)} \tag{2}$$

and
$$A = \left(1 - \frac{2}{R}\right) \frac{dT}{ds} \qquad \text{(conservation of energy)} \tag{3}$$

With the substitution, $u = 1/R$, we get the equation of the orbit:

$$\frac{d^2u}{d\phi^2} + u = \frac{1}{H^2} + 3u^2 \tag{4}$$

The corresponding Newtonian law is

$$\frac{d^2u}{d\phi^2} + u = \frac{1}{H^2} \tag{5}$$

The solution of Eq. (5) is

$$u = \frac{1}{H^2}(1 + \epsilon \cos\phi) \tag{6}$$

This is an ellipse if $\epsilon < 1$. The Einstein term $3u^2$ is very small compared to $1/H^2$ in planetary motion. Therefore Eq. (4) can be solved by the methods of successive approximations. One substitutes on the right-hand side the Newtonian value of u, and obtains the differential equation

$$\frac{d^2u}{d\phi^2} + u = \frac{1}{H^2} + \frac{3}{H^4}\left(1 + \frac{\epsilon^2}{2} + 2\epsilon\cos\phi + \frac{1}{2}\epsilon^2\cos 2\phi\right) \tag{7}$$

The solution is

$$u = \left[\frac{1}{H^2} + \frac{3}{H^4}\left(1 + \frac{\epsilon^2}{2}\right)\right](1 + \epsilon\cos\phi) + \frac{3\epsilon}{H^4}\left(\phi\sin\phi - \frac{1}{6}\epsilon\cos 2\phi\right) \tag{8}$$

If it were not for the term $\phi \sin\phi$, u would have the period 2π. The actual period is $2\pi + \mathcal{E}$, where $\mathcal{E}$ is a small number. If this value is substituted in Eq. (8) and higher powers of $\mathcal{E}$ are neglected, one obtains

$$\mathcal{E} = \frac{6\pi}{H^2} = \frac{6\pi\alpha^2c^2}{h^2} = \frac{6\pi K^2 m^2}{h^2c^2} \tag{9}$$

In the case of the planet Mercury, this formula gives for the rotation of the major axis of the orbit in 100 years approximately 43 seconds of arc. This is in good agreement with observation.

As A. Sommerfeld has pointed out, each deviation from the Newtonian or Coulomb potential causes a perihelion motion of the Kepler ellipse. In the theory of the fine structure of the hydrogen spectrum the deviation results from the mass-variability of the orbital electron. However the effect is much smaller than here. The rate of the precession of the perihelion is only about one-sixth of the rate produced here.

b. *Deflection of light rays by the sun.* For a light ray path $ds = 0$. Therefore $h = \infty$. The differential equation for the path becomes

$$\frac{d^2u}{d\phi^2} + u = 3u^2 \tag{10}$$

If the right-hand side were zero, the solution would be

$$u = \frac{1}{p} \cos \phi \tag{11}$$

where ϕ is the angle between r and the x axis, the equation of a straight-line parallel to the y axis at a distance p. Again, we substitute this value for u into the right-hand side of Eq. (10). This gives

$$\frac{d^2u}{d\phi^2} + u = \frac{3}{p^2} \frac{1 + \cos^2 \phi}{2} \tag{12}$$

The solution is

$$u = \frac{1}{p} \cos \phi + \frac{3}{p^2} \left(\frac{1}{2} - \frac{1}{6} \cos^2 \phi \right) \tag{13}$$

From Eq. (13) $u = 0$, or $r = \infty$ for $\phi = \pm (\pi/2 + \varepsilon)$, where ε is small. To find ε, we have the equation

$$0 = -\frac{\varepsilon}{p} + \frac{3}{p^2} \left(\frac{1}{2} + \frac{1}{6} \right) \tag{14}$$

or

$$\varepsilon = \frac{2}{p} \tag{15}$$

The asymptotic directions of the light ray differ by $2/p$ radians from the undeflected direction. The total deviation of a light ray, starting from infinity where the space is flat, and going off into the opposite direction to infinity, where the space is again flat, is

$$\delta = \frac{4}{p} = \frac{4\alpha}{d} \quad (d = \alpha p) \tag{16}$$

c. *The red shift of spectral lines.* Consider two identical "atomic clocks," one on the surface of the sun and one on the surface of the earth, where the curvature of space may be assumed negligible. If both clocks are at rest, we have

$$dr = d\theta = d\phi = 0 \tag{17}$$

Hence
$$d\tau^2 = c^2\left(1 - \frac{2Km}{c^2r}\right)dt^2 \tag{18}$$

if the proper-time dt/c between two successive signals will be the same on the sun as on the earth. But the observed time intervals dt' (on the sun) and dt'' (on the earth) will have the ratio

$$\frac{dt''}{dt'} = \sqrt{1 - \frac{2Km}{c^2r}} = 1 - \frac{Km}{c^2r} \tag{19}$$

the corresponding frequencies have the ratio

$$\frac{f''}{f'} = 1 + \frac{Km}{c^2r} \tag{20}$$

or
$$\delta f = f'' - f' = \frac{Km}{c^2r}\, f' \tag{21}$$

The frequency of the same spectrum line, emitted by an atom on the surface of the sun, appears shifted to the red as compared with the line emitted by the same atom on the surface of the earth.

Bibliography

See references at end of Chapter 6.

Chapter 8

HYDRODYNAMICS AND AERODYNAMICS

BY MAX M. MUNK

Department of Aeronautical Engineering
Catholic University of America

1. Assumptions and Definitions

1.1. Theoretical fluid dynamics deals with continuous fluid masses. Modern physics interprets the continuous properties of such masses as statistical averages. The passage from the general equations governing the fluid motion to specific solutions often involves the replacement of the correct equations by special and simplified equations approaching merely the exact equations asymptotically as special additional conditions are more and more exactly complied with. These simpler and more special equations again lead to specific formulas. The student of hydrodynamics and of aerodynamics is particularly interested in these formulas. One should also recall that the velocity fields of hydrodynamics and of gas dynamics have much in common with the fields of electric theory and of the theory of elasticity, all being the solutions of the basic equations subject to boundary conditions. The student will therefore find comparison of the formulas of this section with those in other allied sections a useful and often profitable occupation.

1.2. We divide the fluid body into small compact portions called fluid particles, which operate under Newton's law of dynamics, notwithstanding the continual loss of identity because of perpetual diffusion.

1.3. We reconcile the exchange of momentum that arises from diffuse convection, § 1.2, by assuming that viscous stresses act within the fluid.

1.4. A pure pressure p, within the fluid, is the mechanism adapting the fluid density to the physical problem.

1.5. A perfect fluid is one free of diffusion and hence free of viscosity stresses.

1.6. The fluid is called perfect in a stricter sense if its density is constant and unchangeable.

1.7. A perfect gas subordinates itself to § 1.5, it obeys the equation of state of a perfect gas.

$$p/\rho = R\theta \tag{1}$$

where θ = temperature, R = gas constant, ρ = mass density, and its internal energy u per unit mass is proportional to the absolute temperature

$$u = k_V \theta \tag{2}$$

2. Hydrostatics

2.1. The buoyancy of an immersed body is equal to the weight of the displaced fluid.

2.2. The buoyancy force acts along a vertical line passing through the center of gravity (CG) of the displaced fluid.

2.3. The wholly submersed body is in stable equilibrium if it weighs as much as the displaced fluid, and if its center of gravity lies vertically below the center of gravity of the displaced fluid.

2.4. The metacenter of a floating partially submersed body is the point at which the body would have to be suspended to experience moments, when slightly tilted from equilibrium, equal to the hydrostatic moments.

2.5. The distance between the center of gravity of the body and the metacenter is called the metacentric height h.

2.6. Let a denote the vertical distance between the CG of the body and the CG of the displaced fluid, positive if the CG of the body is above the CG of said fluid. Let I denote the axial (not central) moment of inertia

$$I = \int y^2 \, dA \tag{1}$$

of the area A cut by the partially immersed body out of the fluid surface, the CG of said area having the ordinate $y = 0$. The metacentric height is then

$$\text{MCH} = I/V - a \tag{2}$$

where V denotes the volume of the displaced fluid.

2.7. The pressure gradient of standing fluid is vertical and equal to the specific gravity of the fluid.

3. Kinematics

3.1. The bulk or particle velocity $\boldsymbol{v}$ is a vector and a function of the time t and of the space vector $\boldsymbol{r}$. The Cartesian components of $\boldsymbol{r}$ will be designated by x, y, z. The Cartesian components of $\boldsymbol{v}$ parallel to said space vector components respectively will be designated by u, v, w. This way of describing the flow is ordinarily designated as Eulerian.

3.2. The less common Lagrangian description identifies the fluid particles by suitable variables and follows the motion of each individual particle.

3.3. The divergence of a velocity field $\boldsymbol{v}$ is defined by

$$\nabla \cdot \boldsymbol{v} = \partial u/\partial x + \partial v/\partial y + \partial w/\partial z \tag{1}$$

It is a scalar quantity.

3.4. The rotation or " curl " of a velocity field, also called its vorticity, is defined by the vector whose Cartesian components are

$$\nabla \times \boldsymbol{v} = (u_y - v_x, v_z - w_y, w_x - u_z) \tag{1}$$

3.5. The gradient of a scalar field p such as the pressure is defined by

$$\mathbf{grad}\, p = \nabla p = (p_x, p_y, p_z) \tag{1}$$

3.6. The acceleration $\boldsymbol{a}$ of a fluid particle in terms of the Eulerian representation is equal to

$$\boldsymbol{a} = \partial \boldsymbol{v}/\partial t + (\boldsymbol{v} \cdot \nabla)\boldsymbol{v} \tag{1}$$

The first right-hand term is called the local portion of the acceleration, the second term is called the convection portion. The x component of the convective term is

$$uu_x + vu_y + wu_z \tag{2}$$

and similarly the y and z components.

3.7. The convective portion of the acceleration can be transformed as follows.

$$(\boldsymbol{v} \cdot \nabla)\boldsymbol{v} = -\tfrac{1}{2}\nabla v^2 + \boldsymbol{v} \times (\nabla \times \boldsymbol{v}) \tag{1}$$

3.8. A flow is called steady if its velocity and all its other properties are a function of the space coordinates only and not a function of the time t. The local portion of the acceleration $\partial \boldsymbol{v}/\partial t$ is then zero.

3.9. A flow field whose velocity v can be represented as the gradient of a scalar quantity ϕ is called a potential flow, or a nonrotational flow. The quantity ϕ is called the potential of the velocity, or the velocity potential.

3.10. The rotation of a potential flow is zero.

$$\nabla \times \nabla \phi = 0 \tag{1}$$

3.11. A velocity field $\boldsymbol{v}$ that has a potential ϕ is given by the scalar line integral $\phi = -\int d\boldsymbol{r} \cdot \boldsymbol{v}$, taken along any line connecting a reference point with the point in question.

3.12. The scalar line integral $-\int d\boldsymbol{r} \cdot \boldsymbol{v}$ taken along a closed line is called the circulation.

3.13. The scalar integral $\int d\boldsymbol{f} \cdot \boldsymbol{v}$, taken through a surface $\boldsymbol{f}$, bounded or unbounded, is called the flux of $\boldsymbol{v}$ through $\boldsymbol{f}$.

3.14. The flow field of an incompressible fluid has the divergence zero, $\nabla \cdot \boldsymbol{v} = 0$.

3.15. The general equation of continuity for any fluid is

$$\nabla \cdot (\boldsymbol{v}\rho) = -\partial\rho/\partial t \tag{1}$$

3.16. The divergence of the vorticity of any velocity field v is identically zero.

$$\nabla \cdot \nabla \times \boldsymbol{v} = 0 \tag{1}$$

3.17. The substantive rate of change of the circulation of $\boldsymbol{v}$ is that change along a closed line moving with the fluid particles, i.e., along one always occupied by the same fluid particles. This substantive rate of change equals the circulation of the acceleration

$$D/Dt \int d\boldsymbol{r} \cdot \boldsymbol{v} = \int d\boldsymbol{r} \cdot \boldsymbol{a} \tag{1}$$

3.18. In a plane two-dimensional flow of an incompressible fluid, the flux through any line connecting a pair of points is independent of the shape of the line of connection. This flux is called stream function ψ if one of the two points is chosen as a fixed reference point and the other point is varied

$$u = \partial\psi/\partial y, \quad v = -\partial\psi/\partial x \tag{1}$$

The existence of a stream function is equivalent to the absence of divergence.

3.19. Axially symmetric two-dimensional flows have no velocity components at right angles to the meridian plane. They have a stream function ψ if the fluid is incompressible

$$u = \frac{1}{r}\psi_y, \quad v = -\frac{1}{r}\psi_x \tag{1}$$

where r denotes the distance from the axis of symmetry.

3.20. With a steady two-dimensional compressible flow the quantity $\mathbf{v}\rho$, called mass flux density, possesses a stream function ψ equivalent to § 3.18 and § 3.19, respectively. If the flow is plane,

$$u\rho = \partial\psi/\partial y, \quad v\rho = -\partial\psi/\partial x \tag{1}$$

3.21. The plane having the velocity components u and v of a plane two-dimensional flow as Cartesian coordinates is called the hodograph plane.

3.22. The space vector x, y has a potential L in the hodograph plane u, v, provided the velocity vector u, v has a potential ϕ in the physical plane x, y, and conversely.

$$L = -\phi + u \cdot x + v \cdot y \tag{1}$$

3.23. The stream function of (u, v) is the potential of the flow $v, -u$. The theorem of § 3.22 holds therefore also for the stream function.

3.24. In the following, S denotes space, $dS = dx\, dy\, dz$; O denotes a closed surface, $d\sigma$ an element thereof. A bounded surface is denoted by f.

3.25. The kinetic energy K of the flow v within the space S bounded by the surface O is equal to

$$K = \tfrac{1}{2}\int v^2\rho\, dS \tag{1}$$

3.26. If the flow has a potential ϕ, then K is equal to

$$K = \tfrac{1}{2}\int d\sigma \cdot \mathbf{v}\rho\phi \tag{1}$$

3.27. The momentum of the flow is equal to

$$\mathbf{M} = \int dS\, \nabla \cdot (\mathbf{v}\rho)\mathbf{r} + \int d\sigma \cdot \mathbf{v}\rho\mathbf{r} \tag{1}$$

3.28. The moment of momentum of the flow is equal to

$$\int dS\, \mathbf{v}\rho\mathbf{r} = \tfrac{1}{2}\int dS(\nabla \times \mathbf{v}\rho)\mathbf{r} \cdot \mathbf{r} - \tfrac{1}{2}\int d\sigma \times (\rho\mathbf{v})\mathbf{r} \cdot \mathbf{r} \tag{1}$$

3.29. Biot Savart

$$\mathbf{v} = \frac{1}{4\pi}\iint \frac{dS\, dS'(\nabla \times \mathbf{v}) \times (\mathbf{r} - \mathbf{r}')}{[(\mathbf{r} - \mathbf{r}') \cdot (\mathbf{r} - \mathbf{r}')]^{3/2}} \tag{1}$$

4. Thermodynamics

4.1. Notations:

p = pressure	dQ = heat differential
ρ = mass density	u = internal energy
V = specific volume	i = enthalpy, also designated by h
θ = temperature	C_V = specific heat of constant volume
s = entropy	C_p = specific heat of constant pressure

4.2. Notations, perfect gases:

R = gas constant $\quad\gamma$ = isentropic exponent

4.3. $du = -p\,dV + dQ = -p\,dV + \theta\,ds$

4.4. $dh = di = -V\,dp + \theta\,ds$

4.5. Perfect gases

$$p = R\rho\theta, \quad u = C_V\theta, \quad h = C_p\theta$$

$$R = C_p - C_V = C_p\frac{\gamma - 1}{\gamma} = C_V(\gamma - 1)$$

$$\gamma = \frac{n + 2}{n} = \frac{C_p}{C_V} = \frac{C_V + R}{C_V}$$

where n denotes degrees of freedom.

$$s = C_V \ln p - C_p \ln \rho$$

5. Forces and Stresses

5.1. There may act on the fluid external body forces f per unit volume. These forces may have a potential

$$f = -\nabla\phi'$$

5.2. The buoyancy force per unit volume exerted on the fluid particle by a pressure distribution p is $-\nabla p$.

5.3. The force effect per unit volume caused within an incompressible fluid by the simplest type of viscosity is (μ denoting the modulus of viscosity)

$$-\mu\nabla\cdot\nabla\nu = (\mu\nabla\cdot\nabla u,\quad \mu\nabla\cdot\nabla v,\quad \mu\nabla\cdot\nabla w) \tag{1}$$

5.4. The viscosity stress within the incompressible fluid corresponding to the equation in § 5.3 is

$$\Sigma = \mu[\nabla \boldsymbol{v} + \boldsymbol{v}\nabla] =$$

$$\begin{vmatrix} \sigma_{xx} & \sigma_{xy} & \sigma_{xz} \\ \sigma_{yx} & \sigma_{yy} & \sigma_{yz} \\ \sigma_{zx} & \sigma_{zy} & \sigma_{zz} \end{vmatrix} = \mu \begin{vmatrix} u_x & v_x & w_x \\ u_y & v_y & w_y \\ u_z & v_z & w_z \end{vmatrix} + \mu \begin{vmatrix} u_x & u_y & u_z \\ v_x & v_y & v_z \\ w_x & w_y & w_z \end{vmatrix} \quad (1)$$

5.5. The average normal stress $\sigma_{xx} + \sigma_{yy} + \sigma_{zz}$ of the stress § 5.4 is zero.

5.6. The dissipation of kinetic energy into heat in accordance with § 5.4 is, per unit volume,

$$2\mu\nabla \cdot \boldsymbol{a} + \mu(|\nabla \times \boldsymbol{v}|)^2 \quad (1)$$

where $\boldsymbol{a}$ denotes the acceleration of the fluid.

5.7. The same dissipation for a finite portion of space S is

$$2\mu \int d\sigma \cdot \boldsymbol{a} + \mu \int dS(|\nabla \times \boldsymbol{v}|)^2 \quad (1)$$

5.8. If the velocity is, and remains, zero at the boundary of the space portion S, the total dissipation (but not the local dissipation, § 5.4) becomes equal to $\mu/2$ times the integral of the vorticity squared over the space portion.

6. Dynamic Equations

(External forces assumed absent)

6.1. Euler's equation for nonviscous fluids:

$$\partial \boldsymbol{v}/\partial t + (\boldsymbol{v} \cdot \nabla)\boldsymbol{v} = -\frac{\nabla p}{\rho}, \quad \frac{dp}{\rho} + V\,dV = 0 \quad (1)$$

along any streamline.

6.2. Euler's equation for steady potential flows:

$$\frac{\nabla p}{\rho} = -\frac{1}{2}\nabla(u^2 + v^2 + w^2) \quad (1)$$

6.3. For a steady isentropic potential flow of an inviscid gas:

$$h_0 = h + \frac{u^2 + v^2 + w^2}{2} \quad (1)$$

6.4. Stokes-Navier equation for a steady incompressible fluid having viscosity:

$$-\frac{\nabla p}{\rho} = (\boldsymbol{v} \cdot \nabla)\boldsymbol{v} - \frac{\mu}{\rho}\nabla \cdot \nabla \boldsymbol{v} \quad (1)$$

6.5. Helmholtz' theorem: If external body forces and viscosity are absent, individual vortices move with the fluid; i.e., individual vortices always consist of the same fluid particles.

6.6. For plane two-dimensional flow of an incompressible fluid, each fluid particle preserves its vorticity.

6.7. In steady plane two-dimensional flow of a perfect gas along any one streamline, the vorticity is proportional to the pressure.

6.8. Convection and diffusion of vorticity: Plane two-dimensional flow

$$w_T + \boldsymbol{v} \cdot \boldsymbol{\nabla} w = \frac{\mu}{\rho \boldsymbol{\nabla}} \cdot \boldsymbol{\nabla} w \tag{1}$$

6.9. In absence of viscosity and of external body forces, a homogeneous fluid executing a potential flow will continue to do so.

6.10. The continued existence of a potential implies that all dynamic equations and requirements are complied with.

7. Equations of Continuity for Steady Potential Flow of Nonviscous Fluids (see also §§ 3.14, 3.15)

7.1. Laplace's equation for the potential of an incompressible fluid of constant density:

$$\boldsymbol{\nabla} \cdot \boldsymbol{\nabla} \phi = \frac{\partial^2 \phi}{\partial x^2} + \frac{\partial^2 \phi}{\partial y^2} + \frac{\partial^2 \phi}{\partial z^2} = 0 \tag{1}$$

7.2. For plane, two-dimensional flow:

$$\frac{\partial^2 \phi}{\partial x^2} + \frac{\partial^2 \phi}{\partial y^2} = 0 \tag{1}$$

7.3. For polar coordinates, r, θ

$$\phi_{rr} + \frac{1}{r^2} \phi_{\theta\theta} + \frac{\phi_r}{r} = 0 \tag{1}$$

7.4. For spherical polar coordinates:

$$\sin \theta \frac{\partial}{\partial r} (r^2 \phi_r) + \frac{\partial}{\partial \theta} (\sin \theta \phi_\theta) + \frac{1}{\sin \theta} \phi_{\omega\omega} = 0 \tag{1}$$

$$x = r \cos \theta, \quad y = r \sin \theta \cos \omega, \quad z = r \sin \theta \sin \omega \tag{2}$$

7.5. For a plane two-dimensional flow obeying Laplace's equation, the equation for the potential L of (x, y) as a function of (u, v) again obeys Laplace's equation.

$$x = \partial L/\partial u, \qquad y = \partial L/\partial v$$

$$\partial^2 L/\partial u^2 + \partial^2 L/\partial v^2 = 0$$

7.6. The value of a solution ϕ of Laplace's equation (§ 7.1) at any point is the arithmetical mean of its values over any sphere having that point as its center (in three-dimensional flows) or any equivalent circle (in two-dimensional flows).

7.7. Equation of continuity for a steady potential flow of a compressible fluid (see § 6.4).

$$\left.\begin{aligned}\phi_{xx}\left(1-\frac{\phi_x^{\,2}}{c^2}\right)+\phi_{yy}\left(1-\frac{\phi_y^{\,2}}{c^2}\right)+\phi_{zz}\left(1-\frac{\phi_z^{\,2}}{c^2}\right)&\\ -\frac{2}{c^2}\phi_{xy}\phi_x\phi_y-\frac{2}{c^2}\phi_{yz}\phi_y\phi_z-\frac{2}{c^2}\phi_{zx}\phi_z\phi_x=0&\end{aligned}\right\} \qquad (1)$$

where c denotes the local velocity of sound.

7.8. For polytropic expansion (γ = expansion exponent):

$$(1-\phi_{xi}\phi_{xi})\nabla\cdot\nabla\phi=\frac{2}{\gamma-1}\phi_{xi}\phi_{xk}\phi_{xixk} \qquad (1)$$

$$(i,k = 1, 2, 3)$$

7.9. For two-dimensional flow in plane polar coordinates:

$$\left.\begin{aligned}\left(1-\frac{\gamma+1}{\gamma-1}\phi_r^{\,2}-\frac{\phi_\theta^{\,2}}{r^2}\right)\phi_{rr}+\left(1-\phi_r^{\,2}-\frac{\gamma+1}{\gamma-1}\frac{\phi_\theta^{\,2}}{r^2}\right)\left(\frac{\phi_{\theta\theta}}{r^2}+\frac{\phi_r}{r}\right)&\\ -\frac{4}{\gamma-1}\phi_r\frac{\phi_\theta}{r}\left(\frac{\phi_{\theta r}}{r}-\frac{\phi_\theta}{r^2}\right)=0&\end{aligned}\right\} \qquad (1)$$

7.10. In the hodograph plane:

$$\left(1-\frac{w^2}{c^2}\right)w^2\phi_{ww}+w\left(1-2\frac{w^3}{c^3}\frac{dc}{dw}+\frac{w^4}{c^4}\right)\phi_w+\left(1-\frac{w^2}{c^2}\right)^2\phi_{\theta\theta}=0 \qquad (1)$$

7.11. The stream function:

$$w^2\psi_{ww}+w\left(1+\frac{w^2}{c^2}\right)\psi_w+\left(1-\frac{w^2}{c^2}\right)\psi_{\theta\theta}=0 \qquad (1)$$

7.12. Reciprocal potential, L:

$$\left(1-\frac{v^2}{c^2}\right)L_{uu}+\left(1-\frac{u^2}{c^2}\right)L_{vv}+\frac{2uv}{c^2}L_{uv}=0 \qquad (1)$$

8. Particular Solutions of Laplace's Equation

8.1. Sink or source in three dimensions:

$$\phi_1=\frac{1}{4\pi}\frac{1}{r} \qquad (1)$$

8.2. Sink or source in two dimensions:

$$\phi_2=\frac{1}{2\pi}\ln r \qquad (1)$$

8.3. Rankine flow in three or two dimensions:

$$\phi=x+\phi_1, \qquad x+\phi_2 \qquad (1)$$

8.4. Flow past a sphere:

$$\phi=v_0\left(r+\frac{R^3}{2r^2}\right)\cos\theta \qquad (1)$$

8.5. Flow past a circle:

$$\phi=v_0\left(r+\frac{R^2}{r}\right)\cos\theta \qquad (1)$$

8.6. General solution suitable for solving problems relating to a straight line in two-dimensional flow:

$$\phi=\text{real part of }\frac{1}{i(z-\sqrt{z^2-1})^n} \qquad (1)$$

8.7. Complex variables: If $u+iv$ is an analytic function of $x+iy$, then u and v are each solutions of Laplace's equation.

8.8. If $f(x,y,z)$ is a solution of Laplace's equation, then

$$\frac{1}{r}f\left(\frac{x}{r^2},\frac{y}{r^2},\frac{z}{r^2}\right) \qquad (1)$$

is also a solution.

8.9. Degree zero: The following is a solution, F and G denoting any analytic function.

$$u=F\left(\frac{x+iy}{r+z}\right)+G\left(\frac{x-iy}{r+z}\right) \quad (r^2=x^2+y^2+z^2) \qquad (1)$$

8.10. Degree −1: The following are solutions.

$$v = \frac{1}{r} u \tag{1}$$

where u is given in § 8.9.

8.11. A conformal transformation of the system of lines $\phi =$ constant, $\psi =$ constant in a plane leads again to a solution of Laplace's equation if the initial system represents potential lines and streamlines consistent with that equation.

8.12. The lines of equal potential and the lines of equal stream function form an orthogonal system.

8.13. Fundamental solutions of Laplace's equation in three dimensions:

$$\phi = \frac{\partial^n}{\partial x^n} \frac{1}{r} \tag{1}$$

The factor of r^n expressed as a function of cos ϕ is called a Legendre function.

8.14. Kinetic energy of the potential flow of the incompressible fluid

$$T = \frac{\rho}{2} \int dS \quad v^2 = \frac{\rho}{2} \int d\sigma \cdot \mathbf{v}\phi \tag{1}$$

8.15. Variational principle: The potential flow of the incompressible fluid has a smaller kinetic energy than any other thinkable potential flow of the incompressible fluid having the same values of the potential at the boundary of the region considered wherever there is a flux different than zero across the boundary.

8.16. Differentiation: The differential quotient $\partial\phi/\partial x$ of a solution ϕ of Laplace's equation with respect to a Cartesian coordinate x is again a solution.

9. Apparent Additional Mass

9.1. An immersed body, moving in a perfect incompressible fluid otherwise at rest, responds to external forces as if its mass were increased by an apparent additional mass. This mass depends on the shape of the body only and is, in general, different for motions in various directions relative to the body. (Translation.)

9.2. The apparent additional mass is equal to the kinetic energy of the fluid, divided by one-half the square of the velocity of advance.

9.3. The apparent additional mass is equal to the momentum transmitted from the body to the fluid from initial rest, divided by the velocity of advance.

9.4. The apparent mass K of a body having the same density as the fluid is equal to the static moment of the sink source system that would induce the fluid motion, multiplied by the density, or

$$K = \rho \int dS\, \mathbf{r} \nabla \cdot \mathbf{v} \tag{1}$$

9.5. The apparent additional mass is equal to the surface integral of the potential, multiplied by the density, or

$$K = \rho \int d\sigma\, \phi \tag{1}$$

9.6. The apparent additional mass, divided by the mass of the displaced fluid, is called the factor of the apparent mass.

9.7. Factor of apparent additional mass: Circular cylinder moving crosswise, 1; sphere, $\frac{1}{2}$.

9.8. Apparent additional mass: Elliptical cylinder of length $L \to \infty$, moving crosswise at a right angle to major axis, a,

$$La^2 \frac{\pi}{4} \rho \tag{1}$$

Flat plate of width a, moving crosswise,

$$La^2 \frac{\pi}{4} \rho \tag{2}$$

Circular disk of diameter r, moving crosswise,

$$\frac{8a^3\rho}{3} \tag{3}$$

9.10. The additional apparent momentum is a symmetric and linear function of the velocity, i.e., at least three principal directions mutually at right angles to each other exist, for which the momentum is parallel to the velocity. The momentum in any other direction (relative to the body) can be computed from the velocity components in the principal directions by simple superposition.

9.11. The momentum transmitted from a body to a barotropically expanding gas surrounding the body, from the time both, the body and the gas, were at rest until the translational motion of the body as well as the motion of the gas has become steady, does not depend on the time history of the motion.

9.12. The energy for subsonic motion has a lower bound, associated with infinitely gradual setting up of the motion.

9.13. The minimum energy is composed of the kinetic energy and of the elastic energy of the gas.

9.14. We can evaluate the elastic energy if the total mass of the gas involved in the motion remains constant for all velocities of advance.

9.15. For polytropic expansion the elastic energy is the space integral of

$$\left.\begin{aligned}&\frac{[1 - M_0^2(\gamma-1)v^2/2]^{\gamma/(\gamma-1)}}{\gamma(\gamma-1)} + \frac{[1 - M_0^2(\gamma-1)v_0^2/2]^{\gamma/(\gamma-1)}}{\gamma}\\ &\qquad - \frac{[1 - M_0^2(\gamma-1)v^2/2]^{1/(\gamma-1)}[1 - M_0^2(\gamma-1)v_0^2]}{\gamma-1}\end{aligned}\right\} \tag{1}$$

where v_0 = ambient velocity, v = local velocity, M_0 = ambient Mach number, γ = polytropic exponent.

10. Airship Theory

10.1. Assumptions: Perfect incompressible flow, slender airship body with circular section, no fins, small angle of attack, linearization, cross-sectional area, X-axial coordinate.

10.2. Cross force per unit axial length (lift):

$$\frac{dL}{dx} = 2\frac{dS}{dx}\frac{v^2\rho}{2}\sin\alpha\cos\alpha = 2\frac{dS}{dx}\frac{v^2\rho}{2}\alpha \tag{1}$$

10.3. The total air force of streamlined spindle is zero.

10.4. The total unstable moment of the air force (couple) is

$$M = 2\text{ volume}\frac{v^2\rho}{2}\alpha$$

10.5. The total lift of the spindle ending abruptly in a flat base of cross section S is

$$L = 2\frac{Sv^2\rho}{2}\alpha$$

10.6. The pressure around any one cross section is proportional to the distance from the diameter normal to the plane of symmetry.

11. Wing Profile Contours; Two-dimensional Flow with Circulation

11.1. The plane two-dimensional flow past such a contour is determined by the ordinary boundary condition of zero flux in conjunction with Kutta's condition that the separation point coincide with the sharp-pointed trailing edge.

11.2. The resulting air force, or lift, acts at right angles to the translational motion of advance. The lift passes through a point of the contour called center of pressure and is proportional to the circulation about the contour.

$$L = v\rho\Gamma \quad \text{per unit span} \tag{1}$$

11.3. The profile possesses a point called areodynamic center. The moment of the air force with respect to this center is not dependent on the angle of attack but is merely proportional to the density and to the square of the velocity of advance.

11.4. The aerodynamic center of a thin contour extending close to a straight line, called the chord, is positioned near the 25 % point of the chord; that is, it is spaced one-quarter chord length downstream from the leading edge.

11.5. The direction of forward motion for zero lift of a slender profile is nearly parallel to the line connecting the trailing edge with the center of the profile, this center being at the 50 % station and equidistant from the upper and lower camber line, or contour boundary.

11.6. Lift coefficient:

$$C_L = 2\pi\alpha = \frac{L}{cv^2\rho/2} \tag{1}$$

where $L \equiv$ lift per unit span, $v \equiv$ velocity of advance, $c \equiv$ chord, $\rho \equiv$ density, $\alpha \equiv$ angle of attack; $dc_L/d\alpha$ is called the lift slope.

11.7. As the angle of attack increases, the CP (center of pressure) approaches the aerodynamic center (AC) with positively cambered profiles.

11.8. If the CP does not travel but is fixed, it coincides with the AC.

11.9. The linearized theory applies to slender airfoil sections. The boundary condition of zero flux at the contour is replaced by prescribing the flow direction along the chord, namely, to be parallel to the average camber direction at the chord station. Only terms linear in the perturbation velocity components are retained.

11.10. The angle of attack of zero lift is given by the integral

$$\alpha = \frac{1}{\pi}\int_{-1}^{+1} \frac{\xi\, dx}{(1-x)\sqrt{1-x^2}} \tag{1}$$

where chord = 2, x = chord station, ξ = mean camber.

11.11. The angle of attack for zero moment with respect to the 50 % station is given by the integral

$$\alpha = \frac{2}{\pi}\int_{-1}^{+1} \frac{x\xi\, dx}{\sqrt{1-x^2}} \tag{1}$$

11.12. In actual, slightly viscous air the profiles experience, furthermore, a resistance or drag, called the profile drag.

12. Airfoils in Three Dimensions

12.1. Assumptions: The spanwise extension of the airfoil or airfoils is large when compared with the flowwise extension. The ratio of the square of the span, or maximum span, to the total projected airfoil area is called aspect ratio. Viscosity is absent. The air, having been in contact with the upper airfoil surface, slides sideways over the air having been in contact with the lower airfoil surface. The vortices representing this condition coincide with the streamlines and are assumed to extend straight downstream. There occurs a drag called induced drag.

12.2. The spanwise lift distribution for minimum induced drag requires constant downwash velocity along the span or spans, provided all chordwise distances are reduced to zero; that is, the airfoil or airfoils are replaced by the flowwise projection.

12.3. The minimum induced drag is equal to

$$D_i = \frac{L^2}{v^2\rho\, 4k/2} \tag{1}$$

where k denotes the area of apparent additional mass of the front projection of the airfoils.

12.4. The minimum induced drag of a single airfoil, straight in front view, is

$$D_i = \frac{L^2}{v^2 \rho b^2 \pi/2} \tag{1}$$

where b denotes the span. The lift slope of the single airfoil is equal to

$$\frac{dC_L}{d\alpha} = \frac{2\pi}{1 + 2S/b^2}$$

12.5. For the single airfoil the relation of the center-of-pressure travel to the variation of the lift coefficient is the same as with the two-dimensional flow (infinite span).

13. Theory of a Uniformly Loaded Propeller Disk

(Incompressible, nonviscous, infinite number of blades)

13.1. The theoretical efficiency, i.e., the upper bound of the uniformly loaded disk of a propeller is

$$\eta = \frac{2}{1 + \sqrt{1 + 8T/v^2 D^2 \pi \rho}} \tag{1}$$

where T = thrust, D = diameter, v = axial forward velocity.

13.2. The minimum power requirement for a hovering propeller ($v = 0$) is equal to

$$P = \sqrt{2T^3/\rho D^2 \pi} \tag{1}$$

13.3. The local slipstream velocity ratio in the region of a propeller is equal to

$$v_s = \frac{1}{2}\sqrt{[1 + (8T/v^2 D^2 \pi) - 1]}$$

14. Free Surfaces

(Incompressible, nonviscous, plane, two-dimensional)

14.1. The intrinsic equation for flow from all sides into an open channel of constant width 4 is

$$s = -\frac{2}{\pi} \ln \cos \frac{\theta}{2} \tag{1}$$

$$y = \frac{1}{\pi}(\theta - \sin \theta) \tag{2}$$

where s = length of curve, θ = direction of curve bounding stream, coefficient of contraction = $\frac{1}{2}$.

14.2. For a slot in a wall of width $2 + 4/\pi$,

$$s = -\frac{2}{\pi} \ln (-\cos \theta) \tag{1}$$

where final width of the jet is 2. The coefficient of contraction is

$$\frac{\pi}{\pi + 2} = 0.611 \tag{2}$$

14.3. For an infinite stream impinging on a strip of width one, with a cavity pressure equal to undisturbed pressure,

$$s = \frac{1}{1 + \pi} \frac{1}{\cos^2 \theta} \tag{1}$$

$$\frac{\text{force}}{\rho v'^2} = \frac{\pi}{\pi + 4} = 0.440 \tag{2}$$

15. Vortex Motion

15.1. The condition for steady vortex motion in the absence of external forces is

$$\nabla \times [\boldsymbol{v} \times (\nabla \times \boldsymbol{v})] = 0 \tag{1}$$

15.2. Bernoulli's equation holds for steady vortex flow, provided all vortices are parallel to the streamlines.

15.3. The force on a cylinder in laminar straight motion of constant vorticity is

$$f = 2\kappa\rho u \tag{1}$$

where $\kappa =$ circulation of circumference.

15.4. The stagger ratio of stable pairs of vortices (Karman's vortices) is

$$\kappa\pi = b/a, \quad \cosh^2 \kappa\pi = 2, \quad \kappa\pi = 0.8814 \tag{1}$$

where $a =$ distance of vortices in each line, $b =$ distance between the two lines.

15.5. The velocity of translational advance of a circular vortex depends on the local distribution of the vorticity within the region of the ring.

16. Waves
(One-dimensional, weak)

16.1. Group velocity c_g:

$$c_g = c - \lambda \frac{dc}{d\lambda} \qquad (1)$$

where $c =$ wave velocity, $\lambda =$ wavelength, $g =$ acceleration of gravity.

16.2. Surface waves, gravity, infinite depth:

$$c = \sqrt{g\lambda/2\pi} \qquad (1)$$

16.3. Capillary waves:

$$c = \sqrt{2\pi k/\rho\lambda} \qquad (1)$$

where $h =$ surface tension (force/length).

16.4. Gravity and capillarity in unison:

$$c = \sqrt{\frac{g\lambda}{2\pi} + \frac{2\pi k}{\rho\lambda}} \qquad (1)$$

16.5. Tidal waves, shallow depth h:

$$c = \sqrt{gh} \qquad (1)$$

16.6. Water jumps associated with tidal waves, steady:

$$v_1 v_2 = \frac{1}{2}(c_1{}^2 + c_2{}^2) \qquad (1)$$

16.7. Same, relation between heights:

$$h_2 = h_1\left[\sqrt{2M_1{}^2 + \frac{1}{4}} - \frac{1}{2}\right] \qquad (1)$$

where M denotes equivalent Mach number, v/c.

16.8. Progressing wave traveling in one direction only: The kinetic energy is equal to the potential energy.

17. Model Rules

17.1. Froude's rule:

$$Lg/v^2 = \text{constant} \qquad (1)$$

where $L =$ length, $v =$ velocity, $g =$ acceleration of gravity.

17.2. Reynolds' rule

$$R = \text{Reynolds' number} = \frac{vL}{\nu} = \frac{vL}{\mu/\rho} = \text{constant} \tag{1}$$

where ρ = density, μ = modulus of viscosity, $\nu = \mu/\rho$ kinematic modulus of viscosity.

17.3. Mach's rule:

$$\text{Mach number} = M = v/c = \text{constant} \tag{1}$$

where v = velocity, c = velocity of sound (see § 23.2).

17.4. Grashoff's rule:

$$\beta L^3 g \Delta\theta/\nu^2\theta = \text{const.} \tag{1}$$

where θ = temperature, β = thermal expansion.

18. Viscosity
(Incompressible)

18.1. Absence of inertia forces:

$$\mu \nabla^2 v = \nabla p, \qquad \nabla \cdot v = 0 \tag{1}$$

18.2. Sphere with radius a advancing with velocity U:

$$P = 6\pi\mu a U \tag{1}$$

where P = resistance.

18.3. Circular disk having radius r moving broadside on

$$R = 16r/9\pi \tag{1}$$

R denotes the radius of the sphere having the same resistance.

18.4. Circular cylinder, two-dimensional flow:

$$R = \frac{4\pi\mu U}{\frac{1}{2} - \gamma - \ln(\mu a/4v)} \tag{1}$$

where γ = resistance per unit length.

18.5. See §§ 5.6 to 5.8; also Section 24.

18.6. Poiseuille flow through straight vertical circular pipe (laminar):

$$\text{Flux} = \frac{\pi a^4}{8\mu}\left(\frac{p_1 - p_2}{l} + \rho g\right) \tag{1}$$

where a = radius of pipe.

18.7. The velocity profile across the diameter of a Poiseuille flow is a parabola.

18.8. Couette flow between two parallel plane walls sliding relatively to each other, steady flow:

$$\text{Tangential force per unit area} = \sigma = \mu v/a \tag{1}$$

where a = distance of walls, v = difference of wall velocities.

19. Gas Flow, One- and Two-Dimensional (Isentropic)

19.1. Velocity of sound:

$$c^2 = dp/d\rho \tag{1}$$

If p is not a function of ρ, both must be varied for the particle considered. With steady flows they have to be varied along the streamline.

19.2. Velocity of sound for a perfect gas:

$$c^2 = \gamma p/\rho = \gamma g^{R\theta} \tag{1}$$

19.3. General equation for sound wave motion:

$$\partial^2\phi/\partial T^2 = c^2\nabla^2\phi \tag{1}$$

19.4. The kinetic energy of a simple, harmonic sound wave is equal to its elastic energy.

19.5. Force necessary to vibrate a sphere (linearized, amplitude $\ll$ radius):

$$f = \frac{4}{3}\pi\rho a^3\left[\frac{2+k^2a^2}{4+k^4a^4}\frac{du}{dT} + \frac{k^3a^3}{4+k^4a^4}\sigma u\right] \tag{1}$$

where a = radius of cylinder, $k = 2\pi/\text{wavelength}$, $\sigma = 2\pi/\text{period}$.

19.6. Subsonic flow past a circular cylinder (perfect gas):

$$\frac{\phi}{u} = \phi_{00} + M^2\phi_{10} + \ldots = \left(r + \frac{1}{r}\right)\cos\theta \tag{1}$$

$$+ M^2\left[\left(\frac{13}{12}r^{-1} - \frac{1}{2}r^{-3} + \frac{1}{12}r^{-5}\right)\cos\theta + \left(-\frac{1}{4}r^{-1} + \frac{1}{12}r^{-3}\right)\cos 3\theta\right] + \ldots \tag{2}$$

19.7. Characteristic equations for plane two-dimensional potential flows:

$$d\theta \pm dp \frac{\sin \mu \cos \mu}{\rho c^2} = 0 \tag{1}$$

For plane two-dimensional flow:

$$dv/v = \pm \tan \mu \, d\theta, \qquad \sin \mu = 1/M \tag{2}$$

19.8. Same for polytropic expansion, integrated along the Mach line:

$$\mu + \sqrt{\frac{\gamma+1}{\gamma-1}} \tan^{-1} \sqrt{\frac{\gamma+1}{\gamma-1}} \tan \mu \pm \theta = 0 \tag{1}$$

For Prandtl-Meyer flow the characteristic equations (19.7) or (19.8) hold through the entire flow and not merely along any characteristic.

19.9. Characteristic equation for axially symmetric flow:

$$d\theta \pm dp \frac{\sin \mu \cos \mu}{\rho c^2} \mp \frac{dr}{r} \frac{\sin \theta \sin \mu}{\sin (\theta \pm \mu)} = 0 \tag{1}$$

19.10. Characteristic equation for unsteady one-dimensional flow:

$$\frac{2}{\gamma - 1} c \pm v = \text{constant} = \frac{2}{\gamma - 1} c_0 \tag{1}$$

$$\frac{2}{\gamma - 1} c_0 \equiv \text{escape velocity} \tag{2}$$

Equation 19.10 holds throughout centered flow, that is, waves traveling in one direction only.

19.11. Coparallelism: All Mach line systems of plane two-dimensional potential flows of the same polytropic gas are coparallel. That is to say, corresponding sides of the diamonds formed by corresponding pairs of adjacent Mach lines are parallel. Mach lines are corresponding if they are parallel at the same Mach number station. If they are parallel at one Mach number station, they are at all Mach number stations.

20. Gas Flows, Three Dimensional

20.1. Isentropic supersonic flow past a circular cone, symmetric (Taylor-Maccoll):

$$v = u_{\omega} \tag{1}$$

where ω = polar angle, u = radial velocity, v = peripheral velocity.

20.2. Velocity component of general linearized conical flow, $M^2 = 2$:

$$u = F\left(\frac{x + iy}{iz + \sqrt{x^2 + y^2 - z^2}}\right) + F\left(\frac{(x - iy)}{iz + \sqrt{x^2 + y^2 - z^2}}\right) \tag{1}$$

(See § 8.9).

20.3. Throat flow:

$$\xi^2 - \frac{2Rr}{\gamma + 1} \tag{1}$$

where ξ = radius of curvature of sonic line, r = throat radius, R = radius of curvature of nozzle wall, γ = expansion exponent.

20.4. Reversal theorem for three-dimensional airfoil flow (linearized equations). The lift slope of any airfoil remains unchanged if the velocity of advance is reversed.

20.5. Slender delta wing, having span b. The lift coefficient slope computed with $b^2\pi/4$ and $v^2\rho/2$ as reference quantities is equal to 2.

20.6. Initial lift slope of slender, flat-based, finless missile body of circular cross section (linearized equations). The lift coefficient slope computed with $D^2\pi/4$ and $v^2\rho/2$ as reference quantities is equal to 2. (D = base diameter.)

20.7. Canonical variables: Every steady flow of a polytropic gas can be expressed by four dependent variables u/a, v/a, w/a, and p, where a denotes the ultimate velocity of the streamline in question. That is to say, every streamline pattern serves an infinity of flows, in that the density can be prescribed arbitrarily at one point at each streamline. The flow may include shock waves.

21. Hypothetical Gases

21.1. Chaplygin gas:

$$p = -\frac{A}{\rho} + B, \qquad \gamma = -1 \tag{1}$$

This gas can execute a plane, simple, sound wave of finite magnitude with unchangeable contour. In plane two-dimensional potential flows all Mach line elements along any one Mach line of the conjugate family are parallel. The characteristics in the hodograph are straight. In the hodograph, the equation for the stream function is reducible to Laplace's equation or to the simple wave equation.

21.2. Munk's gas:

$$p = A - \tan^{-1}\frac{1}{\rho} - \frac{\rho}{1+\rho^2} \tag{1}$$

All Mach lines in the plane two-dimensional potential flow are straight. In the hodograph the characteristics are circles passing through the center. In the hodograph, the equation for the Legendre reciprocal potential is reducible to Laplace's equation or to the simple wave equation. This is not a polytropic expansion.

21.3. Isothermal expansion:

$$\gamma = +1 \tag{1}$$

Many equations for polytropic gases do not hold for this special case, but must be replaced by special equations containing logarithmic or exponential expressions. The ultimate velocity for this expansion is infinite, and the velocity of sound is constant. This constant velocity of sound is preferably used as reference velocity. The elastic energy of expansion available is unlimited.

21.4. Any expansion law $p = f(\rho)$ may be replaced by

$$p = f(\rho) + \text{constant} \tag{1}$$

without any change of the dynamics of the gas.

22. Shockwaves

22.1. Steady, normal shock: a. *Rankine-Hugoniot relations:*

$$u_1^2 + h_1 = u_2^2 + h_2 \qquad \text{(energy)} \tag{1}$$

$$u_1\rho_1 = u_2\rho_2 \qquad \text{(continuity)} \tag{2}$$

$$p_1 + \rho_1 u_1^2 = p_2 + \rho_2 u_2^2 \qquad \text{(momentum)} \tag{3}$$

where subscript 1 denotes upstream shockwave; subscript 2 denotes downstream shockwave; a denotes ultimate velocity.

b. *Polytropic expansion, perfect gas:*

$$c^{*2} = a_1^2 \frac{\gamma - 1}{\gamma + 1} = a_2^2 \frac{\gamma - 1}{\gamma + 1}$$

$$\theta_{01} = \theta_{02}, \qquad u_1 u_2 = c^{*2}$$

If M_1 is given,

$$M_2^2 = \frac{(\gamma - 1)M_1^2 + 2}{2\gamma M_1^2 - (\gamma - 1)} = \frac{M_1^2 + 5}{7M_1^2 - 1} \quad (\gamma = 1.4) \tag{5}$$

$$\frac{p_2}{p_1} = \frac{2\gamma M_1^2 - (\gamma - 1)}{\gamma + 1} = \frac{7M_1^2 - 1}{6} \quad (\gamma = 1.4) \tag{6}$$

$$\frac{\rho_2}{\rho_1} = \frac{u_1}{u_2} = \frac{(\gamma + 1)M_1^2}{(\gamma - 1)M_1^2 + 2} = \frac{6M_1^2}{M_1^2 + 5} \quad (\gamma = 1.4) \tag{7}$$

$$\frac{\theta_2}{\theta_1} = \frac{p_2 \rho_1}{p_1 \rho_2} \tag{8}$$

$$\frac{\rho_2}{\rho_1} = \frac{(\gamma + 1)(p_2/p_1) + (\gamma - 1)}{(\gamma - 1)(p_2/p_1) + (\gamma + 1)} = \frac{(6p_2/p_1) + 1}{(p_2/p_1) + 6} \quad (\gamma = 1.4) \tag{9}$$

$$\frac{\Delta S}{c_v} = \ln \frac{2\gamma M_1^2 - (\gamma - 1)}{\gamma + 1} - \gamma \ln \frac{(\gamma + 1)M_1^2}{(\gamma - 1)M_1^2 + 2} \tag{10}$$

where $cv = R/\gamma - 1$.

22.2. Oblique shock waves : In equations (22.1*b*-5), (22.1b-6), and (22.1b-7) substitute $M \sin \sigma$ for M. Then

$$\tan \delta = \frac{2M_1^2(\sin \sigma \cos \sigma - \cot \sigma)}{2 + M_1^2(\gamma + 1 - 2 \sin^2 \sigma)} \tag{1}$$

$$v_1 \cos \sigma = v_2 \cos (\sigma - \delta) \tag{2}$$

$$a^2_{\text{oblique}} = a^2_{\text{normal}} + v_1^2 \cos^2 \sigma \tag{3}$$

$$v_1 \sin \sigma \cdot v_2 \sin (\sigma - \delta) = \frac{\gamma - 1}{\gamma + 1} a^2_{\text{normal}} \tag{4}$$

where σ = shock angle, normal shock $\sigma = 90°$, δ = deflection angle.

23. Cooling

23.1. Notation : k = conductivity, c_p = heat capacity (specific heat), μ = modulus of viscosity, $L \equiv$ reference length.

23.2. Prandtl number:

$$\sigma = \mu c/k \tag{1}$$

23.3. Nusselt number:

$$gL/[k(\theta_2 - \theta_1)] \tag{1}$$

23.4. Impact temperature increase (for air) is about

$$\Delta\theta = (MPH/100)^2 \quad \text{centigrade} \tag{1}$$

23.5. Impact temperature increase

$$\Delta\theta = v^2/2c_p, \quad c_p \sim 1000 \text{ m}^2/\text{sec}^2 \quad \text{(air)} \tag{1}$$

24. Boundary layers

24.1. Notations: $u =$ streamwise velocity component, $U =$ undisturbed velocity, $x =$ streamwise coordinate, $X =$ streamwise distance from leading edge, $\rho =$ density, $\mu =$ modulus of viscosity, $\nu = \mu/\rho$, $R =$ Reynolds' number $= XU/\nu$, $T =$ local shear, $\delta =$ boundary layer thickness, δ_{99} same up to where velocity equals 99 % of undisturbed velocity, $\delta_d =$ asymptotic displacement of streamlines, $S =$ " wetted " area.

24.2. Boundary layer equations (incompressible):

$$u_T + uu_x + vu_y = -p_x/\rho + u_{yy}\nu \tag{1}$$

$$u_x + v_y = 0 \tag{2}$$

a. *Boundary layer of flat plate (laminar).* Boundary layer is laminar, for $R = 350{,}000$ to $500{,}000$.

$$T = 0.332u^2\rho\sqrt{\mu/(U\rho X)} = 1.66\mu u/\delta_{99} \tag{1}$$

$$\delta_{99} = 5\sqrt{\nu x/u} \tag{2}$$

$$\delta_{\text{displ}} = 1.73\sqrt{\nu x/u} \tag{3}$$

Total friction

$$D_f = 1.328S\rho u^2/(2\sqrt{xu/\nu}) \tag{4}$$

b. *Boundary layer of flat plate (turbulent).* The following holds for $R = 1{,}000{,}000$ to $10{,}000{,}000$.

$$\delta = \frac{0.37x}{\sqrt[5]{ux/\nu}} \tag{1}$$

$$\tau = 0.0228\rho u^2 \sqrt[4]{\frac{\nu}{u\delta}} \tag{2}$$

$$= 0.0294\rho u^2 \sqrt[5]{\frac{\nu}{ux}} \tag{3}$$

$$D_f = 0.072\rho u^2 \frac{S}{2} \sqrt[5]{\frac{\nu}{ux}} \tag{4}$$

Bibliography

1. Bateman, H., *Partial Differential Equations*, Cambridge University Press, London, 1944.
2. Courant, R. and Friedrichs, R. O., *Supersonic Flow and Shock Waves*, Interscience Publishers, Inc., New York, 1948. Emphasizes the mathematical aspect of gas dynamics.
3. Durand, W. F., *Aerodynamic Theory*, Julius Springer, Berlin, 1935. See especially Vols. 1, 2, 6. (Dover reprint)
4. Goldstein, S., *Modern Developments in Fluid Dynamics*, Clarendon Press, Oxford, 1938.
5. Lamb, H., *Hydrodynamics*, 6th ed., Cambridge University Press, London, 1932. (Also Dover Publications, New York, 1945.) The classical treatise, rich and profound.
6. Sauer, R., *Einführung in die theoretische Gasdynamik*, Julius Springer, Berlin, 1943.
7. Tietjens, O. G., *Applied Hydro- and Aeromechanics* (based on lectures of L. Prandtl; trans. J. P. Den Hartog), McGraw-Hill Book Company, Inc., New York, 1934. Short and clear. (Dover reprint)
8. Tietjens, O. G., *Fundamentals of Hydro- and Aeromechanics* (based on lectures of L. Prandtl; trans. L. Rosenhead), McGraw-Hill Book Company, Inc., New York, 1934. Short and clear. (Dover reprint)

Chapter 9

BOUNDARY VALUE PROBLEMS IN MATHEMATICAL PHYSICS

By Henry Zatzkis

Assistant Professor of Mathematics
Newark College of Engineering

1. The Significance of the Boundary

1.1. Introductory remarks. Boundary value problems occupy a central position in mathematical physics. To give a systematic development of this subject would be well beyond the scope of the book, and we shall therefore confine ourselves to the discussion of some of the equations that occur more frequently. Other specifications will appear in other chapters. The major difficulty is not so much to find a solution of the given differential equation (in fact the equations admit always an infinity of solutions), but to find that solution which fits the given boundary values. If the solution depends on space and time, the latter can be treated formally as an additional coordinate in a space of one extra dimension, and the initial conditions thus also form part of the " boundary " conditions.

We are mainly concerned here with the Laplace equation, wave equation, and heat conduction equation. They are all linear, homogeneous, partial differential equations of second order. We plan first to discuss these equations individually and then to describe some of the features common to all which furnish unifying principles for their solutions. Finally, we shall also add some remarks concerning the corresponding inhomogeneous equations. A powerful tool to solve many important linear, homogeneous, differential equations is the method of separation of variables. However, it is not the only possibility, as will be seen subsequently.

1.2. The Laplace equation. The three-dimensional Laplace equation in Cartesian coordinates is defined by

$$\nabla^2\Phi = \frac{\partial^2\Phi}{\partial x^2} + \frac{\partial\Phi}{\partial y^2} + \frac{\partial^2\Phi}{\partial z^2} = 0 \tag{1}$$

If Φ is independent of z, we obtain the two-dimensional Laplace equation

$$\nabla^2\Phi = \frac{\partial^2\Phi}{\partial x^2} + \frac{\partial^2\Phi}{\partial y^2} = 0 \tag{2}$$

A function Φ satisfying the Laplace equation is called harmonic. An intuitive interpretation of $\nabla^2\Phi$ is the following. Consider a point P_0 at which Φ has the value Φ_0. Draw a little sphere around P_0 with a radius sufficiently small that we can use the Taylor formula to express the values of Φ on the surface of the sphere and neglect third-order and higher-order terms. If we compute the arithmetic mean of Φ on the spherical surface and denote it by $\overline{\Phi}$, then $\overline{\Phi} - \Phi_0 = (\nabla^2\Phi)_0$. We can therefore say that $\nabla^2\Phi$ is a measure for the deviation of the value of Φ at a given point from its average or " equilibrium " value. This interpretation is especially helpful in understanding the nature of the wave equation or heat conduction equation.

The Laplace equation is fundamental for potential theory; Φ then represents the potential produced by gravitational masses (or electrostatic charges) in a region free from mass (or charge). The simplest boundary value

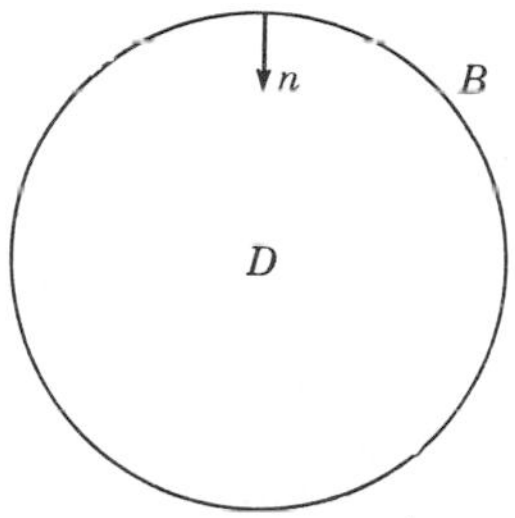

FIGURE 1

problem is the following: Given a simply connected domain D (Fig. 1), bounded by a continuous boundary B, the values of Φ on B are prescribed to be a continuous function $\overline{\Phi}$. We then wish to find the values of Φ in the interior of D, satisfying the Laplace equation and taking on the prescribed boundary values. This is the " inner " Dirichlet problem. In place of the function values $\overline{\Phi}$ themselves, their normal derivatives $\overline{\partial\Phi/\partial n}$ may be prescribed on B. This constitutes the Neumann problem or, finally, a linear combination of the function values and their normal derivatives with constant coefficients may be prescribed on B. This is the " third " problem, which occurs mainly in connection with the heat conduction equation. The condition, imposed on the solution Φ to be harmonic in D, has as a consequence that $\overline{\Phi}$ and $\overline{\partial\Phi/\partial n}$ cannot be prescribed independent of each other and, therefore, cannot be prescribed simultaneously in an arbitrary manner.

If we wanted the solution Φ outside the domain D, with the same boundary conditions, we should speak of the corresponding " outer " problems.

We understand that the positive direction of the normal always extends into the interior of the region D. To give meaning to the concept of " normal, " we must assume that the boundary consists of at least piecewise continuously differentiable sections.

We finally remark that the domain may also be multiply connected.

1.3. Method of separation of variables. Frequently the boundary is of a simple geometrical structure, e.g., a rectangle or a circle. Then we find it useful to employ rectangular or, more generally, curvilinear coordinates in which the boundary consists of sections along which one of the coordinates has a constant value. In case of the rectangle we use of course, the Cartesian coordinates themselves. The boundary consists then of segments, described by equations $x =$ constant and $y =$ constant. In case of a circular boundary of radius a, we use polar coordinates, and the boundary is given by $r = a$. The Laplace equation, after transformation to the new coordinates u, v, w, usually yields to a solution of the form

$$\Phi(u,v,w) = U(u)V(v)W(w) \tag{1}$$

We obtain then a system of ordinary differential equations for the functions U, V, and W. The final step consists of obtaining any appropriate linear combination of the solutions of type Eq. (1) with constant coefficients chosen to satisfy the given boundary conditions. Some examples are the following.

a. *Cartesian coordinates, three-dimensional:*

$$\frac{\partial^2\Phi}{\partial x^2} + \frac{\partial^2\Phi}{\partial y^2} + \frac{\partial^2\Phi}{\partial z^2} = 0 \tag{2}$$

Particular solutions:

$$\Phi_{k,l,m} = (kx + ly + mz), \qquad (k^2 + l^2 + m^2 = 1) \tag{3}$$

b. *Cartesian coordinates, two-dimensional:*

$$\frac{\partial^2\Phi}{\partial x^2} + \frac{\partial^2\Phi}{\partial y^2} = 0 \tag{4}$$

Particular solutions:

$$\Phi_k = e^{\pm k(x \pm iy)} \tag{5}$$

c. *Spherical coordinates* (r,θ,φ):

$$\frac{1}{r^2}\frac{\partial}{\partial r}\left(r^2\frac{\partial\Phi}{\partial r}\right)+\frac{1}{r^2\sin\theta}\frac{\partial}{\partial\theta}\left(\sin\theta\frac{\partial\Phi}{\partial\theta}\right)+\frac{1}{r^2\sin^2\theta}\frac{\partial^2\Phi}{\partial r^2}=0 \tag{6}$$

Particular solutions:

$$\Phi_{m,k}=\left(Ar^m+\frac{B}{r^{m+1}}\right)e^{\pm ik\varphi}P_m{}^k(\cos\theta) \tag{7}$$

where $P_m{}^k(\cos\theta)=$ associated Legendre polynomial.

d. *Cylindrical coordinates, three-dimensional* (ρ,φ,z):

$$\frac{\partial^2\Phi}{\partial\rho^2}+\frac{1}{\rho}\frac{\partial\Phi}{\partial\rho}+\frac{1}{\rho^2}\frac{\partial^2\Phi}{\partial\varphi^2}+\frac{\partial^2\Phi}{\partial z^2}=0 \tag{8}$$

Particular solutions:

$$\Phi_{k,m}=e^{\pm i(kz\pm m\varphi)}J_m(ik\rho) \tag{9}$$

where $J_m=$ Bessel function of order m.

e. *Cylindrical coordinates, two-dimensional* (ρ,φ):

$$\frac{\partial^2\Phi}{\partial\rho^2}+\frac{1}{\rho}\frac{\partial\Phi}{\partial\rho}+\frac{1}{\rho^2}\frac{\partial^2\Phi}{\partial\varphi^2}=0 \tag{10}$$

Particular solutions:

$$\Phi_k=\rho^{\pm k}e^{ik\varphi} \tag{11}$$

f. *Elliptical coordinates* (u_1,u_2,u_3):

The relations between the Cartesian and elliptical coordinates are:

$$\left.\begin{aligned} x&=\sqrt{\frac{(u_1+a^2)(u_2+a^2)(u_3+a^2)}{(b^2-a^2)(c^2-a^2)}}\\ y&=\sqrt{\frac{(u_1+b^2)(u_2+b^2)(u_3+b^2)}{(c^2-b^2)(a^2-b^2)}}\\ z&=\sqrt{\frac{(u_1+c^2)(u_2+c^2)(u_3+c^2)}{(a^2-c^2)(b^2-c^2)}}\end{aligned}\right\} \tag{12}$$

where $a^2>u_3>b^2>u_2>c^2>u_1$. The surfaces $u_1=$ constant, $u_2=$ constant, $u_3=$ constant represent confocal ellipsoids, hyperboloids of one sheet, and hyperboloids of two sheets, respectively.

Then $u_1 = 0$ in the ellipsoid

$$\frac{x^2}{a^2} + \frac{y^2}{b^2} + \frac{z^2}{c^2} = 1$$

If $$R_i = \sqrt{(u_i + a^2)(u_i + b^2)(u_i + c^2)}, \qquad (i = 1,2,3)$$

the Laplacian reads as follows.

$$\left.\begin{aligned}\nabla^2\Phi = \frac{4}{(u_1 - u_2)(u_2 - u_3)(u_3 - u_1)} \bigg[&(u_2 - u_3)R_1 \frac{\partial}{\partial u_1}\left(R_1 \frac{\partial\Phi}{\partial u_1}\right) \\ &+ (u_3 - u_1)R_2 \frac{\partial}{\partial u_2}\left(R_2 \frac{\partial\Phi}{\partial u_2}\right) \\ &+ (u_1 - u_2)R_3 \frac{\partial}{\partial u_3}\left(R_3 \frac{\partial\Phi}{\partial u_3}\right)\bigg] = 0\end{aligned}\right\} \tag{13}$$

As an illustration, let us compute the potential of a charged ellipsoidal conductor, whose surface is given by

$$\frac{x^2}{a^2} + \frac{y^2}{b^2} + \frac{z^2}{c^2} = 1$$

The conductor must be a surface of constant potential. For large distances the potential must fall off as Q/r, where Q is the charge of the conductor. The equipotential surfaces are confocal ellipsoids, i.e., Φ depends only on u_1. The Laplace equation reduces to

$$\frac{d}{du_1}\left(R_1 \frac{d\Phi}{du_1}\right) = 0 \tag{14}$$

The solution is

$$\Phi = k \int_{u_1=0}^{\infty} \frac{d\xi}{R_1(\xi)} \tag{15}$$

where $R_1(\xi) = \sqrt{(\xi + a^2)(\xi + b^2)(\xi + c^2)}$. For large values of $r^2 = x^2 + y^2 + z^2$, the value of ξ approaches r^2, and the potential Φ itself approaches the value

$$\Phi \doteq \frac{2k}{r} \tag{16}$$

Therefore $2k = Q =$ charge of the conductor, and the solution becomes

$$\Phi = \frac{Q}{2} \int_{u_1=0}^{\infty} \frac{d\xi}{R_1(\xi)} \tag{17}$$

1.4. Method of integral equations. We can express the interior Dirichlet problem as the solution of an integral equation. Let Q and T be points on the boundary B and P, and interior point of the domain D; let n be the interior normal of B, and $g(T)$ be the preassigned boundary values of $\Phi(P)$. The solution of the problem is then in the three-dimensional case

$$\Phi(P) = \iint_B f(Q) \frac{\partial}{\partial n_Q} \left(\frac{1}{r_{PQ}}\right) dS_Q \tag{1}$$

where $f(Q)$ is the solution of the integral equation

$$2\pi f(T) + \iint_B f(Q) \frac{\partial}{\partial n_Q} \left(\frac{1}{r_{TQ}}\right) dS_Q = g(T) \tag{2}$$

In the two-dimensional case the corresponding formulas are

$$\Phi(P) = \oint_B f(Q) \frac{\partial}{\partial n_Q} \left(\log \frac{1}{r_{TQ}}\right) dS_Q \tag{3}$$

and

$$\pi f(T) + \oint_B f(Q) \frac{\partial}{\partial n_Q} \left(\log \frac{1}{r_{TQ}}\right) dS_Q = g(T) \tag{4}$$

In the second case B is the curve bounding the area D, dS_Q is a line element, and the line integral has to be taken in the positive sense (the interior of the domain must be to the left of B).

The integral equations for f are Fredholm integral equations of the second kind. For the methods of their solution the reader is referred to any textbook on integral equations or potential theory.

In the special case where the boundary is a sphere or circle of radius 1, an explicit solution can be given. The formulas are known as the Poisson integral formulas:

$$\Phi(P) = \frac{1}{2\pi} \iint_B g(Q) \left[\frac{\partial}{\partial n_Q} \left(\frac{1}{r_{PQ}}\right) - \frac{1}{2lr_{PQ}}\right] d_1S_Q \tag{5}$$

and

$$\Phi(P) = \frac{1}{\pi} \oint g(Q) \left[\frac{\partial}{\partial n_Q} \left(\log \frac{1}{r_{PQ}}\right) - \frac{1}{2l}\right] dS_Q \tag{6}$$

1.5. Method of Green's function. This method is closely related to the foregoing discussion. The Green function G belonging to the domain D with the boundary B is defined in the following way: G is a function of two points $P(x,y,z)$ and $Q(\xi,\eta,\zeta)$, of which the point P varies in the interior of D, and Q varies in $D + B$. Also, G has the property that, as a function of Q, it is harmonic in $D + B$, with the exception of P. In P, the function G becomes singular, like $1/r_{PQ}$.

Therefore, G can be written

$$G(r_{PQ}) = \frac{1}{r_{PQ}} + \omega(r_{PQ}) \tag{1}$$

where r_{PQ} is harmonic in the closed domain $D + B$. Finally, we assume that, on the boundary, ω takes on the value $-1/r_{PQ}$. In the two-dimensional case, the procedure is identical, except that we replace $1/r_{PQ}$ by $\log(1/r_{PQ})$.

The solution of the interior Dirichlet problem then becomes

$$\Phi(P) = \frac{1}{4\pi} \iint_B g(Q) \frac{\partial G}{\partial n} dS_Q \tag{2}$$

in the three-dimensional case, and

$$\Phi(P) = \frac{1}{2\pi} \oint_B g(Q) \frac{\partial G}{\partial n} dS_Q \tag{3}$$

in the two-dimensional case. As in the preceding section, the function $g(Q)$ denotes the prescribed boundary values of Φ.

For a sphere and a circle, Green's function can be stated explicitly. Let Q (Fig. 2) be a point on the boundary, and P an interior point whose

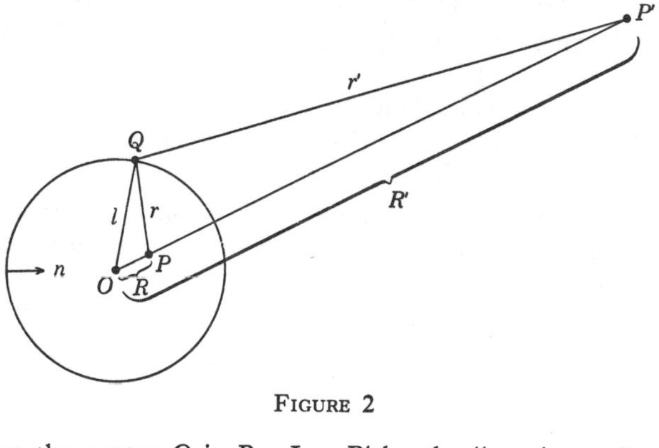

FIGURE 2

distance from the center O is R. Let P' be the " conjugate " point of P with the distance R' from O, i.e., $\overline{OP} \cdot \overline{OP'} = RR' = l^2$ ($l =$ radius). Then

$$G(PQ) = 1/r - l/Rr'$$

for the three-dimensional case, and

$$G(PQ) = \log(1/r) - \log(l/Rr')$$

for the two-dimensional case. W. Thomson's (Lord Kelvin's) method of electrical images is based on these Green functions.

1.6. Additional remarks about the two-dimensional case. The two-dimensional Laplace equation has an important property that any analytic function

$$f(z) = \Phi(x,y) + i\Psi(x,y) \tag{1}$$

is a solution of the Laplace equation. The fact that $f(z)$ is analytic permits us to use all the results of function theory. Especially important is the theorem that an analytic function remains analytic under a conformal mapping. The conjugate functions Φ and Ψ satisfy the Cauchy-Riemann equations

$$\partial\Phi/\partial x = \partial\Psi/\partial y, \qquad \partial\Phi/\partial y = -\partial\Psi/\partial x \tag{2}$$

A consequence of Eq. (2) is that both Φ and Ψ satisfy Laplace's equation. The two families of curves $\Phi =$ constant and $\Psi =$ constant are mutually orthogonal. We can, therefore, interpret them, respectively, as lines of constant potential and lines of force. Under a conformal mapping they will transform into two families of mutually orthogonal curves. We utilize these results in solving problems in electrostatics and in the steady, two-dimensional flow of an incompressible, ideal liquid. If this flow is also irrotational, the velocity distribution can be described by a velocity potential function $\Phi(x,y)$ whose Laplacian is zero. The Cartesian components of the velocity at the point (x,y) are given by

$$u = \partial\Phi/\partial x, \qquad v = \partial\Phi/\partial y \tag{3}$$

As an example let us consider a uniform motion with a constant velocity α in a positive x direction. Then

$$\Phi = \alpha x \tag{4}$$

and, therefore,

$$\Psi = \alpha y \tag{5}$$

or

$$f(z) = \alpha(x + iy) = \alpha z, \qquad (z = x + iy)\ * \tag{6}$$

The mapping $\omega = z^2$ maps the upper half-plane into the upper right-hand quadrant. As we can see from Eq. (8), the new streamlines are hyperbolas with the coordinate axes as asymptotes. The velocity potential takes the form

$$\xi = \alpha^2(x^2 - y^2) \tag{9}$$

The velocity components are

$$u = \partial\xi/\partial x = 2\alpha^2 x, \qquad v = \partial\xi/\partial y = 2\alpha^2 y \tag{10}$$

* The mapping $\omega = z^2$ gives us again a possible flow. If we write

$$\omega = \xi + i\eta \tag{7}$$

the new streamlines are given by

$$\eta = 2\alpha^2 xy = \text{constant} \tag{8}$$

The magnitude of the velocity is

$$\sqrt{u^2 + v^2} = 2\alpha^2 r, \qquad (r^2 = x^2 + y^2) \tag{11}$$

In the plane every analytic function represents a conformal mapping. In three-dimensional space the possibilities of conformal mapping (i.e., angle-preserving mapping) are much more restricted. Aside from the similarity transformations (i.e., rigid rotations, translations, and stretching the x, y, and z axes by the same factor), the only nontrivial conformal mapping of the space upon itself is the inversion at a sphere. If the radius of the sphere is equal to 1, the image point P' of the object point P lies along the infinite half line through the center of the sphere and P, and its distance from the center is given by $\overline{OR} \cdot \overline{OR'} = l^2$. If the center of the sphere is at the origin, the coordinates of the point P$'$ and P are, therefore,

$$\left.\begin{aligned} x' &= \frac{l^2}{R^2}\,x, \qquad y' = \frac{l^2}{R^2}\,y, \qquad z' = \frac{l^2}{R^2}\,z \\ x &= \frac{l^2}{R'^2}\,x', \qquad y = \frac{l^2}{R'^2}\,y', \qquad z = \frac{l^2}{R'^2}\,z' \end{aligned}\right\} \tag{12}$$

As W. Thomson discovered, we have here also a possibility of obtaining new solutions of the Laplace equation from the original one by conformal mapping. However, we find one difference that distinguishes the three-dimensional from the two-dimensional case. If

$$\frac{\partial^2 F}{\partial x^2} + \frac{\partial^2 F}{\partial y^2} + \frac{\partial^2 F}{\partial z^2} = 0 \tag{13}$$

then $F(x',y',z')$ does not satisfy the Laplace equation, but

$$F' = \frac{l}{R'}\,F(x',y',z') \tag{14}$$

does:

$$\frac{\partial^2 F'}{(\partial x')^2} + \frac{\partial^2 F'}{(\partial y')^2} + \frac{\partial^2 F'}{(\partial z')^2} = 0 \tag{15}$$

Green's function is closely related to the problem of conformal mapping in the two-dimensional case. Let $\zeta = f(x + iy)$ be an analytic function that maps conformally the domain D in the z plane ($z = x + iy$) in such a manner into the unit circle in the y plane, such that the point $\zeta(\xi,\eta)$ in D goes over into the center of the unit circle; then $-(1/2\pi)\log|f(x + iy)|$ is the desired Green function. Thus we know the Green functions for all those domains which can be mapped conformally into the unit circle.

1.7. The one-dimensional wave equation. The fundamental equation for the propagation of a wave in an anisotropic and homogeneous medium is

$$\nabla^2\Phi = \frac{1}{c^2}\frac{\partial^2\Phi}{\partial t^2} \tag{1}$$

where c is the velocity at which a disturbance will travel. If the direction of propagation is in the x direction only, the equation above reduces to

$$\frac{\partial^2\Phi}{\partial x^2} = \frac{1}{c^2}\frac{\partial^2\Phi}{\partial t^2} \tag{2}$$

Just as was the case for the two-dimensional Laplace equation, we can write the general solution at once:

$$\Phi(x,t) = F(x - ct) + G(x + ct) \tag{3}$$

where F represents a disturbance traveling in the positive x direction at velocity c, and G is a disturbance traveling in the negative x direction at the velocity c.

We want to apply this solution to find the motion of a vibrating string which is held fixed at its boundary points $x = 0$ and $x = L$, and whose initial displacement is prescribed to be $\Phi(x,0) = f(x)$, and its initial velocity distribution

$$\left.\frac{\partial\Phi}{\partial t}\right|_{t=0} = g(x)$$

From the initial conditions it follows that

$$\left.\begin{aligned} f(x) &= F(x) + G(x) \\ g(x) &= -cF'(x) + cG'(x) \end{aligned}\right\} \quad \text{for} \quad 0 \leqq x \leqq L \tag{4}$$

where primes denote derivatives with respect to the argument. If the second equation is integrated and combined with the first we obtain

$$\left.\begin{aligned} F(x) &= \frac{1}{2}\left(f(x) - \frac{1}{c}\int_{x_0}^{x} g(u)du\right) \\ G(x) &= \frac{1}{2}\left(f(x) + \frac{1}{c}\int_{x_0}^{x} g(u)du\right) \end{aligned}\right\} \quad \text{for} \quad 0 \leqq x \leqq L \tag{5}$$

The lower limit x_0 is arbitrary, and it will be cancelled out if Eq. (5) is substituted into Eq. (3):

$$\Phi(x,t) = \frac{1}{2}\left[f(x - ct) + f(x + ct) + \frac{1}{c}\int_{x-ct}^{x+ct} g(u)du\right] \tag{6}$$

Since f and g are defined only for values between 0 and L, the formula above is valid only for

$$ct \leqq x \leqq L - ct$$

However, the boundary conditions permit us to extend this range, for we must have

$$\Phi(0,t) \equiv F(-ct) + G(ct) = 0$$

for all t. Therefore,

$$F(-x) = -G(x) \tag{7}$$

Hence the second equation (5) permits us to define $F(x)$ also for the range

$$-L \leqq x \leqq 0$$

Using Eq. (7), Eq. (3) can now be written

$$\Phi(x,t) = F(x - ct) - F(-x - ct) \tag{8}$$

The second boundary condition for $x = L$ now yields

$$\Phi(L,t) \equiv F(L - ct) - F(-L - ct) = 0 \tag{9}$$

or

$$F(L - ct) = F(-L - ct) \tag{10}$$

This result shows that F is a periodic function in x with the period $2L$. Thus, F is defined for all values of x. Furthermore, if t increases by $T \equiv 2L/c$, the function will also repeat itself. Therefore F is also defined for all times.

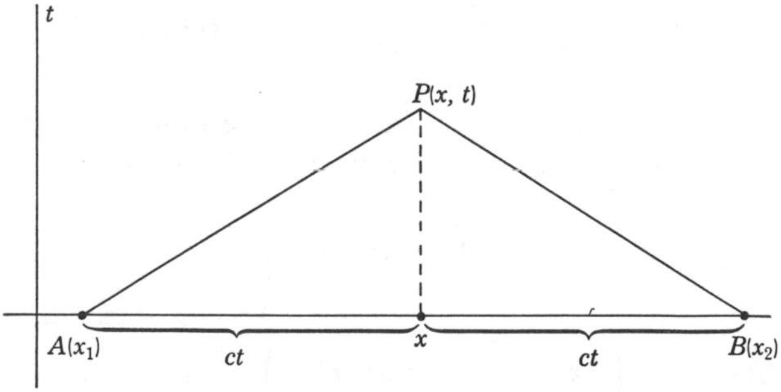

FIGURE 3

Equation (6) has an interesting geometrical interpretation (see Fig. 3). We draw through the point P in the x,t plane the two characteristic lines PA and PB (the characteristic lines are the lines obeying the equations $x = ct =$ constant). It is only those initial data (Cauchy data) which lie on the segment AB that contribute to the disturbance at point P at the time t. In optics the triangular polygon APB is called the retrograde light cone. If the initial velocities are zero, i.e., $g = 0$, then

$$\Phi(x,t) = \tfrac{1}{2}[f(x - ct) + f(x + ct)] \tag{11}$$

This is an expression of Huyghens' principle: The perturbation at x at the time t, originating from the sources at A and B at the time zero, need the retardation time $\tau = (x_2 - x)/c = (x - x_1)/c$ to reach the point x.

1.8. The general eigenvalue problem and the higher-dimensional wave equation. The one-dimensional wave equation is the only one for which the general solution can be given in closed form as in Eq. (3). In all other cases the method of separation of variables must be used. It will, therefore, be useful to consider those features which are common to all these problems and see the common underlying idea. We shall designate the spatial coordinates by x_1, x_2, x_3 (not necessarily Cartesian) and time again by t.

The inhomogeneous wave equation is of the structure

$$L[u] = \rho(x_1,x_2,x_3,\ldots)\frac{\partial^2 u}{\partial t'^2} - F(x_1,\ldots,t), \quad u = u(x_1,\ldots,t) \tag{1}$$

where L is a linear differential operator which operates on the spatial coordinates only. The linearity of L means that L has the property

$$L[au_1 + bu_2] = aL[u_1] + bL[u_2]$$

where a and b are independent of x.

We need also the concept of adjoint and self-adjoint linear differential operators. Let $u = u(x\ldots t)$ and $v = v(x\ldots t)$ be any two sufficiently differentiable functions in x; M is called the adjoint of L if $vL[u] - uM[v]$ can be written as a complete divergence, i.e., there exists a set of functions $f_1(x_1\ldots t)$, $f_2(x_1\ldots t)$, ... such that

$$vL[u] - uM[v] = \frac{\partial f_1}{\partial x_1} + \ldots + \frac{\partial f_n}{\partial x_n}$$

where $n =$ dimensionality of the space, and L is called self-adjoint if $L \equiv M$.

We also need the concept of homogeneous and inhomogeneous boundary conditions. We speak of a problem having a homogeneous character if, with the solution u the function cu is also a solution, where c is a constant. For this to be the case it is necessary not only that the differential equation but also that the boundary conditions, be homogeneous, e.g., $u = 0$ on B, or $\partial u/\partial n = 0$ on B, or $A_1 u + A_2(\partial u/\partial n) = 0$ on B, where A_1 and A_2 are constants.

Now suppose that u satisfies a homogeneous equation but an inhomogeneous boundary condition

$$u = f \quad \text{on } B.$$

It is then possible to continue f in such a manner into the interior of D that $L[f] = g$, where g is a continuous function in D. We can now construct a new function $v \equiv f - u$ which satisfies the inhomogeneous differential equation $L[v] = g$ but satisfies the homogeneous boundary conditions

$$v = 0 \quad \text{on } B.$$

The problem can be reversed in the same manner, and we can state quite generally that a homogeneous differential equation with inhomogeneous boundary conditions is equivalent to an inhomogeneous differential equation with homogeneous boundary conditions.

We shall, from now on, assume L to be self-adjoint (in the case of the wave equation, L is the Laplace operator, which is self-adjoint). If ρ is a function of x only, we can interpret ρ as the mass density, and Eq. (1) represents the dynamical equation for a continuous medium with the restoring force $L[u]$, e.g., an elastic force and an applied, external force F. In the case of equilibrium, after the transient has died and where F is a function of x only, we obtain as a special case Poisson's equation:

$$L[u] = F(x_1,\ldots) \tag{2}$$

The general problem to be solved is *that we assume L, ρ and F are defined in a certain spatial domain D.* On the boundary B of D we prescribe the homogeneous boundary conditions

$$u(x_1,\ldots,0) = \varphi(x_1,\ldots) \quad \text{and} \quad \left.\frac{\partial u}{\partial t}\right|_{t=0} = \psi(x_1,\ldots) \quad \text{on } B$$

We want to find a solution $u(x_1,\ldots,t)$ of Eq. (1) which satisfies these boundary conditions.

We shall first discuss the homogeneous differential equation

$$L[u] = \rho \frac{\partial^2 u}{\partial t^2} \tag{3}$$

i.e., $F \equiv 0$. Among all the possible solutions of Eq. (3) obeying homogeneous boundary conditions, we are especially interested in the synchronous solutions, i.e., those solutions which can be written

$$u = v(x_1,\ldots,x_n)g(t) \tag{4}$$

Since we are interested in periodic phenomena in time, we assume $g(t)$ of the form

$$g(t) = a \cos \lambda t + b \sin \lambda t \tag{5}$$

We obtain then for v the equation

$$L[v] + \lambda^2 \rho v = 0 \tag{6}$$

where v must now satisfy the boundary conditions imposed on u. Eq. (6) is called a Sturm-Liouville equation. We will find that if D is a finite domain, solutions obeying the boundary conditions will exist only for special values of λ: $\lambda_1, \lambda_2, \ldots$, the so-called characteristic values, or eigenvalues. The solutions belonging to them: $v_1(x), v_2(x), \ldots$, are called the characteristic functions or eigenfunctions. The solution of Eq. (3) is given by

$$u - (a \cos \lambda t + b \sin \lambda t) v(x_1, \ldots, x_n) \tag{7}$$

An eigenvalue λ_i can have several eigenfunctions $u_i^{(1)}, \ldots, u_i^{(k)}$. It is then called a k-fold degenerate. The eigenfunctions can be chosen in such a way that they form a " complete orthonormal " set, i.e.,

$$\int_D \rho u_i u_k d\tau = \delta_{ik} \tag{8}$$

Every function $h(x_1, \ldots)$ which has the degree of differentiability required by the operator L and obeys the required boundary conditions can be written in a generalized Fourier series

$$h = \sum_{n=1}^{\infty} c_n v_n(x) \tag{9}$$

where the constant coefficients c_n are given by

$$c_n = \int_D \rho h v_n d\tau \tag{10}$$

The complete, time-dependent solution of Eq. (3) is found by superposition of the particular solutions

$$u = \Sigma(a_n \cos \lambda t + b_n \sin \lambda t) v_n(x) \tag{11}$$

where

$$a_n = \int_D \rho \varphi v_n d\tau, \qquad b_n = \frac{1}{\lambda_n} \int_D \rho \psi v_n d\tau \tag{12}$$

The inhomogeneous Eq. (1) is solved by quite similar methods. But before discussing these we must introduce the concept of homogeneous and inhomogeneous boundary conditions. We speak of a problem having a homogeneous character if with u the function cu is also a solution, where c is a constant. For this to be the case it is necessary not only that the differential equation but also that the boundary conditions be homogeneous, e.g., $u = 0$ on B and $\partial u/\partial n = 0$ on B, or $Au + B\,\partial u/\partial n = 0$ on B, where A and B are constants.

Now suppose that u satisfies an inhomogeneous boundary condition, e.g., $u = f$ on B, but obeys the homogeneous differential equation $L[u] = 0$. It is then possible to continue the values of u on B in such a manner into the interior of D that $L[f] = g$ in D, where g is a continuous function in D. If we wish now to solve the inhomogeneous Eq. (1), we can without loss of generality assume homogeneous boundary conditions. We expand $F(x_1,\ldots,t)$ in terms of the orthonormal functions v_n:

$$F(x_1,\ldots,t) = \sum_{n=1}^{\infty} F_n(t) v_n(x) \tag{13}$$

and similarly

$$u(x_1,\ldots,t) = \sum_{n=1}^{\infty} \gamma_n(t) v_n(x) \tag{14}$$

If these values are substituted into Eq. (1), we obtain the differential equation for $\gamma_n(t)$:

$$\frac{d^2\gamma_n}{dt^2} + \lambda_n{}^2\gamma_n = -F_n \tag{15}$$

A particular solution of Eq. (15) is

$$\gamma_n = \frac{1}{\lambda_n}\int_0^t \sin \lambda_n(t-\tau)F_n(\tau)d\tau \tag{16}$$

Using a series expansion (14) with the coefficients of Eq. (16), we obtain a solution of Eq. (1) and, by a linear superposition, we obtain a solution of Eq. (1) satisfying the prescribed homogeneous boundary conditions.

There exists a close connection between the eigenvalue problem and the calculus of variations. We shall for simplicity state this relation for the one-dimensional case.

Suppose we want to find the function (or functions) $u(x)$ which give the integral

$$\int_a^b \left[A(x)\left(\frac{du}{dx}\right)^2 - B(x)u^2\right]dx = 0 \tag{17}$$

a stationary value. The function u is subject to the auxiliary condition

$$\int_a^b \rho u^2 dx = \text{constant}, \quad [\rho(x) > 0 \quad \text{for} \quad a \leqq x \leqq b] \tag{18}$$

and shall also satisfy the homogeneous boundary conditions

$$x(a) = x(b) = 0 \tag{19}$$

The Euler Lagrange equation belonging to this problem is the Sturm-Liouville equation

$$\frac{d}{dx}\left(A\frac{du}{dx}\right) + Bu + \lambda\rho u = 0 \tag{20}$$

The stationary values of the integral Eq. (17) are achieved if u is one of the characteristic functions with the characteristic value λ. This property of the characteristic values is utilized in the so-called direct methods of the calculus of variations to find the numerical value of the lowest eigenvalue and with it also the eigenfunction itself.

We shall state now the synchronous wave equation in a few typical coordinate systems and its particular solutions. The wave equation is in all cases

$$\nabla^2\Phi + \lambda^2\Phi = 0 \tag{21}$$

a. *Cartesian coordinates, two-dimensional case:*

$$\frac{\partial^2\Phi}{\partial x^2} + \frac{\partial^2\Phi}{\partial y^2} + \lambda^2\Phi = 0 \tag{22}$$

Particular solutions

$$\Phi_{\alpha,\beta} = e^{\pm i(\alpha x + \beta y)} \tag{23}$$

where $\alpha^2 + \beta^2 = \lambda^2$ (plane wave).

$$\Phi_0 = (Ax + B)e^{i\lambda y} \tag{24}$$

b. *Cartesian coordinates, three-dimensional case:*

$$\frac{\partial^2\Phi}{\partial x^2} + \frac{\partial^2\Phi}{\partial y^2} + \frac{\partial^2\Phi}{\partial z^2} + \lambda^2\Phi = 0 \tag{25}$$

Particular solutions

$$\Phi_{k,l,m} = e^{i(kx+ly+mz)} \tag{26}$$

where $k^2 + l^2 + m^2 = \lambda^2$ (plane wave).

c. *Cylindrical coordinates, two-dimensional case:*

$$\frac{\partial^2\Phi}{\partial\rho^2} + \frac{1}{\rho}\frac{\partial\Phi}{\partial\rho} + \frac{1}{\rho^2}\frac{\partial^2\Phi}{\partial\varphi^2} + \lambda^2\Phi = 0 \tag{27}$$

Particular solutions

$$\Phi_k = e^{ik\varphi}\cdot J_k(\lambda\rho), \quad \text{(circular wave)} \tag{28}$$

$$\Phi_0 = (A\varphi + B)J_0(\lambda\rho) \tag{29}$$

d. *Cylindrical coordinates, three-dimensional case:*

$$\frac{\partial^2\Phi}{\partial\rho^2} + \frac{1}{\rho}\frac{\partial\Phi}{\partial\rho} + \frac{1}{\rho^2}\frac{\partial^2\Phi}{\partial\varphi^2} + \frac{\partial^2\Phi}{\partial z^2} + \lambda^2\Phi = 0 \tag{30}$$

Particular solutions (modulated cylindrical wave)

$$\Phi_{k,m} = e^{\pm i(kz+m\varphi)} \cdot J_m(\rho\sqrt{\lambda^2 - k^2}) \tag{31}$$

$$\Phi_{0,0} = (A\varphi + B)(Cz + D)J_0(\lambda\rho) \tag{32}$$

where J_m is the Bessel function of order m.

e. *Spherical coordinates:*

$$\left.\begin{aligned} \frac{1}{r^2}\frac{\partial}{\partial r}\left(r^2\frac{\partial\Phi}{\partial r}\right) + \frac{1}{r^2\sin^2\theta}\frac{\partial}{\partial\theta}\left(\sin\theta\frac{\partial\Phi}{\partial\theta}\right) \\ + \frac{1}{r^2\sin^2\theta}\frac{\partial^2\Phi}{\partial\varphi^2} + \lambda^2\Phi = 0 \end{aligned}\right\} \tag{33}$$

Particular solutions

$$\Phi_{m,k} = \frac{1}{\sqrt{r}} J_{m+(1/2)}(\lambda r)e^{\pm ik\varphi}P_m{}^k(\cos\theta) \tag{34}$$

for a modulated spherical wave, and

$$\Phi_0 = \frac{e^{i\lambda r}}{r} \tag{35}$$

for a pure spherical wave.

By using the Fourier integral it is possible to build up either spherical or cylindrical waves out of plane waves, or, conversely, plane waves out of modulated spherical or cylindrical waves.

a. *Spherical wave as a superposition of plane waves:*

$$\Phi = \frac{i\lambda}{2\pi}\int_0^{(\pi/2)-i\sim} d\theta\sin\theta\int_0^{2\pi} d\varphi e^{i\lambda(x\sin\theta\cos\varphi+y\sin\theta\sin\varphi+|z\cos\theta|)} = \frac{e^{i\lambda r}}{r} \tag{36}$$

b. *Plane wave as a superposition of modulated spherical waves:*

$$\Phi = \sqrt{\frac{2\pi}{\lambda r}}\sum_{n=0}^{\infty}\frac{2n+1}{2}i^n J_{n+(1/2)}(\lambda r)P_n(\cos\theta) = e^{i\lambda r\cos\theta} = e^{i\lambda z} \tag{37}$$

c. *Plane wave as a superposition of circular waves* (two-dimensional case):

$$\sum_{n=-\infty}^{+\infty} i^n J_n(\lambda\rho)e^{in\varphi} = e^{i\lambda\varrho\cos\varphi} = e^{i\lambda x} \tag{38}$$

1.9. Heat conduction equation. Let $T(x,y,z,t)$ be the temperature distribution. In a homogeneous and isotropic medium it obeys the "equation of heat conduction"

$$\nabla^2 T - \frac{1}{a^2}\frac{\partial T}{\partial t} \tag{1}$$

Here $a^2 = k/c\rho$, where $k =$ specific heat conductivity, $c =$ specific heat, $\rho =$ density. If the homogeneous body D with the boundary B is imbedded in an infinite homogeneous medium of constant temperature zero, the boundary condition has the form

$$\frac{\partial T}{\partial n} + \sigma T = 0 \tag{2}$$

where σ is a positive constant and the normal n is directed into the interior of D. We have here again an eigenvalue problem, and the preceeding general principles therefore apply. We write again

$$T = v(x,y,z)g(t) \tag{3}$$

and obtain for v the eigenvalue problem

$$\nabla^2 v + \lambda v = 0 \quad \text{in } D \tag{4}$$

and

$$\frac{\partial v}{\partial n} + \sigma v = 0 \quad \text{in } B \tag{5}$$

The eigenfunction belonging to λ is

$$T = Ave^{-\lambda a^2 t} \tag{6}$$

If $T(x,y,z,0)$ is a prescribed function $\varphi(x,y,z)$, which is twice continuously differentiable and satisfies the boundary conditions, then the solution can be written in terms of the characteristic functions $v_1, v_2, \ldots,$ and their characteristic values $\lambda_1, \lambda_2, \ldots.$

$$T(x,y,z,t) = \sum_{n=1}^{\infty} c_n v_n(x,y,z)e^{-\lambda_n a^2 t} \tag{7}$$

where

$$c_n = \iiint_D \varphi v_n dxdydz \tag{8}$$

If the domain in which the initial temperature $\varphi(x,y,z)$ is prescribed is the entire space, the solution can be written down in closed form:

$$T(x,y,z,t) = \frac{1}{(2a\sqrt{\pi t})^3}\iiint_{-\infty}^{+\infty} \varphi(x',y',z')e^{-\varrho^2/4a^2 t}dx'dy'dz' \tag{9}$$

where

$$\rho^2 = (x-x')^2 + (y-y')^2 + (z-z')^2$$

This solution can be interpreted in the following way. Suppose the initial temperature was zero everywhere, except in a small region around the point, where it had the constant value T_0. Let us assume this region was a small cube of size $\Delta\tau$. We may choose this point P_0 as the origin. The temperature is then

$$T(x,y,z,t) = \frac{T_0\Delta\tau}{(2a\sqrt{\pi t})^3} e^{-\varrho^2/4a^2t} \tag{10}$$

where $$\rho^2 = x^2 + y^2 + z^2$$

At the time zero there exists a sharp maximum at the origin, which gradually becomes less and less defined caused by the diffusion of the heat. The initial accumulation of heat at the origin is called a heat pole. We may therefore interpret Eq. (9) as the result of a continuous set of heat poles having the strength $\varphi(x,y,z)$.

Another solution of the one-dimensional heat equation

$$\frac{\partial^2 T}{\partial x^2} = \frac{1}{a^2}\frac{\partial T}{\partial t} \tag{11}$$

of geophysical interest is the penetration of heat into the interior of a semi-infinite, homogeneous isotropic medium if the temperature on the surface is a prescribed periodic function of time, e.g., the temperature distribution in the interior of the earth caused by the daily or annual temperature oscillations on its surface.

We shall assume the temperature at the surface $x = 0$ to be $A \cos \omega t$. The temperature $T(x,t)$ for $x > 0$ (in the interior of the medium) is given by

$$T = Ae^{-(1/a)\sqrt{\omega/2}\,x} \cos\left(\omega t - \frac{1}{a}\sqrt{\frac{\omega}{2}}\,x\right) \tag{13}$$

This expression shows that temperature distribution has the shape of a damped oscillation, and that the phase is continuously changing. The solution also has another remarkable property; if we write the phase in the form

$$-\frac{1}{a}\sqrt{\frac{\omega}{2}}\,(x - a\sqrt{2\omega}t) \tag{13}$$

we see that the velocity $a\sqrt{2\omega}$ with which the disturbance travels depends on ω. In other words, we have here a dispersion, which was not the case for the one-dimensional wave equation. This is also the reason why we cannot have here a general solution of the type $f(x \pm ct)$.

Another significant difference between the heat and the wave equation is that the heat equation is not isotropic with respect to the direction of time. The solution for $t > 0$ cannot be continued to values $t < 0$. This is connected with the fact that heat flow is an irreversible flow and is connected with an increase in entropy.

1.10. Inhomogeneous differential equations. Suppose we have density distribution ρ of mass in space, and we assume that ρ is independent of time. The potential Φ then obeys Poisson's equation

$$\nabla^2\Phi = -4\pi\rho \tag{1}$$

Its solution is

$$\Phi(P) = \int \frac{\rho_Q}{r_{PQ}}\, d\tau_Q \tag{2}$$

where r_{PQ} is the distance between the " field point " P, where we want to compute the potential, and the " source point " Q with mass $\rho d\tau$.

If we assume that ρ is a function of space and time, we obtain the inhomogeneous wave equation for Φ:

$$\nabla^2\Phi - \frac{1}{c^2}\frac{\partial^2\Phi}{\partial t^2} = -4\pi\rho \tag{3}$$

One solution of this equation is

$$\Phi(P) = \int \frac{\rho(\xi,\eta,\zeta,t - R/c)}{R}\, d\xi d\eta d\zeta \tag{4}$$

where

$$R^2 = (x - \xi)^2 + (y - \eta^2) + (z - \zeta)^2$$

This solution is called the retarded potential, because the contributions to the potential at P are arising from those values of ρ at $Q(\xi,\eta,\zeta)$ which needed the time R/c to reach P.

Bibliography

1. CHURCHILL, R. V., *Fourier Series and Boundary Value Problems*, McGraw-Hill Book Company, Inc., New York, 1941. Emphasizes the expansion in orthogonal functions.
2. COURANT, R. and HILBERT, D., *Methoden der mathematischen Physik*, 2d ed., Vol. 1, 1931; Vol. 2, 1937: Julius Springer, Berlin. (Also Interscience Publishers, Inc., New York, 1943.)
3. FRANK, P. and VON MISES, R., *Die Differential- und Integralgleichungen der Mechanik und Physik*, 2d ed., Vol. 1 (mathematical part), 1930; Vol. 2 (physical part), 1935; F. Viehweg und Sohn, Braunschweig. (Dover reprint)
4. WEBSTER, A. G., *Partial Differential Equations of Mathematical Physics*, 2d ed., G. E. Stechert & Company, New York, 1936. (Dover reprint)

Chapter 10

HEAT AND THERMODYNAMICS

BY P. W. BRIDGMAN

Higgins University Professor
Harvard University

1. Formulas of Thermodynamics

The derivation of the word thermodynamics, which literally means "concerned with the motion of heat," suggests that the science of thermodynamics deals with the details of the interchange of heat between bodies. As a matter of fact, traditional thermodynamics does not deal with the motion of heat at all, but with the equilibrium conditions reached after heat motion has ceased, and should, therefore, more properly be called thermostatics. The results of thermodynamics are derived from two laws: the laws of the conservation and of the degradation of energy. These two laws are, perhaps, the most sweeping generalization from experiment yet achieved. A mechanistic account of these laws can be given by the methods of statistical mechanics, for which the following chapter should be consulted.

1.1. Introduction. The partial derivatives of thermodynamics are somewhat different from the conventional partial derivatives of mathematics in spite of their apparent identity of form. The conventional derivative $(\partial y/\partial x)_z$ implies by its form that y is a function of x and z. In the corresponding thermodynamic derivative the subscript z indicates the *path* along which the derivative is taken, y and x not necessarily being defined off the path. It follows that such derivatives as $(\partial\tau/\partial p)_Q$, $(\partial\tau/\partial p)_W$, $(\partial Q/\partial p)_\tau$, $(\partial W/\partial\tau)_p$ are thermodynamically meaningful, although Q and W are not possible independent variables.

The ordinary mathematical formulas for the manipulation of first derivatives apply to the path derivatives of thermodynamics. In particular,

$$\left(\frac{\partial x}{\partial y}\right)_z = \frac{1}{(\partial y/\partial x)_z} \tag{1}$$

$$\left(\frac{\partial x}{\partial y}\right)_z \left(\frac{\partial y}{\partial z}\right)_x \left(\frac{\partial z}{\partial x}\right)_y = -1 \tag{2}$$

$$\left(\frac{\partial w}{\partial u}\right)_v = \left(\frac{\partial w}{\partial x}\right)_y \left(\frac{\partial x}{\partial u}\right)_v + \left(\frac{\partial w}{\partial y}\right)_x \left(\frac{\partial y}{\partial u}\right)_v \tag{3}$$

The second path derivatives, however, in general require special treatment. In particular,

$$\frac{\partial}{\partial p}\left[\left(\frac{\partial Q}{\partial \tau}\right)_p\right]_\tau = \frac{\partial^2 Q}{\partial p \partial t} \neq \frac{\partial^2 Q}{\partial \tau \partial p}, \quad \text{etc.} \tag{4}$$

1.2. The laws of thermodynamics. The first law is

$$dE = dQ - dW \tag{1}$$

This formula applies to any thermodynamic system for any infinitesimal change of its state. This formula is a definition for dE, it being assumed that dQ (heat absorbed by the system) and dW (work done by the system) have independent instrumental significance. The first law states that the dE so defined is a perfect differential in the independent variables of state. By integration E may be obtained as a function of the state variables. The E so obtained contains an arbitrary constant of integration, without significance for thermodynamics.

The second law is

$$ds = \frac{dQ}{\tau} \tag{2}$$

The second law states that for any *reversible* absorption of heat there is an integrating denominator τ (absolute temperature) such that ds is a perfect differential in the state variables.

Applied to a Carnot engine (engine working on two isothermals and two adiabatics)

$$Q_1/Q_2 = \tau_1/\tau_2 \tag{3}$$

$$\text{efficiency of Carnot engine} \equiv \frac{Q_1 - Q_2}{Q_1} = \frac{\tau_1 - \tau_2}{\tau_1} \tag{4}$$

The two laws apply to any subsystem that can be carved out of the whole system. It follows that there are fluxes of heat $\boldsymbol{q}$, mechanical energy $\boldsymbol{w}$, total energy $\boldsymbol{e}$, and entropy $\boldsymbol{s}$ such that

$$-\operatorname{div} \boldsymbol{q} = dQ/dt \tag{5}$$

$$\operatorname{div} \boldsymbol{w} = dW/dt \tag{6}$$

$$\boldsymbol{e} = \boldsymbol{q} - \boldsymbol{w} \tag{7}$$

$$\boldsymbol{s} = \boldsymbol{q}/\tau \tag{8}$$

where now Q and W are taken for unit volume, and t denotes time.

1.3. The variables. The variables usually associated with thermodynamic systems are p = pressure, τ = absolute temperature, v = volume, s = entropy, E = energy, $G = E + pv - \tau s$ (Gibbs thermodynamic potential), $F = E - \tau s$ (Helmholtz free energy), $H = E + pv$ (total heat or enthalpy), dW = work done by the system, dQ = heat absorbed by the system. The specific heats are defined as

$$C_p \equiv \left(\frac{\partial Q}{\partial \tau}\right)_p, \qquad C_v \equiv \left(\frac{\partial Q}{\partial \tau}\right)_v$$

These variables are all to be taken in consistent units, which means in particular that heat is usually to be measured in mechanical units. The amount of matter in the system to which these quantities refer may be taken arbitrarily, subject to the demands of consistency. Thus, if v represents the volume of one gram, C_p refers conventionally to one gram of matter, but measured in mechanical units. However, if v is taken more ordinarily as the volume of that amount of substance which occupies one cubic centimeter under standard conditions, then C_p has its conventional value multiplied by the density.

1.4. One-component systems. The state is fixed by two independent variables of which p and τ are a possible pair. $dW = pdv$ (one-phase systems). In such a system any three of the first derivatives can, in general, be assigned without restriction from the laws of thermodynamics (there are many exceptions to this). Any fourth first derivative can then, in general, be expressed in terms of the chosen three with the help of the first and second laws. Normally, a thermodynamic relation involving first derivatives is a relation between any four such derivatives. There are approximately 10^{11} such relationships. Various schemes have been proposed by which any desired one of these 10^{11} relationships may be obtained with comparatively little mathematical manipulation *. No attempt is made to reproduce these schemes here, but in the following a few of the most frequently used relations will be given.

$$\left(\frac{\partial \tau}{\partial v}\right)_s = -\left(\frac{\partial p}{\partial s}\right)_v \tag{1}$$

$$\left(\frac{\partial \tau}{\partial p}\right)_s = \left(\frac{\partial v}{\partial s}\right)_p \tag{2}$$

$$\left(\frac{\partial p}{\partial \tau}\right)_v = \left(\frac{\partial s}{\partial v}\right)_\tau \tag{3}$$

* References 1, 2, 3, 4 and 10 in the Bibliography.

$$\left(\frac{\partial v}{\partial \tau}\right)_p = -\left(\frac{\partial s}{\partial p}\right)_\tau \tag{4}$$

Equations (1)-(4) are the four Maxwell relations.

$$\left(\frac{\partial v}{\partial p}\right)_s = \left(\frac{\partial v}{\partial p}\right)_\tau + \frac{\tau}{C_p}\left(\frac{\partial v}{\partial \tau}\right)_p^2 \tag{5}$$

$$C_p = \tau\left(\frac{\partial s}{\partial \tau}\right)_p = \tau\left(\frac{\partial p}{\partial \tau}\right)_s\left(\frac{\partial v}{\partial \tau}\right)_p \tag{6}$$

$$C_v = \tau\left(\frac{\partial s}{\partial \tau}\right)_v = -\tau\left(\frac{\partial \tau}{\partial v}\right)_s\left(\frac{\partial p}{\partial \tau}\right)_v = \left(\frac{\partial E}{\partial \tau}\right)_v \tag{7}$$

$$C_p - C_v = \tau\left(\frac{\partial p}{\partial \tau}\right)_v\left(\frac{\partial v}{\partial \tau}\right)_p = \frac{-\tau(\partial v/\partial \tau)_p{}^2}{(\partial v/\partial p)_\tau} \tag{8}$$

$$C_p/C_v \equiv \gamma = \frac{(\partial p/\partial v)_s}{(\partial p/\partial v)_\tau} \tag{9}$$

$$\frac{(\partial v/\partial \tau)_s}{(\partial v/\partial \tau)_p} = \frac{1}{1-\gamma} \tag{10}$$

$$\left(\frac{\partial Q}{\partial p}\right)_\tau = -\tau\left(\frac{\partial v}{\partial \tau}\right)_p \tag{11}$$

$$\left(\frac{\partial G}{\partial p}\right)_\tau = v, \qquad \left(\frac{\partial G}{\partial \tau}\right)_p = -s \tag{12}$$

$$\left(\frac{\partial F}{\partial v}\right)_\tau = -p, \qquad \left(\frac{\partial F}{\partial \tau}\right)_v = -s \tag{13}$$

$$\left(\frac{\partial H}{\partial s}\right)_p = \tau, \qquad \left(\frac{\partial H}{\partial p}\right)_s = v \tag{14}$$

$$\left(\frac{\partial E}{\partial s}\right)_v = \tau, \qquad \left(\frac{\partial E}{\partial v}\right)_s = -p \tag{15}$$

$$\left(\frac{\partial E}{\partial v}\right)_\tau = -p + \tau\left(\frac{\partial p}{\partial \tau}\right)_v \tag{16}$$

Equation (16) is the so-called " thermodynamic equation of state."

$$\left.\begin{aligned}\mu &\equiv \left(\frac{\partial \tau}{\partial p}\right)_H, \quad \text{(Joule-Thomson coefficient)}\\ &= \frac{1}{C_p}\left[\tau\left(\frac{\partial v}{\partial \tau}\right)_p - v\right] = -\frac{1}{C_p}\left(\frac{\partial H}{\partial p}\right)_\tau\end{aligned}\right\} \tag{17}$$

$$\left(\frac{\partial C_p}{\partial p}\right)_\tau = -\tau\left(\frac{\partial^2 v}{\partial \tau^2}\right)_p \tag{18}$$

$$\left(\frac{\partial C_v}{\partial v}\right)_\tau = \tau\left(\frac{\partial^2 p}{\partial \tau^2}\right)_v \tag{19}$$

1.5. One-component, usually two-phase systems. The state is fixed by two independent variables of which p and τ are *not* a possible pair, $dW = pdv$. A new variable appears, x, the fraction of the total mass of the system present in phase 1. There are now only two independently assignable first derivatives, of which $dp/d\tau$ and $C_v\,[\equiv (\partial Q/\partial \tau)_v]$ form a convenient pair; C_v is expressible in terms of the properties of the separate phases.

$$\left.\begin{aligned}C_v = x&\left[C_{p_1} - 2\tau\frac{dp}{d\tau}\left(\frac{\partial v_1}{\partial \tau}\right)_p - \tau\left(\frac{dp}{d\tau}\right)^2\left(\frac{\partial v_1}{\partial p}\right)_\tau\right]\\ &+ (1-x)\left[C_{p_2} - 2\tau\frac{dp}{d\tau}\left(\frac{\partial v_2}{\partial \tau}\right)_p - \tau\left(\frac{dp}{d\tau}\right)^2\left(\frac{\partial v_2}{\partial p}\right)_\tau\right]\end{aligned}\right\} \tag{1}$$

The general thermodynamic relation between first derivatives is a relation between any three derivatives. In these relations the following function of x occurs frequently :

$$\left.\begin{aligned}f(x, 1-x) \equiv x&\left[\left(\frac{\partial v_1}{\partial \tau}\right)_p + \left(\frac{\partial v_1}{\partial p}\right)_\tau \frac{dp}{d\tau}\right]\\ &+ (1-x)\left[\left(\frac{\partial v_2}{\partial \tau}\right)_p + \left(\frac{\partial v_2}{\partial p}\right)_\tau \frac{dp}{d\tau}\right]\end{aligned}\right\} \tag{2}$$

In addition to the following special formulas, many of the formulas of Section 1.4 also apply.

$$\left(\frac{\partial x}{\partial \tau}\right)_s = -\frac{1}{v_1 - v_2}\left[\frac{C_v}{\tau(dp/d\tau)} + f(x, 1-x)\right] \tag{3}$$

$$\left(\frac{\partial x}{\partial \tau}\right)_v = -\frac{f(x, 1-x)}{v_1 - v_2} \tag{4}$$

$$\left(\frac{\partial x}{\partial v}\right)_\tau = \frac{1}{v_1 - v_2} \tag{5}$$

$$\left(\frac{\partial x}{\partial v}\right)_s = \frac{1}{v_1 - v_2}\left[1 + \frac{\tau}{C_v}\frac{dp}{d\tau} f(x, 1-x)\right] \tag{6}$$

$$\left(\frac{\partial x}{\partial \tau}\right)_E = \frac{1}{v_1 - v_2}\left[\frac{C_v}{p - \tau(dp/d\tau)} - f(x, 1-x)\right] \tag{7}$$

$$\left(\frac{\partial x}{\partial E}\right)_G = \frac{-1}{(v_1 - v_2)\,[p - \tau(dp/d\tau)]} \tag{8}$$

$$\left(\frac{\partial x}{\partial \tau}\right)_H = \frac{-1}{v_1 - v_2}\left[f(x, 1-x) + \frac{C_v + v(dp/d\tau)}{\tau(dp/d\tau)}\right] \tag{9}$$

$$\left(\frac{\partial x}{\partial \tau}\right)_F = \frac{-1}{v_1 - v_2}\left[f(x, 1-x) + \frac{s}{p}\right] \tag{10}$$

$$\frac{d\tau}{dp} = \frac{\tau(v_1 - v_2)}{L}, \quad \text{(Clapeyron's equation)} \tag{11}$$

where L is the heat absorbed by the system when it passes from phase 2 to phase 1 (latent heat).

$$\left.\begin{aligned} \frac{d^2\tau}{dp^2} = \frac{-1}{v_1 - v_2}\frac{d\tau}{dp}\Bigg\{ & \frac{C_{p_1} - C_{p_2}}{\tau}\left(\frac{d\tau}{dp}\right)^2 \\ & -2\left[\left(\frac{\partial v_1}{\partial \tau}\right)_p - \left(\frac{\partial v_2}{\partial \tau}\right)_p\right]\frac{d\tau}{dp} - \left[\left(\frac{\partial v_1}{\partial p}\right)_\tau - \left(\frac{\partial v_2}{\partial p}\right)_\tau\right]\Bigg\} \end{aligned}\right\} \tag{12}$$

$$\frac{dL}{dp} = \frac{d\tau}{dp}(C_{p_1} - C_{p_2}) + (v_1 - v_2) - \tau\left[\left(\frac{\partial v_1}{\partial \tau}\right)_p - \left(\frac{\partial v_2}{\partial \tau}\right)_p\right] \tag{13}$$

If the phase 1 is a perfect gas,

$$\frac{d \log p}{d\tau} = \frac{L}{R\tau^2} \tag{14}$$

1.6. Transitions of higher orders

$$\frac{dp}{d\tau} = -\frac{\Delta(\partial z/\partial \tau)_p}{\Delta(\partial z/\partial p)_\tau} \tag{1}$$

Here z is any continuous thermodynamic function with discontinuous first derivatives along a line in the p,τ plane. The formula gives the slope of the line of discontinuity. By specializing z, various formulas may be obtained. By identifying z successively with s and v and eliminating $dp/d\tau$, Ehrenfest's formula for a transition of the second kind is obtained:

$$\Delta C_p = -\tau \left[\frac{\Delta(\partial v/\partial \tau)_p{}^2}{\Delta(\partial v/\partial p)_\tau} \right] \tag{2}$$

1.7. Equations of state

$$pv = R\tau, \quad \text{(perfect gas)} \tag{1}$$

If p is measured in dynes/cm^2, v is the volume in cm^3 of one gram molecule, and τ is absolute Kelvin degrees, and $R = 8.3144 \times 10^7$ergs/g-mole-deg.

$$\left(p + \frac{a}{v^2}\right)(v - b) = R\tau, \qquad \text{(van der Waals)} \tag{2}$$

$$\left(p + \frac{a}{\tau v^2}\right)(v - b) = R\tau, \qquad \text{(Berthelot)} \tag{3}$$

$$p(v - b) = R\tau \exp \frac{-C}{R\tau v}, \qquad \text{(Dieterici)} \tag{4}$$

$$p = \frac{R\tau(1 - \epsilon)}{v^2}(v + B) - \frac{A}{v^2}, \quad \text{(Beattie-Bridgeman)} \tag{5}$$

where $$A = A_0\left(1 - \frac{a}{v}\right), \quad B = B_0\left(1 - \frac{b}{v}\right), \quad \epsilon = C/v\tau^3$$

and A_0, a, B_0, b, and c are constants.

$$pv = A + \frac{B}{v} + \frac{C}{v^2} \cdots, \quad \text{(" virial equation of state ")} \tag{6}$$

The coefficients A, B, etc., which are functions of temperature, are called the first, second, etc. virial coefficients.

Excepting the perfect gas equation, all these equations have the property of indicating the " critical point " at which vapor and liquid phases are identical. The parameters of the system at the critical point are designated by p_C, τ_C, v_C. The " reduced coordinates " of a system are defined as $p_r = p/p_C$, $\tau_r = \tau/\tau_C$, $v_\tau = v/v_C$. If the various equations of state are expressed in terms of reduced coordinates, a functional form will be obtained, the same for all substances. In particular, the reduced van der Waals equation of state is

$$p_r = \frac{8\tau_r}{3v_r - 1} - \frac{3R}{(v_r)^2} \tag{7}$$

1.8. One-component, two-variable systems, with $dW = Xdy$. Here X is "generalized force" and y is "generalized displacement." Possible pair of variables, τ and X. The simplest examples of such systems are electrical or magnetic systems, or simple elastic systems with a single stress component not a hydrostatic pressure. All the formulas of § 1.4 apply, replacing p by X and v by y.

1.9. One-component, two-variable systems, with $dW = Xdy$. Here X is generalized force and y is generalized displacement, τ and X *not* a possible pair of variables, X being fixed as a function of τ by the physics of the system. The formulas of § 1.5 apply, replacing p by X and v by y. Special examples are

$$u = \text{constant } \tau^4 \quad \text{(Stefan's law)} \tag{1}$$

where u is the density of black-body radiation.

$$\frac{d\,\text{emf}}{d\tau} = \frac{1}{\tau}\left(\frac{\partial Q}{\partial q}\right)_\tau, \quad \text{(Helmholtz)} \tag{2}$$

the equation for the temperature change of the emf of a reversible cell in terms of the heat absorbed when unit quantity of electricity flows through the cell isothermally.

$$\frac{d\sigma}{d\tau} = -\frac{1}{\tau}\left(\frac{\partial Q}{\partial A}\right)_\tau \tag{3}$$

the temperature derivative of surface tension in terms of the heat absorbed when the surface increases isothermally.

1.10. One-component, multivariable systems, with $dW = \Sigma X_i dy_i$. Here the X_i's are generalized forces and the y_i's are generalized displacements. Much variation is possible in the treatment, depending on the choice of independent variables. The generalized equation of state must be known for such systems. A general method of getting information is to put ds a perfect differential and write the relations on the cross derivatives of the coefficients. The various potential functions may be generalized for such systems. For example,

$$G_{\text{gen}} = E - \tau s + \Sigma Xy \tag{1}$$

This potential function is a minimum at equilibrium for changes at constant τ and X.

1.11. Multicomponent, multivariable systems, $dW = pdv$. A system fixed by composition variables n_i, usually given in gram molecules, plus two others, for example p and τ. In general a composition variable will be needed for each independent component in each homogeneous part (phase) of the system. A complete scheme for obtaining all the possible relations in such a system has been given by Goranson (see Bibliography), analogous to the schemes for dealing with § 1.4. Only a few of the more important relations will be given here. For arbitrary values of the independent variables in multicomponent systems, the system is generally not in equilibrium, and the spontaneous changes of variable which occur as the system settles toward equilibrium are irreversible. At equilibrium

$$G \text{ is a minimum for changes at constant } p \text{ and } \tau \tag{1}$$

$$F \text{ is a minimum for changes at constant } v \text{ and } \tau \tag{2}$$

$$H \text{ is a minimum for changes at constant } s \text{ and } p \tag{3}$$

$$E \text{ is a minimum for changes at constant } s \text{ and } v \tag{4}$$

In general all the formulas of § 1.4 apply with the addition of all the n_i's as subscripts. For example,

$$\left(\frac{\partial G}{\partial \tau}\right)_{p_i n_i} = v \tag{5}$$

The new formulas involve derivatives with respect to the n_i's. These derivatives define various " partial " quantities. In particular the " partial chemical potentials " are defined by

$$\mu_i \equiv \left(\frac{\partial G}{\partial n_i}\right)_{\tau_i p_i n_j} \equiv \left(\frac{\partial F}{\partial n_i}\right)_{\tau_i v_i n_j} \equiv \left(\frac{\partial H}{\partial n_i}\right)_{p_i s_i n_j} \equiv \left(\frac{\partial E}{\partial n_i}\right)_{s_i v_i n_j} \tag{6}$$

Here the subscript n_j indicates that all the n's except n_i are held constant during the differentiation. Other typical partial quantities are

$$\bar{v}_i \equiv \left(\frac{\partial v}{\partial n_i}\right) \tag{7}$$

$$\bar{s}_i \equiv \left(\frac{\partial s}{\partial n_i}\right) \tag{8}$$

$$C_{pi} \equiv \left(\frac{\partial C_p}{\partial n_i}\right) \tag{9}$$

No subscripts are indicated in these derivatives. The understanding is that all the other independent variables except n_i, whatever they may be, are held constant during the differentiation.

For the various *extensive* properties of the system there are linear relations on the n's, such as

$$G = \Sigma\, n_i\mu_i \tag{10}$$

$$F = \Sigma\, n_i\mu_i \tag{11}$$

$$v = \Sigma\, n_i\bar{v}_i \tag{12}$$

$$s = \Sigma\, n_i\bar{s}_i \tag{13}$$

Formulas for the various partial quantities hold similar to those for the total quantities. For example,

$$\left(\frac{\partial\mu_i}{\partial\tau}\right)_{p_i n_i} = -\,\bar{s}_i \tag{14}$$

$$\left(\frac{\partial\mu_i}{\partial p}\right)_{\tau_i n_i} = \bar{v}_i \tag{15}$$

1.12. Homogeneous systems. a. *Gaseous mixtures.* The mole fractions are defined by

$$N_i = \frac{n_i}{\sum\limits_k n_k} \tag{1}$$

$$\Sigma\, N_i = 1 \tag{2}$$

The partial vapor pressures are defined by

$$\bar{p}_i = N_i p \tag{3}$$

$$\Sigma\, \bar{p}_i = p \tag{4}$$

Perfect gaseous mixtures:

$$\bar{v}_i = \left(\frac{\partial\mu_i}{\partial p}\right)_{\tau_i n_i} = \frac{R\tau}{p}\, N_i \tag{5}$$

$$p = \sum \frac{n_i}{v}\, R\tau \tag{6}$$

$$\bar{p}_i = \frac{n_i}{v}\, R\tau \tag{7}$$

$$\left[\frac{\partial(\mu_i/\tau)}{\partial\tau}\right]_{p_i n_i} = -\,\frac{\bar{H}_i}{\tau^2} \tag{8}$$

If the perfect gases react according to the equation

$$lL + mM \ldots \rightleftarrows qQ + rR \ldots \tag{9}$$

where L, M, etc. represent moles of the reacting species, and l, m, etc. are the numerical coefficients in the reaction equation,

$$\frac{(\bar{p}_Q)^q(\bar{p}_R)^r \cdots}{(\bar{p}_L)^l(\bar{p}_M)^m \cdots} = \bar{K}(\tau), \quad \text{(mass action law for perfect gases)} \tag{10}$$

Here $\bar{K}(\tau)$ is a function of temperature only and is called the *equilibrium constant.*

$$\frac{d \log \bar{K}}{d\tau} = \frac{L_p}{R\tau^2} \tag{11}$$

where L_p is the latent heat at constant pressure of the reaction from left to right.

Imperfect gaseous mixtures: The " fugacity " $(\equiv \bar{p}_i^*)$ of the ith component is a quantity closely related to the partial vapor pressure and is defined by the equations

$$\mu_i = \mu_i^0(\tau) + R\tau \log \bar{p}_i^* \tag{12}$$

$$\lim_{p\to 0} \bar{p}_i^* = \bar{p}_i \tag{13}$$

$$\left(\frac{\partial \log \bar{p}_i^*}{\partial \tau}\right)_{p_i n_i} = -\frac{\bar{H}_i - \bar{H}_i^0}{R\tau^2} \tag{14}$$

where μ_i^0 and $\bar{H}_i^0$ are the limiting values at infinite volume.

If the imperfect gases react according to Eq. (9)

$$\frac{(\bar{p}_Q^*)^q(\bar{p}_R^*)^r \cdots}{(\bar{p}_L^*)^l(\bar{p}_M^*)^m \cdots} = K(\tau) \tag{15}$$

$$\frac{d \log K(\tau)}{d\tau} = \frac{L_p^0}{R\tau^2} \tag{16}$$

where L_p^0 is the heat of reaction at infinite volume.

b. *Ideal solutions:*

$$\mu_i = \mu_i^0 + R\tau \log N_i \tag{17}$$

where μ_i^0 is a function of pressure and temperature but not of composition.

$$\frac{\partial \mu_i}{\partial p} = \frac{\partial \mu_i^0}{\partial p} = \bar{v}_i \tag{18}$$

$$\frac{\partial(\mu_i/\tau)}{\partial \tau} = \frac{\partial(\mu_i^0/\tau)}{\partial \tau} = -\frac{\bar{H}_i}{\tau^2} \tag{19}$$

Ideal solutions of the same solvent mix with no change in volume or total heat content.

1.13. Heterogeneous systems. The phase rule of Gibbs is

$$f = c - n + 2 \tag{1}$$

where f is the number of degrees of freedom (counting intensive variables only), c the number of components, and n the number of phases.

For a liquid or solid solution in equilibrium with a vapor which behaves as a perfect gas

$$\Sigma\, N_i d(\log \bar{p}_i) = 0, \quad \text{(Duhem-Margules)} \tag{2}$$

for changes in composition at constant p and τ, where N_i refers to the condensed phase and $\bar{p}_i$ to the vapor phase in equilibrium with it.

$$\bar{p}_i = k_i N_i, \quad \text{(Henry's law for ideal solutions)} \tag{3}$$

where k_i is independent of N_i.

$$\bar{p}_0 = \bar{p}_0{}^0 N_0, \quad \text{(Raoult's law for ideal solutions)} \tag{4}$$

Here the subscript 0 denotes the solvent, and $\bar{p}_0{}^0$ the vapor pressure of the pure solvent

$$\Pi = \frac{R\tau}{\bar{v}_0} \log \frac{1}{N_0} = \frac{R\tau}{\bar{v}_0} N_1, \quad \text{(osmotic pressure)} \tag{5}$$

Here Π is the osmotic pressure of the ideal solution of the component N_1 in the solvent N_0.

Depression of the freezing point of an ideal dilute solution

$$\theta = \frac{R\tau_f{}^2}{L_f{}^0} N_1, \quad \text{(van 't Hoff)} \tag{6}$$

where $L_f{}^0$ is the latent heat of freezing of the pure solvent. Raising of the boiling point of the ideal dilute solution

$$\theta = \frac{R\tau_b{}^2}{L_v{}^0} N_1, \quad \text{(van 't Hoff)} \tag{7}$$

where $L_v{}^0$ is the latent heat of vaporization of the pure solvent at the normal boiling point τ_b;

$$\frac{N_i{}^\alpha}{N_i{}^\beta} = K, \quad \text{(Nernst's distribution law for ideal solutions)} \tag{8}$$

where α and β denote different condensed phases, and K depends on solvent, p, and τ, but not on N_i.

Bibliography

1. Bridgman, P. W., *Phys. Rev.*, April, 1914.
2. Bridgman, P. W., *A Condensed Collection of Thermodynamic Formulas*, Harvard University Press, Cambridge, Mass., 1925.
3. Crawford, F. H., *Am. J. Phys.*, **17**, 1-5 (1949).
4. Crawford, F. H., *Proc. Am. Acad. Arts Sci.*, **78**, 165-184 (1950).
5. Epstein, P. S., *Textbook of Thermodynamics*, John Wiley & Sons, Inc., New York, 1937. A comprehensive book for the more advanced student.
6. Goranson, R. W., *Publication 408*, Carnegie Institution of Washington, Washington, 1930.
7. Guggenheim, E. A., *Thermodynamics*, North Holland Publishing Company, Amsterdam, 1949. Features especially the use of thermodynamic potentials.
8. Prigogine, I. and Defay, R., *Thermodynamique chimique*, Éditions Desoer, Liège, 1950. Deals with " open " systems and irreversible processes.
9. Schottky, W., *Thermodynamik*, Julius Springer, Berlin, 1929. Notable for rigorous examination of fundamentals.
10. Shaw, N., *Trans. Roy. Soc.* (London), 1935, pp. 234, 299.
11. Zemansky, M. W., *Heat and Thermodynamics*, 3d ed., McGraw-Hill Book Company, Inc., New York, 1951. A comprehensive treatment from the elementary point of view.

Chapter 11

STATISTICAL MECHANICS

BY DONALD H. MENZEL

Professor of Astrophysics at Harvard College
Director of Harvard College Observatory

1. Statistics of Molecular Assemblies

1.1. Partition functions. Thermodynamics and statistical mechanics constitute independent approaches to essentially the same problem: the physical properties of assemblies of atoms or molecules. Thermodynamics depends on certain postulates concerning the behavior of an artificially defined quantity called entropy. Statistical mechanics concerns itself with averages over an assembly wherein a certain amount of energy E has been partitioned among the various atomic or molecular systems.

Fowler defines a quantity known as a "partition function," for each component of the assembly. By component we mean each kind of atom, molecule, particle, or "system," whether interacting or not. Thus all atoms of neutral iron comprise one component; those of singly ionized iron a second, etc. In its most general form the partition function, $f(T)$, reduces to a sum:

$$f(T) = \sum_i e^{-\varepsilon_i/kT} \tag{1}$$

where ϵ_i is the energy of state i, k is Boltzmann's constant, and T is the absolute temperature. If a given level is degenerate, consisting of $\tilde{\omega}_i$ states, we sum by multiplying the summand by $\tilde{\omega}_i$, which we term the statistical weight.

For a molecule, we often can break up the total energy of a given state into a number of independent energies, viz., $\epsilon_V =$ potential energy, $\epsilon_T =$ kinetic energy, $\epsilon_r =$ rotational energy, $\epsilon_v =$ vibrational energy, $\epsilon_e =$ internal electronic energy.

Of these energies the last three are usually quantized and, for classical assemblies, the first two are nonquantized. The potential energy is a function of the coordinates q_1, q_2, q_3; the kinetic energy, in turn, depends on the momenta p_1, p_2, p_3.

Writing

$$\epsilon = \epsilon_V + \epsilon_T + \epsilon_r + \epsilon_v + \epsilon_e \tag{2}$$

we see that the partition function factors whenever the energies are independent as above. Then

$$\left.\begin{aligned} f(T) &= \iiint e^{-\varepsilon_V/kT} dq_1 dq_2 dq_3 \iiint e^{-\varepsilon_T/kT} \frac{dp_1 dp_2 dp_3}{h^3} \\ &\quad \times \sum_r e^{-\varepsilon_r/kT} \sum_v e^{-\varepsilon_v/kT} \sum_e e^{-\varepsilon_e/kT} \\ &= V(T)H(T)B_r(T)B_v(T)B_e(T) \end{aligned}\right\} \tag{3}$$

The quantity h is Planck's constant, which we include so that the product dp_1dq_1/h will be dimensionless. We arbitrarily associate h with the kinetic rather than with the potential energy.

$$V(T) = \iiint e^{-\varepsilon_V/kT} dq_1 dq_2 dq_3 = V \tag{4}$$

the volume, in the absence of an external potential field. Under a constant gravitational acceleration g, the potential is $\epsilon_V = mgz$, where m is the mass of the particle and z the vertical coordinate. Then

$$V(T) = e^{-mgz/kT} V \tag{5}$$

Usually we integrate over only a small volume V of the assembly, wherein we can take ϵ_V constant or zero.

The classical partition function for the kinetic energy becomes

$$H(T) = \int_0^\infty \int_0^\infty \int_0^\infty e^{-(p_1^2+p_2^2+p_3^2)/2mkT} \frac{dp_1 dp_2 dp_3}{h^3} = \frac{(2\pi mkT)^{3/2}}{h^3} \tag{6}$$

A factor $H^{1/3}$ belongs to each degree of freedom of kinetic energy.

For a diatomic molecule, we can write

$$\epsilon_r = hcBJ(J+1) \tag{7}$$

where c is the velocity of light, B the quantity

$$B = h/8\pi^2 cI \tag{8}$$

with I the moment of inertia, and J the quantum number, an integer: 1,2,.... For a given J, we get $2J + 1$ states of identical energy. Thus

$$B_r(T) = \sum_{J=1}^{\infty} \frac{2J+1}{\sigma} e^{-J(J+1)hcB/kT} \tag{9}$$

When $hcB/kt \gg 1$ we cannot easily simplify the expression. But when this quantity is small, we can approximate to $B_r(T)$ by an integral

$$B_r(T) \sim \int_0^\infty \frac{2J+1}{\sigma} e^{-J(J+1)hcB/kT} dJ$$
$$\sim \frac{1}{B\sigma} \cdot \frac{kT}{hc} = \frac{8\pi^2 IkT}{h^2\sigma} \tag{10}$$

We call σ the symmetry factor. It is unity for diatomic molecules whose components are unlike, e.g., C^{12}-C^{13}. It is 2 for diatomic molecules whose constituents are identical, e.g., C^{12}-C^{12}. The number of rotational states is halved for a symmetrical molecule.

The spacing of the vibrational levels, if independent of rotation, will usually follow a law similar to

$$\epsilon_v = (v + \tfrac{1}{2})hc\omega_e \tag{11}$$

where ω_e is a basic vibrational wave number and v an integral vibrational quantum number.

Equation (11) measures ϵ_v from the lowest point of the potential energy curve. If, instead, we measure it from the condition of complete dissociation, we get

$$\epsilon_v = \left(v + \frac{1}{2}\right)hc\omega_e - D - \frac{hc\omega_e}{2} = hc\omega_e v - D \tag{12}$$

with D equal to the dissociation energy. Then

$$B_v(T) = \sum_0^\infty e^{D - hc\omega_e v/kT} = \frac{e^{D/kT}}{1 - e^{-hc\omega_e/kT}} \tag{13}$$

We cannot usually express the electronic energy in any simple form. We must therefore merely write

$$B_e(T) = \sum_j \tilde{\omega}_j e^{-\epsilon_j/kT} \tag{14}$$

with the sum taken over all relevant levels of electronic excitation. For atoms we take the statistical weight of a single state to be unity. If, as in common practice, we represent a level by its inner quantum number, J,

$$\tilde{\omega}_j = 2J + 1 \tag{15}$$

If a complete term, denoted by the orbital and spin quantum numbers L and S, is degenerate,

$$\tilde{\omega}_j = (2S + 1)(2L + 1) \tag{16}$$

etc. (see Chapter 19).

A spinning electron has the invariant weight:

$$\tilde{\omega}_j = 2 \tag{17}$$

For molecules we encounter electronic states characterized by the quantum numbers S and Λ. $\Lambda = 0$ corresponds to Σ levels; $\Lambda = 1$ to Π; $\Lambda = 2$ to Δ, etc. If we set $r = 2S + 1$, where r is the multiplicity, we get

$$\left.\begin{aligned} \tilde{\omega}_j &= 2S + 1 = r && \text{for } {}^r\Sigma \text{ levels} \\ \tilde{\omega}_j &= 2(2S + 1) = 2r, && \text{for } {}^r\Pi,\ {}^r\Delta, \text{ etc. levels} \end{aligned}\right\} \tag{18}$$

The complete partition function for a diatomic molecule whose components possess masses m_1 and m_2 becomes

$$f(T) = \frac{[2\pi(m_1 + m_2)kT]^{3/2}}{h^3} V(T) \frac{1}{B\sigma} \cdot \frac{kT}{hc} \cdot \frac{e^{D/kT}}{(1 - e^{-hc\omega_e/kT})} B_e(T) \tag{19}$$

1.2. Equations of state. For a perfect gas, the equation of state becomes

$$p = \frac{2}{3} \cdot \frac{\bar{E}_{\text{kin}}}{V} = NkT \frac{\partial}{\partial V} \ln f(T) \tag{1}$$

Since V usually enters into $f(T)$ only as a multiplicative parameter, we get

$$p = \frac{N}{V} kT = nkT \tag{2}$$

where n is the number of systems per unit volume. If we take V as the volume of a gram molecule and set N equal to the number of atoms per mol,

$$pV = NkT = R_0 T \tag{3}$$

If ρ is the density,

$$p = \rho kT/m_0\mu \tag{4}$$

where m is the mass of an atom of unit atomic weight, and μ the atomic weight of the gas.

For an imperfect gas the best-known equation is that of van der Waals:

$$(p + a/V^2)(V - b) = NkT \tag{5}$$

Berthelot gives an alternative empirical equation:

$$(p + a'/TV^2)(V - b) = NkT \tag{6}$$

and Dieterici still another, viz:

$$p(V - b) = NkTe^{-a'/NkT^sV} \tag{7}$$

where a, a', b, and s are all constants, the last being an exponent.

To the first order in $1/V$ these three equations assume the respective forms,

$$pV = NkT + \frac{(NkTb - a)}{V} \tag{8}$$

$$pV = NkT + \frac{(NkTb - a'/T)}{V} \tag{9}$$

$$pV = NkT + \frac{(NkTb - a'/T^{s-1})}{V} \tag{10}$$

1.3. Energies and specific heats of a one-component assembly. If E is the total energy assigned to the N systems of a one-component assembly,

$$E = NkT^2 \frac{d}{dT} \ln f(T) \tag{1}$$

The mean kinetic energy per particle is

$$\bar{\epsilon} = \frac{E_{\text{kin}}}{N} = kT^2 \frac{d}{dT} \ln H(T) = \frac{3}{2} kT \tag{2}$$

The specific heats per mol at constant volume C_v, and at constant pressure C_p are

$$\left.\begin{aligned} C_V &= \left(\frac{\partial E}{\partial T}\right)_V = \left[\frac{\partial}{\partial T} M_0 kT^2 \frac{\partial}{\partial T} \ln f(T)\right]_V \\ C_p &= \left[\frac{\partial}{\partial T}(E + pV)\right]_p = C_V + \left[\frac{\partial}{\partial T} pV\right]_p \end{aligned}\right\} \tag{3}$$

where the subscripts indicate that V and p are held constant during the respective differentiations; M_0 is the number of particles per mol.

1.4. Adiabatic processes. Let γ be the ratio

$$\gamma = C_p/C_V \tag{1}$$

The so-called adiabatic law, wherein no heat is added or subtracted from a gas is

$$pV^\gamma = \text{constant} \tag{2}$$

$$pT^{\gamma/(1-\gamma)} = \text{constant} \tag{3}$$

$$\rho T^{1/(1-\gamma)} = \text{constant} \tag{4}$$

In each of these equations we regard V as the volume of a mol (or some other conveniently defined mass of gas.)

1.5. Maxwell's and Boltzmann's laws. The number of systems in a given quantum state is

$$N_i = N \frac{\tilde{\omega}_i e^{-\varepsilon_i/kT}}{B(T)} \tag{1}$$

which is Boltzmann's law. Also

$$\frac{N_i}{N_j} = \frac{\tilde{\omega}_i}{\tilde{\omega}_j} e^{-(\varepsilon_i - \varepsilon_j)/kT} \tag{2}$$

In a classical assembly, the number, ϕ of systems in a volume element dV, with components of momenta in dp_1, dp_2, dp_3, becomes

$$\left.\begin{aligned} \phi dp_1 dp_2 dp_3 dV &= \frac{N}{H(T)V} e^{-(p_1^2+p_2^2+p_3^2)/2mkT} \frac{dp_1 dp_2 dp_3 dV}{h^3} \\ &= \frac{N}{V} \cdot \frac{1}{(2\pi mkT)^{3/2}} e^{-(p_1^2+p_2^2+p_3^2)/kT} dp_1 dp_2 dp_3 dV \end{aligned}\right\} \tag{3}$$

normalized to give

$$\int \phi dp_1 dp_2 dp_3 dV = N/V \tag{4}$$

where the integral is over a unit volume, with the respective momenta limits $-\infty$ to $+\infty$.

In terms of Cartesian velocity components, u, v, w, we have the alternative form of Maxwell's equation :

$$\phi du dv dw dV = \frac{N}{V}\left(\frac{m}{2\pi kT}\right)^{3/2} e^{-m(u^2+v^2+w^2)/2kT} du dv dw dV \tag{5}$$

If c is the actual space velocity, such that

$$c^2 = u^2 + v^2 + w^2 \tag{6}$$

the number of systems moving with velocities in dc and within solid angle $d\omega$ are

$$f dc d\omega = \frac{N}{V}\left(\frac{m}{2\pi kT}\right)^{3/2} c^2 e^{-mc^2/2kT} dc d\omega \tag{7}$$

To get the total number per cm^3, with velocities in dc, replace $d\omega$ by 4π.

For additional theorems on kinetic theory of gases see Chapter 20, pp. 288-302.

1.6. Compound and dissociating assemblies. When an assembly consists of more than one component, interacting or not, we write, instead of Eq. (1) of § 1.3,

$$E = \sum_j N_j k T^2 \frac{d}{d\pi} \ln f_j(T) \tag{1}$$

where the subscript j refers to a given component. The partition function is the same as before. For noninteracting components N_j is constant. Otherwise, N_j will depend on T and V according to the dissociation equation. For example, when N_m atoms of component m interact with N_n atoms of component n to form N_s compound molecules of component s, in accord with the symbolic chemical equation

$$M + N \rightleftarrows S \tag{2}$$

where S is a molecule of type MN, the equilibrium conforms to the law

$$\frac{N_m N_n}{N_s} = \frac{f_m(T) f_n(T)}{f_s(T)} \tag{3}$$

In a more general case, when N_m, N_n, N_p, ..., etc., interact to form a compound molecule, so that α atoms of type m, β of type n, γ of type p, etc., are required to form μ compound molecules of type s, according to the reaction

$$\alpha M + \beta N + \gamma P + \ldots = \mu S = \mu[M_{\alpha/\mu} N_{\beta/\mu} P_{\gamma/\mu} \cdots] \tag{4}$$

in ordinary chemical notation. The general dissociation equation becomes

$$\frac{N_m{}^{\alpha} N_n{}^{\beta} N_p{}^{\gamma} \ldots}{N_s{}^{\mu}} = \frac{(f_m)^{\alpha}(f_n)^{\beta}(f_p)^{\gamma} \ldots}{(f_s)^{\mu}} \tag{5}$$

in which the superscripts α, β, γ, ..., μ are true algebraic exponents. For more detailed forms of these equations, including the Saha "ionization formula," see Chapter 21.

1.7. Vapor pressure. The vapor pressure of a gas in equilibrium with its solid or liquid phase is

$$p = \frac{kT}{V} \cdot \frac{f(T)}{\kappa(T)} \tag{1}$$

For structureless systems, the partition function

$$f(T) = \frac{(2\pi mkT)^{3/2}}{h^3} V e^{-\chi/kT} \tag{2}$$

where χ is the energy required to evaporate an atom from the liquid or solid to the gaseous phase at absolute zero. The partition function for the crystal is

$$\ln \kappa(T) = \int_0^{T/V} \frac{dT}{RT^2} \int_0^T (C_p)_{\text{sol}} dT' \tag{3}$$

where $(C_p)_{\text{sol}}$ is the specific heat of the solid or condensed phase, calculated at the vapor pressure of the gas for the value of T in the integrand; $(C_p)_{\text{sol}}$ also includes heats of transition (melting, evaporation, etc.)

For molecular gases or other gases having internal structure, the appropriate factors relation to rotations, vibrations, electronic energies, etc., must be included in $f(T)$. For many applications we can set, approximately,

$$\kappa(T) \sim H(T) \tag{4}$$

see Eq. (6) of § 1.1.

More generally we note that the empirical relation,

$$p = CT^{\alpha}e^{-X/T} \tag{5}$$

with three disposable constants, C, α, and X usually gives a very accurate representation between any two given transition points.

1.8. Convergence of partition functions. The partition function $B_e(T)$, Eq. (14) of § 1.1, does not converge if we sum over all quantum numbers to $n = \infty$. The presence of neighboring atoms fixes an effective upper limit to the summation, beyond which we regard the electron as free rather than bound. Thus we excluded the volume assigned to other atoms. Denote by the subscript s the lower electronic states of a given atom, and by r the lower electron states of the same atom in the next higher ionization stage. Approximately

$$B(T) \sim \sum_s \tilde{\omega}_s e^{-\varepsilon_s/kT} + \tilde{\omega}_\varepsilon \left[\frac{64}{9\pi}\left(\frac{3}{4\pi}\right)^{1/2} \frac{(Rhc)^{3/2}}{\epsilon^3}\right] \frac{Z^{3/2}}{n_0^{1/2}} e^{-\chi_0/kT} \sum \tilde{\omega}_r e^{-\varepsilon_r/kT} \tag{1}$$

The bracketed constant, we shall term Q. Its value is

$$Q \sim 1.02 \times 10^{12} \text{ cm}^{-3/2} \tag{2}$$

In this equation, $\tilde{\omega}_\varepsilon$ is the electron spin, Z is the effective charge of the higher ionization state involved, χ_0 is the ionization potential, and n_0 is the number of systems per cm^3. Usually only the first or the second term of (1) will dominate.

1.9. Fermi-Dirac and Bose-Einstein statistics. The foregoing formulas, especially those relating to equations of state and partition of kinetic energy, involve the tacit assumption that the energies of such a system are not quantized. This assumption is sufficiently correct at low densities and high temperatures. However, at high densities and low temperatures, we find that even the so-called " continuous states " are quantized and, furthermore, that we can assign only one particle to each state. The Pauli exclusion principle is a consequence of this general law.

For systems whose wave functions are antisymmetric, we get the Fermi-Dirac statistics (indicated by the upper sign in the following formulas). When the wave functions are symmetric, we get the Bose-Einstein statistics (lower signature).

In our given volume V, which contains N systems in all, we find that N_s are assigned to the energy level ϵ_s, whose statistical weight is $\tilde{\omega}_s$. Then

$$N_s = \pm \tilde{\omega}_s \lambda \frac{\partial}{\partial \lambda} \ln (1 \pm \lambda e^{-\varepsilon_s/kT}) = \frac{\tilde{\omega}_s}{\lambda^{-1} e^{\varepsilon_s} \pm 1} \tag{1}$$

The parameter λ follows from the condition that

$$\sum_s N_s = N = \sum_s \frac{\tilde{\omega}_s}{\lambda^{-1} e^{\varepsilon_s/kT} \pm 1} \tag{2}$$

For Bose-Einstein statistics we must have

$$0 < \lambda \leqslant 1 \tag{3}$$

For Fermi-Dirac, we have the condition

$$\lambda > 0 \tag{4}$$

To complete these equations we must have formulas for the separations ϵ_s. For kinetic energy in a cubical volume bounded by a side of length l,

$$\epsilon_s = \frac{p^2}{2m} = \frac{h^2}{8l^2 m}(x^2 + y^2 + z^2) \tag{5}$$

where x, y, and z are integers $0 < x,y,z \leqslant \infty$. Then

$$N = \sum_{x=1}^{\infty} \sum_{y=1}^{\infty} \sum_{z=1}^{\infty} \frac{\tilde{\omega}}{\lambda^{-1} e^{h^2(x^2+y^2+z^2)/8l^2 mkT} \pm 1} \tag{6}$$

where $\tilde{\omega}$ is a weight factor set equal to unity for longitudinal and equal to 2 for transverse waves. The quantity $h^2/8l^2 mkT = 4.4 \times 10^{-11}$, if $l = 1$ cm and $T = 1^\circ$ K. Hence we can approximate to the sum by an integral

$$\left. \begin{aligned} N &= \int_0^\infty \int_0^\infty \int_0^\infty \frac{\tilde{\omega}\, dx\, dy\, dz}{\lambda^{-1} e^{h^2(x^2+y^2+z^2)/8l^2 mkT} \pm 1} \\ &= \int_0^{\pi/2} \int_0^{\pi/2} \int_0^\infty \frac{\tilde{\omega} r^2 \sin\theta \, dr\, d\theta\, d\varphi}{\lambda^{-1} e^{h^2 r^2/8l^2 mkT} \pm 1} \\ &= \frac{\pi}{2} \int_0^\infty \frac{\tilde{\omega} r^2 dr}{\lambda^{-1} e^{hr^2/8l^2 mkT} \pm 1} = \frac{2\pi\tilde{\omega}(2m)^{3/2} V}{h^3} \int_0^\infty \frac{\epsilon_s^{1/2}}{\lambda^{-1} e^{\epsilon_s/kT} \pm 1} d\epsilon_s \end{aligned} \right\} \tag{7}$$

where we set

$$x^2 + y^2 + z^2 = r^2 \tag{8}$$

and transform from Cartesian to spherical coordinates. We have set $l^3 = V$.

We cannot integrate again in finite terms. However, when $\lambda \ll 1$,

$$N \sim \frac{2\pi\tilde{\omega}(2m)^{3/2}V\lambda}{h^3} \int_0^\infty e^{-\varepsilon_s/kT}\epsilon_s^{1/2}d\epsilon_s = \frac{\tilde{\omega}(2\pi mkT)^{3/2}V\lambda}{h^3} \tag{9}$$

Thus, when

$$\lambda = \frac{N}{\tilde{\omega}V} \cdot \frac{h^3}{(2\pi mkT)^{3/2}} \ll 1 \tag{10}$$

$$N_s = N \frac{2\pi(2m)^{1/2}}{(2\pi mkT)^{3/2}} e^{-\varepsilon_s/kT}\epsilon_s^{1/2}d\epsilon_s \tag{11}$$

a form of Maxwell's law, expressed in terms of energies instead of momenta or velocities as before. Thus, when $\lambda \ll 1$ we recover the classical formulas.

When $\lambda \gg 1$ (for the Fermi-Dirac case only)

$$N \sim \frac{2\pi\tilde{\omega}(2m)^{3/2}V}{h^3} \int_0^{\varepsilon'} \epsilon_s^{1/2}\, d\epsilon_s = \frac{4\pi}{3\pi^{1/2}} \frac{(2\pi mkT)^{3/2}V}{h^3} (\ln \lambda)^{3/2} \tag{12}$$

where ϵ' is defined by

$$\lambda^{-1}e^{\varepsilon'/kT} = 1 \tag{13}$$

For large λ, then, the Fermi-Dirac distribution becomes

$$N_s = 2\pi\tilde{\omega}(2m)^{3/2}V\epsilon_s^{1/2}d\epsilon_s \tag{14}$$

for the energy range below ϵ'. For higher energies, the formula grades into the classical expression.

For the Bose-Einstein case, $\lambda \leqslant 1$, we get

$$N \sim \frac{\tilde{\omega}(2\pi mkT)^{3/2}V}{h^3} \sum_{j=1}^{\infty} \frac{\lambda^j}{j^{3/2}} \tag{15}$$

1.10. Relativistic degeneracy. At extremely high temperatures we must allow for relativistic change of mass with velocity. The deBroglie wavelength of a particle moving with velocity v is

$$\lambda_v = \frac{h}{m_0 v}\left(1 - \frac{v^2}{c^2}\right)^{1/2} \tag{1}$$

where m_0 is the " rest mass." The kinetic energy is

$$\epsilon_s = m_0c^2 \left[\left(1 - \frac{v^2}{c^2}\right)^{-1/2} - 1\right] \tag{2}$$

or

$$\frac{1}{\lambda_v^2} = \frac{2\epsilon_s m_0(1 + \epsilon_s/2m_0c^2)}{h^2} \tag{3}$$

when relativity effects are predominant,

$$\epsilon_s \gg 2m_0c^2 \tag{4}$$

and

$$\epsilon_s = \frac{hc}{\lambda v} = h\nu_s \tag{5}$$

Planck's relation. The energy is independent of m_0.

For Fermi-Dirac statistics we get, for a relativistically degenerate gas,

$$N = \frac{4\pi\tilde{\omega}V}{h^3c^2}\int_0^\infty \frac{\epsilon_s^{\,2}}{\lambda^{-1}e^{\epsilon_s/kT}+1}\,d\epsilon_s \tag{6}$$

When $\lambda \ll 1$,

$$N \sim 8\pi\tilde{\omega}V\lambda\left(\frac{kT}{hc}\right)^3 \tag{7}$$

When $\lambda \gg 1$,

$$N \sim \frac{4\pi\tilde{\omega}V}{3}\left(\frac{kT}{hc}\ln\lambda\right)^3 \tag{8}$$

1.11. Dissociation laws for new statistics. Let

$$\beta_m = m_m/m\epsilon \tag{1}$$

the ratio of the mass of the system to that of the electron.

For each component m of an assembly find the appropriate value of λ_m according to the conditions of temperature and density.

Classical nonrelativistic:

$$\lambda = \frac{Nm}{\tilde{\omega}_m V}\cdot\frac{h^3}{(2\pi m_m kT)^{3/2}} \tag{2}$$

conditions: $\left\{\begin{array}{l} N_m/V \ll \tilde{\omega}_m/h^3(2\pi m_m kT)^{3/2} = 4.87 \times 10^{15}(\beta_m T)^{3/2} \\ T \ll m_m c^2/k = 5.9 \times 10^9\beta_m \end{array}\right\}$

Classical, relativistic:

$$\lambda_m = \frac{N_m}{8\pi\tilde{\omega}_m V}\left(\frac{hc}{kT}\right)^3 \tag{3}$$

conditions: $\left\{\begin{array}{l} N_m/V \ll 4.87 \times 10^{15}(\beta_m T)^{3/2} \\ T \gg 5.9 \times 10^9\beta_m \end{array}\right\}$

Fermi-Dirac degeneracy, nonrelativistic:

$$\ln\lambda_m = \frac{h^2}{2m_m kT}\left(\frac{3}{4\pi\tilde{\omega}_m}\frac{N_m}{V}\right)^{2/3} \tag{4}$$

conditions: $\left\{\begin{array}{l} N_m/V \gg 4.87 \times 10^{15}(\beta_m T)^{3/2} \\ T \ll 5.9 \times 10^9\beta_m \end{array}\right\}$

Bose-Einstein degeneracy, nonrelativistic:

$$\sum_{j=1}^{\infty} \frac{\lambda_m{}^j}{j^{3/2}} = \frac{N_m}{\tilde{\omega}_m V} \frac{h^3}{(2\pi m_m kT)^{3/2}} \tag{5}$$

Conditions : same as for Fermi-Dirac, above.

Fermi-Dirac degeneracy, relativistic (electron gas):

$$\ln \lambda_m = \left(\frac{3N_m}{4\pi\tilde{\omega}_m V}\right)^{1/3} \frac{hc}{kT} \tag{6}$$

conditions : $\begin{cases} N_m/V \gg (kT/hc)^3\,(4\pi\tilde{\omega}_m/3) = 2.86T^3 \\ T \gg 5.9 \times 10^9/\ln \lambda_m \end{cases}$

These equations ignore internal excitation energies. Now in a reaction such as

$$M + N \rightleftarrows S \tag{7}$$

where S is a molecule of type MN, of dissociation energy Q_s, we can determine the equilibrium from the equation

$$\lambda_s = \lambda_m \lambda_n e^{Q_s/kT} \tag{8}$$

For the more complex reaction

$$\alpha M + \beta N \rightleftarrows \gamma S \tag{9}$$

with α, β, and γ integers, in which the chemical formula for S is $M_{\alpha/\gamma}N_{\beta/\gamma}$, we get

$$\lambda_s{}^\gamma = \lambda_m{}^\alpha \lambda_n{}^\beta e^{Q_s/kT} \tag{10}$$

1.12. Pressure of a degenerate gas. The general equation of state still holds for nonrelativistic statistics :

$$p = \frac{2}{3} \cdot \frac{\bar{E}_{\text{kin}}}{V} \tag{1}$$

$$\bar{E}_{\text{kin}} = \frac{2\pi\tilde{\omega}(2m)^{3/2}V}{h^3} \int_0^\infty \frac{\epsilon_s{}^{3/2}}{\lambda^{-1}e^{\epsilon_s/kT} \pm 1}\, d\epsilon_s \tag{2}$$

For small λ, we recover the classical expression

$$p = NkT/V \tag{3}$$

For $\lambda \gg 1$, for Fermi-Dirac statistics,

$$p = \frac{4\pi}{15} \frac{h^2\tilde{\omega}}{m} \left(\frac{3}{4\pi\tilde{\omega}} \cdot \frac{N}{V}\right)^{5/3} + \frac{2\pi^3}{9} \frac{m(kT)^2}{h^2} \left(\frac{3}{4\pi\tilde{\omega}} \frac{N}{V}\right)^{1/3} + \ldots \tag{4}$$

Note that the leading term gives a pressure independent of T.

For Bose-Einstein statistics

$$p = \tilde{\omega} \frac{(2\pi m)^{3/2}(kT)^{5/2}}{h^3} \sum_1^\infty \frac{\lambda^j}{j^{3/2}} \tag{5}$$

For relativistic statistics,

$$p = \frac{\bar{E}_{\text{kin}}}{3V} \tag{6}$$

$$\bar{E}_{\text{kin}} = \frac{4\pi\tilde{\omega}V}{h^3c^3} \int_0^\infty \frac{\epsilon_s^3}{\lambda^{-1}e^{\epsilon_s/kT} \pm 1} d\epsilon_s \tag{7}$$

When $\lambda \ll 1$,

$$p = NkT/V \tag{8}$$

When $\lambda \gg 1$, Fermi-Dirac,

$$p = \frac{hc}{4} \left(\frac{3}{4\pi\tilde{\omega}}\right)^{1/3} \left(\frac{N}{V}\right)^{4/3} \tag{9}$$

1.13. Statistics of light quanta. Quanta obey the Bose-Einstein relativistic statistics with $\lambda = 1$. This condition is equivalent to the assumption that total number of quanta are not conserved.

$$\bar{E}_{\text{kin}} = \frac{4\pi\tilde{\omega}Vh}{c^3} \int_0^\infty \frac{\nu^3}{e^{h\nu/kT} - 1} d\nu \tag{1}$$

Set $\tilde{\omega} = 2$ for transverse waves. Then the energy density of ν radiation becomes

$$\rho_\nu d\nu = \frac{8\pi h\nu^3}{c^3} \frac{1}{e^{h\nu/kT} - 1} d\nu \tag{2}$$

For additional radiation formulas, see Chapter 21.

Bibliography

1. CHANDRASEKHAR, S., *Monthly Notices Roy. Astronom. Soc.*, **95**, 225 (1935).
2. FOWLER, R. H. and GUGGENHEIM, E. A., *Statistical Mechanics*, Cambridge University Press, London, 1939.

Chapter 12

KINETIC THEORY OF GASES: VISCOSITY, THERMAL CONDUCTION, AND DIFFUSION

BY SYDNEY CHAPMAN

Oxford University

The kinetic theory of gases was the pioneer branch of statistical mechanics, applied to the motion of gas molecules moving freely except during the brief fraction of time occupied by collisions. Kinetic theory thus provides the necessary connection between the microscopic viewpoint of the molecules on one hand and the macroscopic viewpoint of hydrodynamics on the other. The formulas here given were chosen as those most likely to be useful to the working physicist, especially those who like to know the steps leading the mathematicians to the results as well as the results themselves. Some of the following formulas, as, for example, those for diffusion, are not so generally well known as they should be.

1. Preliminary Definitions and Equations for a Mixed Gas, Not in Equilibrium

1.1. $\boldsymbol{r}, x, y, z, t, d\boldsymbol{r}$. Position is indicated by the position vector $\boldsymbol{r}$ or by orthogonal coordinates x, y, z; and time by t. A small volume element enclosing the point $\boldsymbol{r}$ is denoted by $d\boldsymbol{r}$.

1.2. $m, n, \rho, n_{10}, n_{12}, m_0$. A mixed gas of two constituents is considered; suffixes 1,2 refer to the two constituents, e.g., m_1, m_2 denote the molecular masses, n_1, n_2 the number densities (or numbers per cc), ρ_1, ρ_2 the mass densities n_1m_1, n_2m_2. The total number or mass density is denoted by n or ρ.

$$n \equiv n_1 + n_2, \quad \rho = \rho_1 + \rho_2 \equiv n_1m_1 + n_2m_2 \tag{1}$$

Also let

$$n_{10} = n_1/n, \quad n_{20} = n_2/n, \quad n_{12} = n_1/n_2 \tag{2}$$

so that

$$n_{10} + n_{20} = 1 \tag{3}$$

and let

$$m_0 = m_1 + m_2, \quad m = \rho/n \tag{4}$$

so that m_0 is the combined, and m the mean, molecular mass.

Thus the suffix 1 or 2 refers to one constituent only, and without the suffix the reference is to the whole gas (or to a simple gas); this is here called the suffix convention and is to be applied in each subsequent section concluded by (S.C.) meaning *suffix convention.*

1.3. The external forces $\boldsymbol{F}$, X, Y, Z, Ψ. Every molecule is supposed to be subject to an external force $m\boldsymbol{F}$, with components mX, mY, mZ, the same for all those of one kind ($m_1\boldsymbol{F}_1$ or $m_2\boldsymbol{F}_2$). Moreover it is supposed that these are conservative forces with potentials Ψ_1, Ψ_2, so that

$$\boldsymbol{F}_1 = -\operatorname{grad}\Psi_1, \qquad \boldsymbol{F}_2 = -\operatorname{grad}\Psi_2, \qquad \text{(S.C.)} \tag{1}$$

1.4. $\boldsymbol{c}$, u, v, w, $d\boldsymbol{c}$. The velocity of a typical molecule is denoted by $\boldsymbol{c}$, and its components by u, v, w; these may be regarded as specifying a point in a *velocity space*, referring to molecules of one kind only, if suffix 1 or 2 is added, otherwise to molecules of both kinds (or to those of a simple gas). A small volume element in the velocity space, enclosing the *velocity point* $\boldsymbol{c}$, is denoted by $d\boldsymbol{c}$. (S.C.)

1.5. The velocity distribution function f. Each of the $nd\boldsymbol{r}$ molecules in the space volume element $d\boldsymbol{r}$ at time t has its own velocity point; the number of these within the velocity volume element $d\boldsymbol{c}$ is denoted by $fd\boldsymbol{r}d\boldsymbol{c}$, that is, $fd\boldsymbol{r}$ is the number density at $\boldsymbol{c}$ in the velocity space; f is a function of $\boldsymbol{c}$, $\boldsymbol{r}$, t in general, and to indicate this it is also written $f(\boldsymbol{c},\boldsymbol{r},t)$; it is called the *velocity distribution function.* (S.C.)

1.6. Mean values of velocity functions. The average of any function $F(\boldsymbol{c})$, vector or scalar, e.g., $\boldsymbol{c}$ or u or uv or c^2, over all the molecules (of either or both kinds) in $d\boldsymbol{r}$ at $\boldsymbol{r}$, t, is indicated by the function symbol with an overline, e.g., $\bar{\boldsymbol{c}}$, $\overline{F(\boldsymbol{c})}$, $\bar{u}$, $\overline{uv}$, $\bar{c}$. Clearly

$$n\overline{F(\boldsymbol{c})} = \int F(\boldsymbol{c})f(\boldsymbol{c},\boldsymbol{r},t)d\boldsymbol{c} \tag{1}$$

the integral being taken over the whole velocity space. (S.C.)

In consequence of this definition,

$$n\bar{\boldsymbol{c}} = n_1\bar{\boldsymbol{c}}_1 + n_2\bar{\boldsymbol{c}}_2, \quad n\bar{u} = n_1\bar{u}_1 + n_2\bar{u}_2, \quad \text{etc.} \tag{2}$$

1.7. The mean mass velocity $\boldsymbol{c}_0$; u_0, v_0, w_0. The mean momentum per molecule, for the whole gas, is $(n_1m_1\bar{\boldsymbol{c}}_1 + n_2m_2\bar{\boldsymbol{c}}_2)/n$; this is denoted by $m\boldsymbol{c}_0$; and u_0, v_0, w_0 denote the components of $\boldsymbol{c}_0$. Clearly

$$\rho\boldsymbol{c}_0 = \rho_1\bar{\boldsymbol{c}}_1 + \rho_2\bar{\boldsymbol{c}}_2 \tag{1}$$

1.8. The random velocity $\boldsymbol{C}$; U, V, W; $d\boldsymbol{C}$. The *random* velocity $\boldsymbol{C}$ of a typical molecule, and its components U, V, W, are defined by

$$\boldsymbol{C} = \boldsymbol{c} - \boldsymbol{c}_0, \quad U = u - u_0, \quad \text{etc.} \tag{1}$$

Clearly

$$\bar{\boldsymbol{C}}_1 = \bar{\boldsymbol{c}}_1 - \boldsymbol{c}_0, \quad \bar{\boldsymbol{C}}_2 = \bar{\boldsymbol{c}}_2 - \boldsymbol{c}_0, \quad \bar{\boldsymbol{C}} = \bar{\boldsymbol{c}} - \boldsymbol{c}_0 \tag{2}$$

If the origin in the velocity space is moved to the velocity point $\boldsymbol{c}_0$, the velocity position vector of a typical molecule is changed to $\boldsymbol{C}$, and the distribution function f can be regarded alternatively as a function of $\boldsymbol{C}$, $\boldsymbol{r}$, t, namely $f(\boldsymbol{C},\boldsymbol{r},t)$; if $d\boldsymbol{C}$ denotes a velocity volume element enclosing the velocity point $\boldsymbol{C}$, the equation for $\overline{F(\boldsymbol{c})}$ in § 1.5 can alternatively be written

$$n\overline{F(\boldsymbol{c})} = \int F(\boldsymbol{c})f(\boldsymbol{C},\boldsymbol{r},t)d\boldsymbol{C}, \quad \text{(S.C.)} \tag{3}$$

1.9. The "heat" energy of a molecule E: its mean value $\bar{E}$. The heat energy E of a molecule is defined to be the energy $\frac{1}{2}mC^2$ of its random translatory energy, together with any additional energy, either rotatory or internal (e.g., vibratory), *which is communicable from one molecule to another on encounter*; for example, this would not include any rotatory energy of a smooth, rigid, elastic, spherical molecule. Its value at $\boldsymbol{r}$, t is denoted by $\bar{E}$. Naturally $\bar{E}$ is a function of the (absolute or Kelvin) temperature T.

1.10. The molecular weight W and the constants m_u, N_L. A molecule of mass m is said to have molecular "weight" W equal to m/m_u, where m_u ($= 1.6603 \times 10^{-24}$ gram), the unit atomic mass, is one-sixteenth the mass of the O^{16} isotope of oxygen. A mass of W grams of a gas of (mean) molecular weight W is called a gram molecule or mole; it contains N_L molecules, where

$$N_L = 1/m_u = 6.023 \times 10^{23} \tag{1}$$

N_L is called Loschmidt's number (or sometimes, less appropriately, Avogadro's number; see § 2.3).

1.11. The constants J, k, R. The following constants are used in later sections, e.g., 1.12.

$J = 4.185 \times 10^7$ ergs per calorie (Joule's mechanical equivalent of heat).
$k = 1.3805 \times 10^{-16}$ erg per degree C (Boltzmann's constant).
$R = kN_L = 8.314 \times 10^7$ ergs per degree C per mole (the gas constant in mechanical units).
$R = kN_L/J = 1.9865$ calorie per degree C per mole (the gas constant in heat units).

1.12. The kinetic theory temperature T is defined by

$$kT = \tfrac{1}{3}m\overline{C^2} \tag{1}$$

whether or not the gas be in thermodynamic equilibrium (see § 2.3).

1.13. The symbols c_v, C_v denote the specific heat at constant volume, per gram (c_v) or per mole (C_v) : that is, the heat required to raise the temperature of this mass of gas by 1° C, at constant volume; this is indicated formally by

$$c_v = (1/m)\,(d\bar{E}/dT)_v, \quad C_v = Wc_v = (1/m_u)\,(d\bar{E}/dT)_v = N_L(d\bar{E}/dT)_v \tag{1}$$

If the molecules have no communicable rotatory or internal energy, $\bar{E} = \tfrac{1}{2}m\overline{C^2} = \tfrac{3}{2}kT$, and $c_v = 3k/2m$; and $C_v = 3k/2m_u$, the same for all gases. If E is a constant multiple (s) of $\tfrac{1}{2}m\overline{C^2}$, so that $E = \tfrac{3}{2}skT$, $c_v = 3ks/2m$, $C_v = 3ks/2m_u$; for many diatomic molecules at ordinary temperatures $s = \tfrac{5}{3}$, as they possess rotatory energy of amount two-thirds the random translatory kinetic energy (see § 2.4 for c_p, C_p, γ).

1.14. The stress distribution, p_{xx}, p_{xy}, ... The x, y, z components of the stress (force per unit area) exerted at $\boldsymbol{r}$, t, *upon* the gas on the positive side of a plane through $\boldsymbol{r}$ normal to Ox, *by* the gas on the negative side, are denoted by p_{xx}, p_{xy}, p_{xz}; similarly p_{yx}, p_{yy}, p_{yz}, and p_{zx}, p_{zy}, p_{zz} denote the stress components across planes normal to Oy and Oz. These nine components are the elements of the stress *tensor*; p_{xx}, p_{yy}, p_{zz} are *normal* stresses; the other six are *tangential* stresses.

In the interior of a gas of ordinary low density (e.g., normal air) these stresses arise mainly from the transfer of momentum by molecules crossing the planes concerned; this produces the *kinetic stresses*. In addition there is a much smaller part due to intermolecular forces; a gas in which this part is negligible is called a perfect gas. The gas here considered will be supposed perfect, but the equation of state will be given also for a slightly imperfect gas (§ 2.5).

In a perfect mixed gas the constituents contribute independently to the stresses :

$$\left.\begin{aligned} p_{xx} &= (p_{xx})_1 + (p_{xx})_2, & p_{xy} &= (p_{xy})_1 + (p_{xy})_2, \quad \text{etc.} \\ (p_{xx})_1 &= \rho_1\overline{U_1^{\,2}}, & (p_{xy})_1 &= \rho_1\overline{U_1V_1}, \quad \text{etc.} \end{aligned}\right\} \tag{1}$$

Hence (see § 1.6)

$$p_{xx} = \rho_1\overline{U_1^{\,2}} + \rho_2\overline{U_2^{\,2}} = \rho\overline{U^2}, \quad p_{xy} = \rho\overline{UV}, \quad \text{etc.} \tag{2}$$

1.15. The hydrostatic pressure p; the partial pressures p_1, p_2. The hydrostatic pressure p is defined as the mean of the three normal stresses; that is,

$$p = \tfrac{1}{3}(p_{xx} + p_{yy} + p_{zz}) = \tfrac{1}{3}\rho\overline{C^2} = knT \tag{1}$$

using § 1.12. The " partial " pressures p_1, p_2 caused by the separate constituents are similarly given by

$$p_1 = \tfrac{1}{3}\rho_1\overline{C_1^2}, \quad p_2 = \tfrac{1}{3}\rho_2\overline{C_2^2}, \quad p = p_1 + p_2 \tag{2}$$

1.16. Boltzmann's equation for f. Boltzmann showed that each velocity distribution function f_1, f_2, satisfies an integro-differential equation. This (e.g., for f_1) equates the following combination of partial derivatives of f_1,

$$\frac{\partial f_1}{\partial t} + u_1\frac{\partial f_1}{\partial x} + v_1\frac{\partial f_1}{\partial y} + w_1\frac{\partial f_1}{\partial z} + X_1\frac{\partial f_1}{\partial u_1} + Y_1\frac{\partial f_1}{\partial v_1} + Z_1\frac{\partial f_1}{\partial w_1} \tag{1}$$

to a multiple integral (not given here), which involves both f_1 and f_2, and represents the rate of change of f_1 caused by molecular encounters. In a perfect gas only simple encounters, each involving but two molecules, need be considered. The evaluation of this integral naturally, in general, involves knowledge as to the properties of the molecules, that is, as to their modes of interaction in encounters.

1.17. Summational invariants. Certain properties are unchanged by an encounter. These are (1) the number of molecules involved, namely two, (2) their combined momentum, namely $m_1\mathbf{c}_1 + m_1\mathbf{c}_1'$, $m_2\mathbf{c}_2 + m_2\mathbf{c}_2'$, or $m_1\mathbf{c}_1 + m_2\mathbf{c}_2$, according as the encounter is between like or unlike molecules, and (3) their combined energy, $E + E'$, or $\tfrac{1}{2}mC^2 + \tfrac{1}{2}m'C'^2$ (suffix convention) if the molecules have no rotatory or internal energy. These properties (number, $m\mathbf{c}$, and E) are called summational invariants of an encounter.

1.18. Boltzmann's H theorem. For a uniform gas whose molecules are spherical and possess only translatory kinetic energy, and are subject to no external forces, the ' differential ' side of Boltzmann's equation reduces to $\partial f_1/\partial t$. Boltzmann also considered in this case the variation of the function H defined by $\int f \ln f \, d\mathbf{c}$, integrated over the whole velocity space. He showed that $\partial H/\partial t$ is negative or zero, and that in the steady state, when it must be zero, $\ln f$ must be a summational invariant, and consequently (see § 1.16) of the form $a_1 + \mathbf{a}_1' . \ m\mathbf{c} + a_1''mC^2$, where a_1, $\mathbf{a}_1'$, a_1'' are arbitrary (scalar or vector) constants; the middle term is the scalar product of $\mathbf{a}_1'$ and $m\mathbf{c}$.

2. Results for a Gas in Equilibrium*

2.1. Maxwell's steady-state solutions. Maxwell determined f in three important special cases, of steady or quasi-steady states; Boltzmann improved the proofs by means of his H theorem (§ 1.18). Later he showed that H_0, the space integral of H over a given volume of gas in such a state, satisfies the equation

$$S = -kH_0 + \text{constant}$$

where S denotes the *entropy* of the gas. In all three of Maxwell's special states,

$$f_1 = n_1\left(\frac{m_1}{2\pi kT}\right)^{3/2} e^{-m_1C_1^2/kT} \tag{1}$$

which is known as the Maxwellian velocity distribution. His three cases are:

a. Uniform steady state of rest or uniform motion under no external forces; n_1, n_2, T, and $\boldsymbol{c}_0$ are all uniform.

b. Steady state at rest under external forces derived from potentials Ψ_1, Ψ_2 (1.3); in this case T and $\boldsymbol{c}_0$ are uniform, but not n_1 and n_2, which are given by

$$n_1 = (n_1)_0 e^{-m_1\Psi_1/kT}, \qquad n_2 = (n_2)_0 e^{-m_2\Psi_2/kT} \tag{2}$$

where $(n_1)_0$, $(n_2)_0$ are constants (the values of n_1 and n_2 at the points, if any, where Ψ_1 or Ψ_2 is zero).

c. A gas in uniform rotation or screw-motion subject to no external force (or to forces whose potentials are constant along the circles or spirals traced out by any point fixed relative to the gas). In this case T must be uniform, but not $\boldsymbol{c}_0$, which is given by

$$\boldsymbol{c}_0 = (\boldsymbol{c}_0)_0 + \boldsymbol{\omega x r}$$

(where ω denotes the angular velocity and $(\boldsymbol{c}_0)_0$ the constant velocity at the origin, and $\boldsymbol{\omega x r}$ denotes the *vector* product of and $\boldsymbol{r}$), nor n_1, n_2, which are given by

$$n_1 = (n_1)_0 e^{-m_1\Psi_0/kT} \tag{3}$$

where $\Psi_0 = -\frac{1}{2}\omega^2 d^2$, and d denotes the distance of the point $\boldsymbol{r}$ from the axis of the rotation or screw. Thus the density distribution is the same as if the gas were at rest in a field of (centrifugal) force with this potential.

* See also Chapters 10 and 11.

2.2. Mean values when f is Maxwellian. In this case the following mean values (§ 1.6) of various functions of $\boldsymbol{C}$ are readily found.

$$\bar{C}_1 = \left(\frac{8kT}{\pi m_1}\right)^{1/2}, \qquad \bar{U}_{1+} = \frac{1}{2}\bar{C}_1 \tag{1}$$

where $\bar{U}_{1+}$ denotes the mean value of any component of $\boldsymbol{C}_1$, e.g., $\boldsymbol{U}_1$, averaged over those molecules 1 for which this component is positive. Also

$$\overline{U_1^2} = \overline{V_1^2} = \overline{W_1^2} = \frac{1}{3}\overline{C_1^2} = \frac{kT}{m_1}, \quad \overline{U_1V_1} = \overline{V_1W_1} = \overline{W_1U_1} = 0 \tag{2}$$

so that

$$\left.\begin{aligned} (p_{xx})_1 &= (p_{yy})_1 = (p_{zz})_1 = p_1 = kn_1T \\ (p_{xy})_1 &= (p_{yz})_1 = (p_{zx})_1 = 0 \end{aligned}\right\} \tag{3}$$

Thus the stress distribution is hydrostatic, that is, normal and equal across all planes through $\boldsymbol{r}$.

The " root-mean-square " speed $\sqrt{\overline{C_1^2}}$ is given by

$$\sqrt{\overline{C_1^2}} = \sqrt{(3kT/m_1)} = 1.086\bar{C}_1 \tag{4}$$

Also

$$\tfrac{1}{2}m_1\overline{C_1^2} = \tfrac{1}{2}m_2\overline{C_2^2} = \tfrac{3}{2}kT \tag{5}$$

thus the mean random kinetic energy is the same for both kinds of molecule; this is a particular case of a very general theorem of *equipartition of energy*.

2.3. The equation of state for a perfect gas. The equation $p = knT$ may be written, in reference to a mass M of the gas, occupying a volume V, so that $n = M/mV$, in the form

$$pV = (kM/m)T \tag{1}$$

This is the *equation of state* for a perfect gas, that is, the equation connecting p, V, and T in the equilibrium state; it includes Boyle's law ($pV =$ constant at any given temperature), and Charles' law (V proportional to T at a given pressure). These laws, however, refer to the thermodynamic temperature, and it is necessary to show that in equilibrium (for which alone the thermodynamic temperature is defined) this is the same as the kinetic theory temperature T. This can be achieved by considering Carnot's cycle for a perfect gas.

The equation $p = knT$ then confirms, for a perfect gas, Avogadro's hypothesis that all gases in equilibrium, at the same p and T, contain the same number N_A of molecules in unit volume. For 1 cc at N.T.P. (0° C and

1 atmosphere pressure namely, 1.013×10^6 dynes/cm²) this number N_A is 2.687×10^{19}; it is known as Avogadro's number (or sometimes, less appropriately, as Loschmidt's number. See § 1.10).

2.4. Specific heats. The symbols c_p and $C_p\,(= Wc_p)$ denote the specific heat at constant pressure, per gram (c_p) or per mole (C_p); see § 1.13. That is, c_p is the heat required to raise 1 gram (containing $1/m$ molecules) by 1° C ($= dT$), at constant pressure. It includes, besides the heat c_v required to increase the mean molecular energy of the molecules, the mechanical energy (pdV/J in heat units) required to increase the volume by dV, against the constant pressure p, according to Charles' law. The equation of state (§ 2.3) gives, for $M = 1$ and $dT = 1°$ C,

$$pdV/J = k/Jm \tag{1}$$

Thus

$$c_p = c_v + \frac{k}{Jm} \tag{2}$$

$$C_p = C_v + \frac{Wk}{Jm} = C_v + \frac{N_L k}{J} = C_v + R \tag{3}$$

or $C_p - C_v = R$ (the same for all perfect gases); R is here expressed in heat units (see § 1.11).

The ratio c_p/c_v or C_p/C_v is denoted by γ; by § 1.13,

$$\gamma = 1 + \frac{k}{Jmc_v} = 1 + \frac{2}{3}s \tag{4}$$

Thus $\gamma = \frac{5}{3}$ for molecules with no rotatory or internal communicable energy, for which $s = 1$; for the diatomic molecules for which (at ordinary temperatures) $s = \frac{5}{3}$, $\gamma = 1.4$.

2.5. Equation of state for an imperfect gas. The simplest generalization of the equation of state $p = knT$ or $pV = (kM/m)T$ for a perfect gas is

$$p + a\rho^2 = knT(1 + b\rho\chi) \tag{1}$$

or, less accurately,

$$(p + a\rho^2)(1 - b\rho) = knT \tag{2}$$

which is van der Waals' equation. These equations allow for both the intermolecular forces and the finite size of molecules. For a simple gas in which the molecules are rigid, elastic, attracting spheres of diameter σ, exerting a force $F(r)$ at distance r,

$$b = \frac{2}{3}\cdot\frac{\pi\sigma^3}{m}, \qquad a = \frac{2}{3}\cdot\frac{\pi}{m^2}\int r^3F(r)dr \tag{3}$$

and to a first approximation

$$\chi = 1 + \tfrac{5}{8}\rho b \tag{4}$$

2.6. The free path, collision frequency, collision interval, and collision energy (perfect gas). For a mixed gas composed of rigid, elastic, spherical molecules of diameters σ_1, σ_2, (for kinds 1 and 2), the following results are obtained by use of Maxwell's velocity distribution function (§ 2.1).

The number of collisions N_{11}, N_{22}, N_{12} per unit volume per unit time, between like molecules (1 or 2) or unlike molecules, are given by

$$N_{11} = 2n_1^2\sigma_1^2\left(\frac{\pi kT}{m_1}\right)^{1/2}, \quad N_{12} = 2n_1n_2\sigma_{12}^2\left(\frac{2\pi m_0 kT}{m_1m_2}\right)^{1/2} \tag{1}$$

where $\sigma_{12} = \frac{1}{2}(\sigma_1 + \sigma_2)$.

The *collision frequencies*, ν_1, ν_2, or average number of collisions for a molecule 1 or 2 per unit time, are given by

$$\nu_1 = \frac{2N_{11} + N_{12}}{n_1}, \qquad \nu_2 = \frac{2N_{22} + N_{12}}{n_2} \tag{2}$$

The *collision intervals*, τ_1, τ_2, or mean time interval between collisions for a molecule of kind 1 or 2, are

$$\tau_1 = 1/\nu_1, \qquad \tau_2 = 1/\nu_2 \tag{3}$$

The mean free paths l_1, l_2, between collisions for a molecule of kind 1 or 2, are

$$\frac{1}{l_1} = \pi\left[2^{1/2}n_1\sigma_1^2 + \left(\frac{1 + m_1}{m_2}\right)^{1/2} n_2\sigma_{12}^2\right] \tag{4}$$

If the gas is simple,

$$l = 0.707/\pi n\sigma^2 \tag{5}$$

The mean free path $l_1(C_1)$ of a molecule of kind 1, moving with random speed C_1 is, for various values of $C_1/\overline{C_1}$, as follows :

$C_1/\overline{C_1}$:	0	0.25	0.50	1	2	3	4	5	6	∞
$l_1(C_1)/l_1$:	0	0.344	0.641	1.026	1.288	1.355	1.380	1.392	1.399	1.414

The probability that a molecule 1 with a particular speed C_1 shall travel a distance at least equal to l is $e^{-l/l_1(C_1)}$.

The probability that a molecule 1 moving with any speed should describe a free path at least equal to l is approximately $e^{-1.04l/l_1}$.

The fraction of the N_{12} collisions per unit volume and unit time, between unlike molecules whose collisional energy (namely, their combined trans-

latory kinetic energy relative to their mass center, before collision) is $x_0 kT$, is $(1 + x_0^2)e^{-x_0^2}$.

The mean value of the collisional energy is $2kT$.

3. Nonuniform Gas

3.1. The second approximation to f. Maxwell's velocity distribution is not correct when T is nonuniform, nor when $\boldsymbol{c}_0$ and the concentration ratio n_1/n_2 vary from point to point, except for his special states (§§ 2.1a and 2.1b). Even so, however, it is a good first approximation to f, if n is 10^{13} or more and if the gradients of T and $\boldsymbol{c}_0$ do not exceed 1°/cm and 1 sec^{-1}, respectively. Also, for these conditions, the second approximation to f gives reasonably accurate expressions for the stress distribution, thermal conduction, and diffusion.

The second approximation has the form

$$f_1 = f_1^{(0)}(1 + \Phi_1^{(1)}) \tag{1}$$

where $f_1^{(0)}$ denotes the Maxwellian expression (2.1), and

$$\Phi_1^{(1)} = -A_1(C_1)\boldsymbol{C}_1 \cdot \textbf{grad} \ln T - D_1(C_1)\boldsymbol{C} \cdot \boldsymbol{d}_{12} - B_1(C_1)B(\boldsymbol{C}_1\boldsymbol{c}_0) \tag{2}$$

where

$$\boldsymbol{d}_{12} = \textbf{grad}\, n_{10} + \frac{n_1 n_2}{n^2 p\rho}(m_1 - m_2)\boldsymbol{f} - \frac{n_1 n_2}{np}(m_1 \boldsymbol{F}_1 - m_2 \boldsymbol{F}_2) \tag{3}$$

$$\left.\begin{aligned} B(C_1, c_0) = \frac{1}{3} U_1^2\left(2\frac{\partial u_0}{\partial x} - \frac{\partial v_0}{\partial y} - \frac{\partial w_0}{\partial z}\right) + \ldots \\ + V_1 W_1\left(\frac{\partial v_0}{\partial z} + \frac{\partial w_0}{\partial y}\right) + \ldots \end{aligned}\right\} \tag{4}$$

and A_1, D_1, B_1, are scalar functions of the random speed C_1, determinable from Boltzmann's equation (§ 1.16) and involving molecular interactions. In $\boldsymbol{d}_{12}$ the vector $\boldsymbol{f}$ denotes the acceleration of the gas at $\boldsymbol{r}$, namely $D\boldsymbol{c}_0/Dt$, where D/Dt denotes the hydrodynamic " mobile operator "

$$\frac{\partial}{\partial t} + \frac{u_0 \partial}{\partial x} + \frac{v_0 \partial}{\partial y} + \frac{w_0 \partial}{\partial z}$$

$\boldsymbol{f}$ is given by

$$\rho \boldsymbol{f} = \rho_1 \boldsymbol{F}_1 + \rho_2 \boldsymbol{F}_2 - \textbf{grad}\, p \tag{5}$$

3.2. The stress distribution. From this expression for f_1, and the corresponding one for f_2, the components of the stress distribution are found to be

$$p_{xx} = knT - \frac{2}{3}\mu\left(\frac{2\partial u_0}{\partial x} - \frac{\partial v_0}{\partial y} - \frac{\partial w_0}{\partial z}\right), \quad p_{yz} = \mu\left(\frac{\partial v_0}{\partial z} + \frac{\partial w_0}{\partial y}\right), \quad \text{etc.} \tag{1}$$

where μ is expressible in terms of integrals involving the functions B_1 and B_2 and depends on the mode of molecular interaction.

These expressions for the stress agree with those of the formal theory of viscous stress, so that μ is identified with the coefficient of viscosity. In laminar motion, such that $u_0 = w_0 = 0$, $v_0 = az$, where a is a constant $(= \partial v_0/\partial z)$,

$$p_{xx} = p_{yy} = p_{zz} = knT, \quad p_{xy} = p_{xz} = 0, \quad p_{yz} = -\mu a \tag{2}$$

which is the simplest formal definition of μ.

3.3. Diffusion. In general $\overline{C_1}$ and $\overline{C_2}$ are unequal in a nonuniform mixed gas; that is, the two sets of molecules diffuse through each other. The rate of diffusion is expressible by $\overline{C_1} - \overline{C_2}$, which is independent of any uniform motion of the axes of reference. It is convenient to consider the flux of either type of molecule relative to axes moving with the mean speed $\overline{\mathbf{c}}$ (rather than $\mathbf{c}_0$), in which case

$$n_1\overline{C_1} = -n_2\overline{C_2} = \frac{n_1 n_2}{n}(\overline{C_1} - \overline{C_2}) \tag{1}$$

In this case the expression for $\overline{C_1} - \overline{C_2}$ derived from f_1 and f_2 gives

$$n_1\overline{C_1} = -nD_{12}\mathbf{d}_{12} - nD_T \,\mathbf{grad} \ln T \tag{2}$$

where D_{12} and D_T are expressible in terms of integrals involving the functions A_1 and D_1 and depend on the law of molecular interaction.

This expression shows that diffusion is set up by any of the four following causes : (a) a concentration gradient, (b) acceleration of the gas, (c) unequal external forces on the molecules 1, 2, and (d) a temperature gradient. Diffusion resulting from this last cause is called thermal diffusion. When T is uniform and there is no external force and no acceleration, p and n are uniform (see §§ 3.1-3.5), and (2) reduces to

$$n_1\overline{C_1} = -nD_{12}\,\mathbf{grad}\, n_{10} = -D_{12}\,\mathbf{grad}\, n_1 \tag{3}$$

This is the formal definition of D_{12} as the coefficient of diffusion. This coefficient also governs accelerative and forced diffusion, due to the causes (b) and (c).

Thermal diffusion is governed by a different factor D_T, called the coefficient of thermal diffusion; the ratio D_T/D_{12} is called the thermal diffusion ratio, and denoted by k_T: it has $n_{10}n_{20}$ as a factor, and therefore varies greatly with the concentration ratio. The thermal diffusion factor α, defined by $k_T/n_{10}n_{20}$, is much less dependent on n_1/n_2. Thus

$$n\overline{\boldsymbol{C}_1} = -D_{12}\boldsymbol{d}_{12}' \tag{4}$$

where

$$\boldsymbol{d}_{12}' = n\,\textbf{grad}\,n_{10} + \frac{n_1 n_2}{np\rho}(m_1 - m_2)\boldsymbol{f}$$

$$-\frac{n_1 n_2}{p}(m_1\boldsymbol{F}_1 - m_2\boldsymbol{F}_2) + nk_T\,\textbf{grad}\ln T$$

If T is uniform and $\boldsymbol{f}$ zero, and $\boldsymbol{F}_1$ and $\boldsymbol{F}_2$ are derived from potentials Ψ_1 and Ψ_2, diffusion proceeds in the sense which decreases $\boldsymbol{d}_{12}$, and if in the limit diffusion ceases, the condition $\boldsymbol{d}_{12}' = 0$ leads to

$$\textbf{grad}\ln\frac{n_1}{n_2} = -\,\textbf{grad}\,\frac{m_1\Psi_1 - m_2\Psi_2}{kT} \tag{5}$$

in accordance with Eq. (3) of § 2.1.

3.4. Thermal conduction. Heat conduction in a diffusing mixed gas proves to be expressible most simply by considering the flux relative to axes moving with the molecular mean speed $\bar{\boldsymbol{c}}$ instead of $\boldsymbol{c}_0$ (when there is no diffusion $\bar{\boldsymbol{c}}$ and $\boldsymbol{c}_0$ are identical). In this case the expressions for f_1 and f_2 lead to the equation

$$\boldsymbol{q} = -\lambda\,\textbf{grad}\,T - np\alpha D_{12}\boldsymbol{d}_{12}' \tag{1}$$

where λ depends on the mode of molecular interaction, through the functions A_1 and D_1, and is identified with the coefficient of thermal conduction in accordance with the formal definition for a nondiffusing gas. The second term in $\boldsymbol{q}$ is proportional to the rate of diffusion; it is an effect inverse to thermal diffusion, and is called the diffusion thermo-effect.

4. The Gas Coefficients for Particular Molecular Models

4.1. Models *a* to *d*. The simplest molecular models are:

a. *Smooth rigid elastic spheres* of diameters σ_1, σ_2; let

$$\sigma_{12} = \tfrac{1}{2}(\sigma_1 + \sigma_2)$$

This is the model already considered in 2.6. For such molecules $E = \tfrac{1}{2}mC^2$.

b. *Smooth rigid elastic attracting spheres.* This (Sutherland's) model has already been considered in § 2.5. The mutual force is $F(r)$ at distance r between two molecules.

c. *Point centers of repulsive force*, given at distance r by $\kappa r^{-\nu}$; κ is called the force constant and ν the force index. Suffixes 11, 22, or 12 are to be added to κ and ν according as the interacting molecules are both m_1, both m_2, or unlike.

d. *Point centers of repulsive and attractive force*, expressed by $\kappa r^{-\nu} - \kappa' r^{-\nu'}$, where κ' and ν' refer to the attraction. Alternatively the interaction may be expressed by the formula $4\epsilon[(\sigma'/r)^{\nu-1} - (\sigma'/r)^{\nu'-1}]$ for the *mutual potential energy* of the two molecules; σ is the distance at which the force changes sign, and ϵ the potential energy due to either the attractive or repulsive field at this mutual distance. Suffixes 11, 22, or 12 must be added to κ, ν, κ', ν'.

4.2. Viscosity μ and thermal conductivity λ for a simple gas

Model a:

$$\mu = 1.016 \frac{5}{16\sigma^2}\left(\frac{kmT}{\pi}\right)^{1/2} = 0.1792\left(\frac{kmT}{\pi}\right)^{1/2} \propto (mT)^{1/2}$$

$$\lambda = 2.522 \mu c_v$$

Model b (Sutherland's). In this case the expressions for μ and λ are those for Model a, divided by $1 + S/T$. The term S, known as Sutherland's constant, is proportional to the potential energy of attraction of the molecules at contact. The term S/T is only the first in a series of descending powers of T, and when S/T is not small, the neglect of the latter terms impairs this interpretation of S, as found empirically from the variation of μ or λ with T, as the potential energy.

Model c:

$$\mu = \frac{5}{8} C_\nu \left(\frac{kmT}{\pi}\right)^{1/2} \frac{(2kT/\kappa)^s}{A_2(\nu)\Gamma(4-s)} \propto \frac{m^{1/2}T^{1/2+s}}{\kappa^s}$$

where $$s = \frac{2}{\nu - 1}, \quad \text{and} \quad C_\nu = 1 + \frac{3(\nu-5)^2}{2(\nu-1)(101\nu - 113)} + \cdots$$

and $A_2(\nu)$ is a function of ν here given for several values of ν.

ν:	2	3	5	7	9	11	15	21	25	∞
s:	2	1	0.5	0.33	0.25	0.2	0.14	0.1	0.08	0
$A_2(\nu)$:			0.436	0.357	0.332	0.319	0.309	0.307	0.306	0.333
$A_1(\nu)$:			0.422	0.385	0.382	0.383	0.393			0.5

Also $\lambda = f\mu c_v$

where $$f = \frac{5}{2}\left(1 + \frac{(\nu - 5)^2}{4(\nu - 1)(11\nu - 13)} + \dots\right)/C_\nu$$

and f declines from 2.522 for $\nu = \infty$ (equivalent to Model a) to 2.5 for $\nu = 5$.

Model d. In this case, if the attractive field is weak, the expressions for μ and λ are those for Model c, divided by $1 + S/T^a$, where $a = (\nu - \nu')/(\nu - 1)$, and S is a function of κ, κ', ν, ν'.

In terms of ϵ and σ, an alternative expression for μ is

$$\mu = \frac{5}{16(\sigma')^2}\left(\frac{kmT}{\pi}\right)^{1/2} \chi \frac{(kT)}{\epsilon};$$

if T is large, the repulsive part of the field is dominant, and χ is proportional to $(kT/\epsilon)^s$. If T is small, the attractive field is dominant, and χ is proportional to $(kT/\epsilon)^{s'}$, where $s' = 2/(\nu' - 1)$.

The values $\nu = 13$, $\nu' = 7$ are found to be specially appropriate for many gases. A graph of $\log_{10} \chi$ is given for these values and also for $\nu = 9$, $\nu' = 5$ in the second item of the bibliography.

The preceding formulas all refer to spherically symmetrical models, for which $\gamma = \frac{5}{3}$; this applies to monatomic molecules and also to some diatomic molecules (e.g., H_2) at very low temperatures. For other values of γ, Eucken has given the following empirical formula for f, which is fairly satisfactory.

$$f = \tfrac{1}{4}(9\gamma - 5).$$

4.3. The first approximation to D_{12}. This is independent of the concentration ratio n_1/n_2. Its values for particular molecular models are

Model a:

$$\frac{3}{8n\sigma_{12}^2}\left(\frac{kT(m_1 + m_2)}{2\pi m_1 m_2}\right)^{1/2} \propto \frac{(T/m')^{1/2}}{n\sigma_{12}^2} \tag{1}$$

where $m' = m_1 m_2/(m_1 + m_2)$; m' is called the "reduced" mass of a pair of unlike molecules.

Model b:

$$\frac{3}{8n}\left(\frac{kT}{m_1}\right)^{1/2} \frac{(2kT/\kappa_{12})^s}{A_1(\nu)\Gamma(3 - s)} \propto \frac{T^{1/2+s}}{n\kappa_{12}^s (m_1)^{1/2}} \tag{2}$$

where $A_1(\nu)$ has values as in the preceding table.

The addition of an attracting force to Model a adds a Sutherland factor $1/(1 + S_{12}/T)$ to the D_{12} expression for this model.

The correction to the foregoing first approximations to D_{12} may amount to a few per cent, and it introduces a small dependence of D_{12} upon n_1/n_2.

4.4. Thermal diffusion. The formula for even the first approximation to k_T or α is very complicated, and this approximation is less close to the true value than in the case of μ, λ, and D_{12}. The correction (which is additive) may be as great as 23 per cent for Model a, when m_1/m_2 is very large. For other models it is less. And for Model b, when $\nu = 5$, k_T is zero.

The end values of α, for $n_{10} = 1$ and $n_{10} = 0$, will be denoted by α_1 and α_2. Values of α for $0 < n_{10} < 1$ are generally of similar magnitude, though they do not necessarily lie between α_1 and α_2. For molecules of very nearly equal mass, and with very different values of σ, it is possible for α_1 and α_2 to be of different sign, in which case α is always very small (e.g., 0.1).

Here only the first approximation to α_1 will be given for Model a (that for α_2 is obtained by changing suffix 2 to 1 except in σ_{12}). It is

$$\frac{2^{3/2}m_1(m_0{}^3/m_2)^{1/2}(\sigma_1/\sigma_{12})^2 - m_2(15m_2 - 7m_1)}{(2/5)(\sigma_1/\sigma_{12})^2(2m_0/m_2)^{1/2}} \tag{1}$$

$$= \frac{5m_1m_0 + (5/2^{3/2})\,(\sigma_{12}/\sigma_1)^2(m_2{}^3/m_0)^{1/2}(7m_1 - 15m_2)}{13m_1{}^2 + 16m_1m_2 + 30m_2{}^2}$$

Unless σ_1 and σ_2 differ greatly, α_1 and α_2 are positive if $m_1 > m_2$. If there is a steady temperature gradient in a gas mixture, and no external forces or acceleration, diffusion will tend to set up a steady state in which

$$\log n_{12} = -\alpha \log T + \text{constant}$$

(treating α as independent of n_{12} and T, which is usually approximately correct). If α is positive, and the suffix 1 refers to the heavier molecules ($m_1 > m_2$), the proportion of gas 1 is enriched in the colder and reduced in the hotter regions, relative to gas 2. Thus the heavier molecules are usually more numerous in the cooler region.

5. Electrical Conductivity in a Neutral Ionized Gas with or without a Magnetic Field

5.1. Definitions and symbols. The gas is supposed to consist of neutral particles, ions (positive and negative) and electrons. These constituents are distinguished by suffixes 1, 2, ... attached to the symbols for their properties; mass m; number density n; charge e; diameter σ; etc. The suffix 1 will refer to the neutral molecules, so that $e_1 = 0$.

The electric intensity is denoted by $\boldsymbol{E}$, and the magnetic by $\boldsymbol{H}$, in the direction of the unit vector $\boldsymbol{h}$. In the presence of a uniform magnetic field alone, a charge e_s will spiral round the lines of force with angular velocity ω_s given by

$$\omega_s = e_s H / m_s$$

The current intensity is denoted by $\boldsymbol{j}$.

5.2. The diameters σ_s. The diameter σ_1 for the neutral molecules, and also the "joint diameter" or effective distance between centers, σ_{1s}, at a collision between a neutral and charged particle ($s \neq 1$), are a moderate multiple of 10^{-8} cm (the factor being 1 to 5, diminishing slightly as T increases). The joint diameter σ_{ei} for a collision between an electron and an ion is of order $10^{-5}\, Z_i\, (300/T)$, where Z_i is the number of unit charges on the ion. For a collision between two ions, it is similarly about $10^{-5} Z_i Z_i'(300/T)$, where Z_i and Z_i' are the charge numbers of the two ions.

Owing to the large magnitude of the effective diameters for collisions between charged particles (due to the slow inverse-square decrease of electrostatic force with increasing distance), the collision frequencies of the charged particles are little affected by the presence of neutral molecules unless these are in large excess, e.g., n_1 more than a hundredfold as great as the number density of all the charged particles, as in a slightly ionized gas (such as the E and F layers of the ionosphere).

5.3. Slightly ionized gas. When n_1/n_s ($s \neq 1$) is large, the different kinds of charged particles each contribute independently to the current and the electric conductivity, their mutual collisions being negligible. If $\boldsymbol{E}$ is *perpendicular* to $\boldsymbol{H}$,

$$\boldsymbol{j} = K\boldsymbol{E} + K'\boldsymbol{h} \times \boldsymbol{E} \tag{1}$$

where
$$K + iK' = \sum_{s \neq 1} \frac{n_s e_s^2 \tau_{1s}}{m_s(1 + i\omega_s \tau_{1s})}$$

or
$$\boldsymbol{j} = \sum \frac{n_s e_s^2 \tau_{1s}}{m_s(1 + \omega_s^2 \tau_{1s}^2)} \quad (\boldsymbol{E} + \omega_s \tau_{1s} \boldsymbol{E} \times \boldsymbol{h})$$

An electric field perpendicular to $\boldsymbol{H}$ hence produces a current in its own direction, for which the conductivity is K, and also a transverse current perpendicular both to $\boldsymbol{E}$ and $\boldsymbol{H}$, for which the conductivity is K'.

If $\boldsymbol{E}$ is along $\boldsymbol{H}$, there is no transverse current, and K is increased to $\sum n_s e_s^2 \tau_{1s}/m_s$.

5.4. Strongly ionized gas. Consider a gas in which n_1 is zero or negligible, and the charged particles are electrons, and ions of two kinds, with very different masses. Let suffixes e, i, and j refer to the electrons and the heavier and lighter ions, and take m_e/m_j, m_j/m_i to be negligible. The electronic contribution to the conductivity is

$$\mathrm{K}_e + i\mathrm{K}_e' = \frac{n_e e^2 \tau_{ie}}{m_e(1 + \tau_{ie}/\tau_{je} + i\omega_e \tau_{ie})} \tag{1}$$

independent of the mutual ionic collisions.

The ion conductivity is similarly given by

$$\frac{n_i n_j (e_i m_j - e_j m_i)^2 \tau_{ij}}{(n_i m_i + n_j m_j) m_i m_j (1 + i\omega_{ij} \tau_{ij})} \tag{2}$$

where

$$\omega_{ij} = eH(m_i + m_j)/m_i m_j$$

Bibliography

1. Boltzmann, L., *Vorlesungen über die Gas Theorie* (2 vols.), J. A. Barth, Leipzig, 1910. This pioneer work well shows the nascent ideas of the kinetic theory in their first detailed mathematical development.
2. Chapman, S. and Cowling, T. G., *Mathematical Theory of Non-Uniform Gases*, 2d ed., Cambridge University Press, London, 1953. This severely mathematical work gives the classical kinetic theory in detail, but sets out and explains the formulas in a way designed for the use of the physicist and chemist.
3. Jeans, J. H., *Kinetic Theory of Gases*, Cambridge University Press, London, 1925. This book in its earlier editions stresses the difficulties later met by the quantum theory, and has interesting passages on the statistical mechanical aspects of the kinetic theory. (Dover reprint)
4. Kennard, E. H., *Kinetic Theory of Gases*, McGraw-Hill Book Company, New York, Inc., 1938. This book is written for the physicist and does not contain the full mathematical theory.
5. Loeb, L. B., *Kinetic Theory of Gases*, 2d ed., McGraw-Hill Book Company, Inc., New York, 1934. See remarks on the Kennard book above.
6. Maxwell, J. C., *Scientific Papers* (2 vols.), Cambridge University Press, London, 1890. See remarks on the Boltzmann book above.

Chapter 13

ELECTROMAGNETIC THEORY

By Nathaniel H. Frank

Professor of Physics, Massachusetts Institute of Technology

and William Tobocman

Research Assistant, Department of Physics, Massachusetts Institute of Technology

1. Definitions and Fundamental Laws

1.1. Primary definitions. The study of mechanics is founded on three basic concepts—space, time, and mass. In the theory of electromagnetism, the additional concept of charge is introduced. Quantity of electric charge is denoted by q, electric current by i, and electric current density by $\boldsymbol{J}$. The mutual interactions of charges and currents are described in terms of fields of electric intensity $\boldsymbol{E}$ and magnetic induction $\boldsymbol{B}$. The force on an element of charge caused by all other charges is represented as the interaction of the element of charge with the field ($\boldsymbol{E}$ and $\boldsymbol{B}$) in its vicinity, the sources of the field being all the other charges. These fields are defined by

mks	Gaussian	
$\dfrac{d\boldsymbol{F}}{dq} = \boldsymbol{E} + \boldsymbol{V} \times \boldsymbol{B},$	$\dfrac{d\boldsymbol{F}}{dq} - \boldsymbol{E} + \dfrac{\boldsymbol{V} \times \boldsymbol{B}}{c}$	(1)

where $d\boldsymbol{F}$ is force experienced by an element dq of charge, $\boldsymbol{V}$ is the velocity of the element of charge, and c is the velocity of light in vacuum.

The electromagnetic field in a given region of space depends on the kind of matter occupying the region as well as on the distribution of charge giving rise to the field. This phenomenon can be described in nonconducting, nonferromagnetic media by writing the field vectors as products of two quantitities:

mks	Gaussian	
$\boldsymbol{E} = \dfrac{\boldsymbol{D}}{\epsilon} = \dfrac{\boldsymbol{D}}{\kappa_\varepsilon \epsilon_0},$	$\boldsymbol{E} = \dfrac{\boldsymbol{D}}{\epsilon} = \dfrac{\boldsymbol{D}}{\kappa_\varepsilon}$	(2)
$\boldsymbol{B} = \mu \boldsymbol{H} = \mu_0 \kappa_m \boldsymbol{H},$	$\boldsymbol{B} = \mu \boldsymbol{H} = \kappa_m \boldsymbol{H}$	(3)

where $\boldsymbol{D}/\epsilon_0$ and $\mu_0\boldsymbol{H}$ are what the electric field $\boldsymbol{E}$ and the magnetic induction $\boldsymbol{B}$ would be in vacuum, and the κ's are quantities characteristic of the medium. In isotropic media the κ's are scalars. $\boldsymbol{D}$ is the electric displacement; $\boldsymbol{H}$ is the magnetic field intensity; κ_ε is the dielectric constant; κ_m is the relative permeability; ϵ_0 and μ_0 are universal constants

mks	Gaussian
$\epsilon_0 = 8.854 \times 10^{-12} \dfrac{\text{farad}}{\text{meter}},$	$\epsilon_0 = 1$
$\mu_0 = 4\pi \times 10^{-7} \dfrac{\text{henry}}{\text{meter}},$	$\mu_0 = 1$

In terms of the above, the following quantities are defined

	mks	Gaussian	
polarization :	$\boldsymbol{P} = \boldsymbol{D} - \epsilon_0 \boldsymbol{E},$	$\boldsymbol{P} = \dfrac{1}{4\pi}(\boldsymbol{D} - \epsilon_0 \boldsymbol{E})$	(4)
magnetization :	$\boldsymbol{M} = \dfrac{\boldsymbol{B}}{\mu_0} - \boldsymbol{H},$	$\boldsymbol{M} = \dfrac{1}{4\pi}\left(\dfrac{\boldsymbol{B}}{\mu_0} - \boldsymbol{H}\right)$	(5)
magnetic susceptibility :	$\chi_m = \dfrac{\lvert \boldsymbol{M} \rvert}{\lvert \boldsymbol{H} \rvert},$	$\chi_m = \dfrac{\lvert \boldsymbol{M} \rvert}{\lvert \boldsymbol{H} \rvert}$	(6)
electric susceptibility :	$\chi_\varepsilon = \dfrac{\lvert \boldsymbol{P} \rvert}{\epsilon_0 \lvert \boldsymbol{E} \rvert},$	$\chi_\varepsilon = \dfrac{\lvert \boldsymbol{P} \rvert}{\epsilon_0 \lvert \boldsymbol{E} \rvert}$	(7)

1.2. Conductors. A conductor is a material within which charge is free to move. Metals and electrolytes are conductors; the flow of charge within a metal or electrolyte is governed by Ohm's law :

$$\boldsymbol{J} = \sigma \boldsymbol{E} \tag{1}$$

where σ (conductivity) is a constant characteristic of the medium, and $\rho = 1/\sigma$ is called the resistivity. The resistance of a conductor of length l and constant cross section A is defined

$$R = \frac{l}{\sigma A} = \frac{l\rho}{A} \tag{2}$$

Clearly it is impossible to maintain a static potential difference between two points of a conductor without flow of charge, so that under electrostatic conditions a conductor is an equipotential. This implies that the tangential component of $\boldsymbol{E}$ must vanish at the surface of a conductor for static situations.

1.3. Ferromagnetic materials. The magnetic induction $\boldsymbol{B}$ within a ferromagnetic medium is not a linear function of the magnetic intensity $\boldsymbol{H}$. The magnetic induction depends not only on the value of the magnetic intensity but also on the previous history of the medium, so that it is possible to have a finite value for $\boldsymbol{B}$ when $\boldsymbol{H} = 0$. In such a case the body has a residual magnetization $M_1 = B/\mu_0$ (in mks units) and acts as a source of magnetic field.

1.4. Fundamental laws. Electromagnetic theory can be based on the definitions above and the following three fundamental laws:

conservation of charge:

mks Gaussian

$$\nabla \cdot \boldsymbol{J} = -\frac{\partial \rho}{\partial t}, \qquad \nabla \cdot \boldsymbol{J} = -\frac{\partial \rho}{\partial t} \tag{1}$$

Faraday's law:

$$\nabla \times \boldsymbol{E} = -\frac{\partial \boldsymbol{B}}{\partial t}, \qquad \nabla \times \boldsymbol{E} = -\frac{1}{c} \cdot \frac{\partial \boldsymbol{B}}{\partial t} \tag{2}$$

Ampère's law:

$$\nabla \times \boldsymbol{H} = \boldsymbol{J} + \frac{\partial \boldsymbol{D}}{\partial t}, \qquad \nabla \times \boldsymbol{H} = \frac{4\pi \boldsymbol{J}}{c} + \frac{1}{c} \cdot \frac{\partial \boldsymbol{D}}{\partial t} \tag{3}$$

Taking the divergence of both sides of Eq. (2) and (3) and using Eq. (1) there follow Gauss's law:

mks Gaussian

$$\nabla \cdot \boldsymbol{D} = \rho, \qquad \nabla \cdot \boldsymbol{D} = 4\pi\rho \tag{4}$$

and

$$\nabla \cdot \boldsymbol{B} = 0, \qquad \nabla \cdot \boldsymbol{B} = 0 \tag{5}$$

Equations (2), (3), (4), and (5) are called Maxwell's equations.

1.5. Boundary conditions. Application of Maxwell's equations to infinitesimal contours and volumes at the boundary between two regions in space yield the following boundary conditions for electromagnetic fields.

mks Gaussian

$$\boldsymbol{n} \cdot (\boldsymbol{B}_2 - \boldsymbol{B}_1) = 0, \qquad \boldsymbol{n} \cdot (\boldsymbol{B}_2 - \boldsymbol{B}_1) = 0 \tag{1}$$

$$\boldsymbol{n} \times (\boldsymbol{H}_2 - \boldsymbol{H}_1) = \boldsymbol{J}_s, \qquad \boldsymbol{n} \times (\boldsymbol{H}_2 - \boldsymbol{H}_1) = \frac{4\pi}{c} \boldsymbol{J}_s \tag{2}$$

$$\boldsymbol{n} \times (\boldsymbol{E}_2 - \boldsymbol{E}_1) = 0, \qquad \boldsymbol{n} \times (\boldsymbol{E}_2 - \boldsymbol{E}_1) = 0 \tag{3}$$

$$\boldsymbol{n} \cdot (\boldsymbol{D}_2 - \boldsymbol{D}_1) = \sigma_s, \qquad \boldsymbol{n} \cdot (\boldsymbol{D}_2 - \boldsymbol{D}_1) = 4\pi\sigma_s \tag{4}$$

where $\boldsymbol{n}$ is a unit vector normal to the boundary and pointing from region 1 into region 2, σ_s is surface charge density and $\boldsymbol{J}_s$ is the surface current density.

1.6. Vector and scalar potentials. From (5) of § 1.4 it follows that

$$B = \nabla \times A, \quad (A = \text{vector potential}) \tag{1}$$

and from (2) of § 1.4 that

mks Gaussian

$$E = -\nabla\varphi - \frac{\partial A}{\partial t}, \quad E = -\nabla\varphi - \frac{1}{c}\cdot\frac{\partial}{\partial t}A, \quad (\varphi = \text{scalar potential}) \tag{2}$$

These definitions leave $\nabla \cdot A$ (the gauge) undetermined, and the potential φ contains an arbitrary additive constant.

From the fact that the fields must always be finite in regions devoid of charge or current, it follows that the potentials must be continuous functions within such regions.

Using the definition

$$\nabla^2 V = \nabla\nabla \cdot V - \nabla \times \nabla \times V, \tag{3}$$

the differential equations relating the potential functions to their sources follow from Eqs. (2), (3) of § 1.4, and (1), (2) of § 1.6; for homogeneous, isotropic media these equations are

mks

$$\nabla^2\varphi = \epsilon\mu\frac{\partial^2\varphi}{\partial t^2} - \frac{\partial}{\partial t}\left(\nabla \cdot A + \epsilon\mu\frac{\partial\varphi}{\partial t}\right) - \frac{\rho}{\epsilon},$$

Gaussian

$$\nabla^2\varphi = \frac{\epsilon\mu}{c^2}\cdot\frac{\partial^2\varphi}{\partial t^2} - \frac{1}{c}\cdot\frac{\partial}{\partial t}\left(\nabla \cdot A + \frac{\epsilon\mu}{c}\cdot\frac{\partial\varphi}{\partial t}\right) - \frac{4\pi\rho}{\epsilon}, \tag{4}$$

mks

$$\nabla^2 A = \mu\epsilon\frac{\partial^2}{\partial t^2}A + \nabla\left(\nabla \cdot A + \epsilon\mu\frac{\partial\varphi}{\partial t}\right) - \mu J,$$

Gaussian

$$\nabla^2 A = \frac{\mu\epsilon}{c^2}\cdot\frac{\partial^2}{\partial t^2}A + \nabla\left(\nabla \cdot A + \frac{\epsilon\mu}{c}\cdot\frac{\partial\varphi}{\partial t}\right) - \frac{4\pi\mu}{c}J, \tag{5}$$

1.7. Bound charge. Combining Eqs. (4) of § 1.1 and (4) of § 1.4 gives for isotropic media

mks Gaussian

$$\nabla \cdot E = \frac{1}{\epsilon_0}(\rho - \nabla \cdot P), \quad \nabla \cdot E = \frac{4\pi}{\epsilon_0}(\rho - \nabla \cdot P) \tag{1}$$

This result has led to the following nomenclature

$-\nabla \cdot P =$ bound charge density

$P \cdot n =$ bound surface charge density

where n is the unit outward normal to surface.

1.8. Amperian currents. Combining Eqs. (5) of § 1.1 and (3) of § 1.4 for a nonferromagnetic isotropic medium gives

$$\text{mks} \qquad\qquad \text{Gaussian}$$

$$\nabla\times \boldsymbol{B} = \mu_0(\boldsymbol{J} + \nabla\times \boldsymbol{M}), \quad \nabla\times \boldsymbol{B} = 4\pi\mu_0(\boldsymbol{J} + \nabla\times \boldsymbol{M}) \tag{1}$$

when $\partial \boldsymbol{D}/\partial t = 0$.

This result provides the justification for the following nomenclature:

$$\nabla\times \boldsymbol{M} = \text{amperian current density}$$

$$\boldsymbol{M}\times \boldsymbol{n} = \text{amperian surface current density}$$

where $\boldsymbol{n}$ is the outward normal to the surface.

2. Electrostatics

2.1. Fundamental laws. For electrostatic situations, Maxwell's equations for the electric field reduce to

$$\text{mks} \qquad\qquad \text{Gaussian}$$

$$\nabla\times \boldsymbol{E} = 0, \qquad \nabla\times \boldsymbol{E} = 0 \tag{1}$$

$$\nabla\cdot \boldsymbol{D} = \rho, \qquad \nabla\cdot \boldsymbol{D} = 4\pi\rho \tag{2}$$

Since $\boldsymbol{E}$ is now irrotational, it can be derived from a scalar potential:

$$\boldsymbol{E} = -\nabla\varphi \tag{3}$$

Combining this result with Eqs. (2) of § 1.1 and (4) of § 1.4, there follows Poisson's equation:

$$\text{mks} \qquad\qquad \text{Gaussian}$$

$$\nabla^2\varphi = -\frac{\rho}{\epsilon}, \qquad \nabla^2\varphi = -\frac{4\pi\rho}{\epsilon} \tag{4}$$

Integration of Poisson's equation by the Green function method gives

$$\varphi = -\frac{1}{4\pi}\int_V \frac{\nabla^2\varphi}{r}\,dV + \frac{1}{4\pi}\int_S \left[\frac{\nabla\varphi}{r} - \varphi\nabla\left(\frac{1}{r}\right)\right]\cdot d\boldsymbol{S} \tag{5}$$

where φ is the potential at a point inside volume V, which is bounded by surface S. Taking S at infinity, where we assume φ vanishes, reduces Eq. (5) to

$$\text{mks} \qquad\qquad \text{Gaussian}$$

$$\varphi = \frac{1}{4\pi\epsilon}\int \frac{\rho\,dV}{r}, \qquad \varphi = \frac{1}{\epsilon}\int \frac{\rho\,dV}{r} \tag{6}$$

For a point charge, i.e.,

$$\rho = q\delta(x)\delta(y)\delta(z)$$

equation (6) becomes

$$\text{mks} \qquad \text{Gaussian}$$

$$\varphi = \frac{q}{4\pi\epsilon r}, \qquad \varphi = \frac{q}{\epsilon r} \tag{7}$$

which is called Coulomb's law.

2.2. Fields of some simple charge distributions. Equations (7) of § 2.1 and (2) of § 2.1 can be used to give the fields due to various simple charge distributions. Let $\zeta = 1$ for mks units, $\zeta = 4\pi$ for Gaussian units.

a. *Line of charge* (τ = linear charge density) :

$$\varphi = -\frac{\zeta\tau}{2\pi\epsilon} \ln R \tag{1}$$

b. *Sphere of charge* (q = total charge; a = radius of the sphere) :

$$\varphi = \frac{\zeta q}{4\pi\epsilon r}, \quad (r \geq a); \qquad \varphi = \frac{\zeta q}{4\pi\epsilon a}, \quad (r \leq a) \tag{2}$$

c. *Ring of charge* Fig. 1 (q = total charge) :

$$\left.\begin{aligned} \varphi &= \frac{\zeta q}{4\pi\epsilon d} \sum_{n=0}^{\infty} \left(\frac{d}{r}\right)^{n+1} P_n(\cos\alpha) P_n(\cos\theta), \quad (r > d \text{ or } \theta \neq \alpha) \\ \varphi &= \frac{\zeta q}{4\pi\epsilon d} \sum_{n=0}^{\infty} \left(\frac{r}{d}\right)^{n} P_n(\cos\alpha) P_n(\cos\theta), \quad (r < d \text{ or } \theta \neq \alpha) \end{aligned}\right\} \tag{3}$$

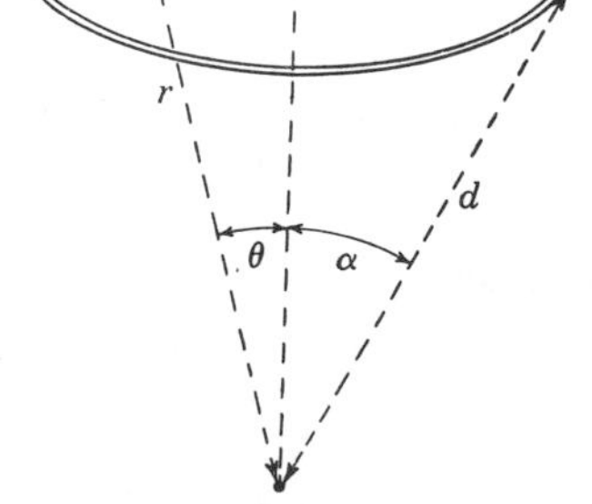

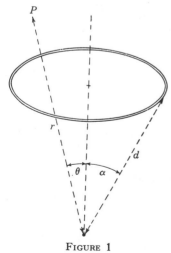

FIGURE 1

where P_n is the Legendre polynomial.

2.3. Electric multipoles; the double layer. A point charge is regarded as a monopole (2^0 pole). A 2^{n+1} pole is defined in terms of a 2^n pole in the following manner. Let

$$\boldsymbol{l}_n = l_n \boldsymbol{u}_n, \qquad |\boldsymbol{u}_n| = 1$$

be a vector at the origin. At the tip of $\boldsymbol{l}_n$ place the charge singularity characterized by 2^n pole of moment p_n; at the back end of $\boldsymbol{l}_n$ place a 2^n pole of moment $-p_n$; let $l_n \to 0$ and $p_n \to \infty$ in such a way as to keep the product $l_n p_n$ constant.

The resulting charge singularity is called a 2^{n+1} pole of moment

$$p_{n+1} = (n+1)l_n p_n \tag{1}$$

The potential resulting from such a singularity is

$$\varphi_{n+1} = \frac{(-1)^{n+1}}{1/\zeta} \frac{p_{n+1}}{(n+1)!} \boldsymbol{u}_n \cdot \nabla \boldsymbol{u}_{n-1} \cdot \nabla \boldsymbol{u}_{n-2} \cdot \ldots \cdot \boldsymbol{u}_0 \cdot \nabla\left(\frac{1}{r}\right) \tag{2}$$

Accordingly,

$$\varphi_0 = \frac{\zeta}{4\pi\epsilon} \cdot \frac{p_0}{r}, \quad [p_0 = q \quad \text{(monopole)}] \tag{3}$$

$$\varphi_1 = -\frac{\zeta}{4\pi\epsilon} p_1 \boldsymbol{u}_0 \cdot \nabla\left(\frac{1}{r}\right), \quad [p_1 = p_0 l_0 \quad \text{(dipole)}] \tag{4}$$

$$\left.\begin{aligned} &\varphi_2 = \frac{\zeta}{4\pi\epsilon} \cdot \frac{p_2}{2!} \boldsymbol{u}_1 \cdot \nabla\left[\boldsymbol{u}_0 \cdot \nabla\left(\frac{1}{r}\right)\right] \\ &\qquad [p_2 = 2l_1 p_1 = 2l_1 p_0 l_0 \quad \text{(quadrupole)}] \end{aligned}\right\} \tag{5}$$

where $\zeta = 1$ in mks units, $\zeta = 4\pi$ in Gaussian units.

The potential at point

$$\boldsymbol{r} = x\boldsymbol{u}_x + y\boldsymbol{u}_y + z\boldsymbol{u}_z$$

of an element of charge $dq = \rho dV$ located at point

$$\boldsymbol{r}_1 = x_1\boldsymbol{i} + y_1\boldsymbol{j} + z_1\boldsymbol{k}$$

when $|\boldsymbol{r}| > |\boldsymbol{r}_1|$ is

$$d\varphi = \frac{\zeta}{4\pi\epsilon} \frac{dq}{|\boldsymbol{r} - \boldsymbol{r}_1|} = \sum_{n=0} d\varphi_n \tag{6}$$

where

$$d\varphi_n = \frac{\zeta(-1)^n}{4\pi\epsilon n!} dq \left[(\boldsymbol{r}_1 \cdot \nabla)_n (\boldsymbol{r}_1 \cdot \nabla)_{n-1} \ldots (\boldsymbol{r}_1 \cdot \nabla)_2 (\boldsymbol{r}_1 \cdot \nabla)_1 \left(\frac{1}{r}\right)\right]$$

$$d\varphi_0 = \frac{\zeta dq}{4\pi\epsilon}\left(\frac{1}{r}\right)$$

by Taylor's theorem. The subscripts simply serve to count the number of factors.

A double layer is a surface distribution of dipoles. Let τ be the dipole moment per unit area; then the potential due to the double layer is

$$\varphi = -\frac{\zeta}{4\pi\epsilon} \int \tau d\Omega \tag{7}$$

where $d\Omega$ is the element of solid angle intercepted by an element of area of the double layer.

2.4. Electrostatic boundary value problems. The electrostatic field potential in a given region must satisfy the following conditions :

a. φ must be a solution of Poisson's equation, Eq. (4) of § 2.1.

b. The change in $(\epsilon\nabla\varphi \cdot \boldsymbol{n})$ must be equal to the surface charge density across any boundary.

c. φ must be constant in a conductor.

d. φ is continuous everywhere (except across a double layer).

e. φ vanishes at least as fast as $1/r$ at infinity if all sources are within a finite region. These conditions determine the field uniquely.

2.5. Solutions of simple electrostatic boundary value problems. Let

$$\zeta = 1 \text{ for the mks system of units,}$$

$$\zeta = 4\pi \text{ for the Gaussian system of units.}$$

a. *Conducting sphere in the field of a point charge*, center at the origin, where

r_1 = radius of sphere
q_1 = net charge on sphere
charge $= q$ at $z = Z > r_1,\ y = 0,\ x = 0$

$r_2 = \sqrt{r^2 + Z^2 - 2rZ\cos\theta}$
Φ = potential at (r,θ,φ)

$$\left.\begin{aligned} \Phi &= \frac{\zeta}{4\pi\epsilon}\left[\frac{q}{r_2} + \frac{r_1\varphi_s}{r} - q\sum_{n=0}^{\infty}\frac{r_1^{2n+1}}{Z^{n+1}}\frac{P_n(\cos\theta)}{r^{n+1}}\right], \quad (r > r_1) \\ \Phi &= \varphi_s = \left(\frac{q_1}{r_1} + \frac{q}{Z}\right)\frac{\zeta}{4\pi\epsilon}, \quad (r < r_1) \end{aligned}\right\} \tag{1}$$

b. *Dielectric sphere of inductive capacity* ϵ_1 *in the field of a point charge* q, both immersed in a medium of inductive capacity ϵ :

$$\Phi = \frac{\zeta}{4\pi\epsilon}\left\{\begin{aligned} &\frac{q}{r_2} + q\sum_{n=0}^{\infty}\left[\frac{n(\epsilon - \epsilon_1)}{n\epsilon_1 + (n+1)\epsilon}\right]\frac{r_1^{2n+1}}{Z^{n+1}}\frac{P_n(\cos\theta)}{r^{n+1}}, \quad (r > r_1) \\ &q\sum_{n=0}^{\infty}\left[\frac{(2n+1)\epsilon}{n\epsilon_1 + (n+1)\epsilon}\right]\frac{r^n}{Z^{n+1}}P_n(\cos\theta), \quad (r < r_1) \end{aligned}\right. \tag{2}$$

c. *Conducting sphere of radius r_1 in a uniform electric field:*

$$\left.\begin{aligned}\Phi &= -E_0 r\cos\theta + E_0\frac{r_1^3\cos\theta}{r^2} + \frac{q_1\zeta}{r4\pi\epsilon}, \quad (r > r_1)\\ \Phi &= \frac{\zeta q_1}{4\pi\epsilon r_1} \quad (r < r_1)\end{aligned}\right\} \tag{3}$$

where the external field has the potential

$$\Phi_0 = -E_0 r\cos\theta$$

d. *Dielectric sphere of radius r_1 in the uniform electric field:*

$$\left.\begin{aligned}\Phi_0 &= -E_0 r\cos\theta\\ \Phi &= -E_0 r\cos\theta + \frac{\epsilon_1-\epsilon}{\epsilon_1+2\epsilon}r_1^3 E_0\frac{\cos\theta}{r^2}, \quad (r > r_1)\\ \Phi &= -\frac{3\epsilon}{\epsilon_1+2\epsilon}E_0 r\cos\theta \quad (r < r_1)\end{aligned}\right\} \tag{4}$$

Let $\boldsymbol{E}_1$ be the field within the sphere, and let $\boldsymbol{P}_1$ be the polarization of the dielectric sphere (example d). The depolarizing factor L, which is defined by the equation

$$LP_1 = E_0 - E_1 \tag{5}$$

is then

$$L = \frac{\zeta}{3\epsilon_0}\frac{\kappa_1-\kappa}{\kappa(\kappa_1-1)} \tag{6}$$

for a sphere.

2.6. The method of images. For the case of a regularly shaped conductor or dielectric in the field of a simple distribution of charge, it is often possible to construct a potential distribution which satisfies the boundary conditions outside the conductor or dielectric by replacing the dielectric or conductor by a distribution of charges located inside of its boundaries. Since the field so constructed satisfies Poisson's equation as well as the boundary conditions, it must represent the actual field outside the conductor. Such fictitious charge distributions are called image charges.

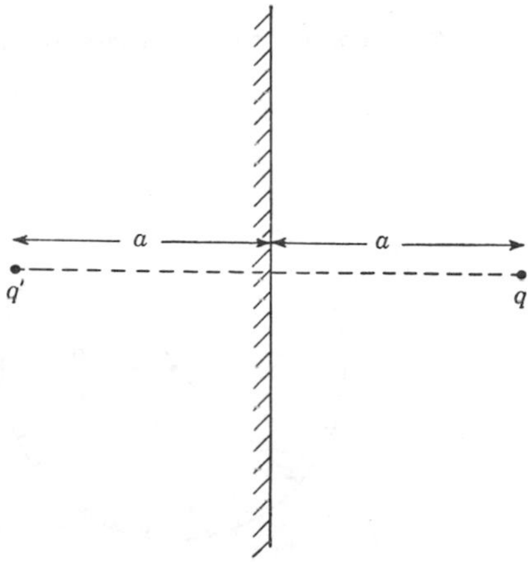

FIGURE 2

Examples :

1. Point charge outside an infinite grounded plane conductor, Fig. 2, p. 315 (q' = image charge = $-q$).

2. Point charge outside a pair of intersecting grounded plane conductors, Fig. 3; (n = integer):

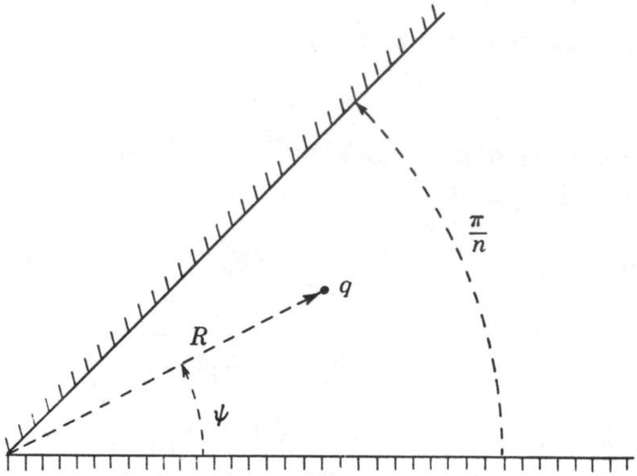

FIGURE 3

Image charges of charge q at

$$\left(R, \psi + \frac{2\pi}{n}\right), \quad \left(R, \psi + \frac{4\pi}{n}\right), \quad \ldots$$

Image charges of charge $-q$ at

$$(R, -\psi), \quad \left(R, -\psi - \frac{2\pi}{n}\right), \quad \left(R, -\psi - \frac{4\pi}{n}\right), \quad \ldots$$

3. Point charge outside a conducting sphere :

$$q' = \text{image charge} = -\frac{qa}{\overline{OQ}}, \quad OR = \frac{a^2}{\overline{OQ}} \quad \text{(Fig. 4)}$$

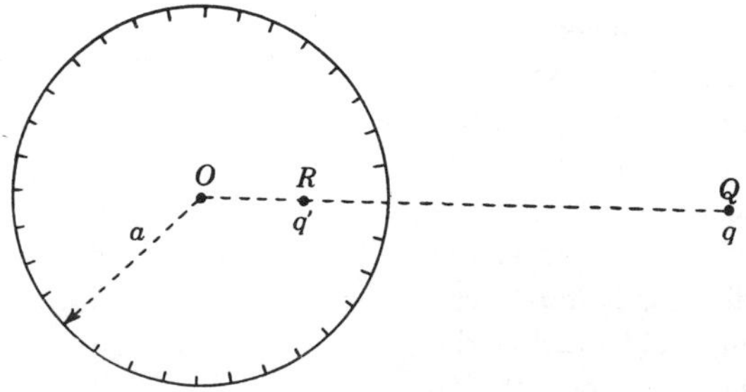

FIGURE 4

4. Line of charge outside a conducting cylinder, Fig. 5; (τ = charge per unit length):

$$\tau' = -\tau, \qquad \overline{OR} = \frac{a^2}{\overline{OQ}}$$

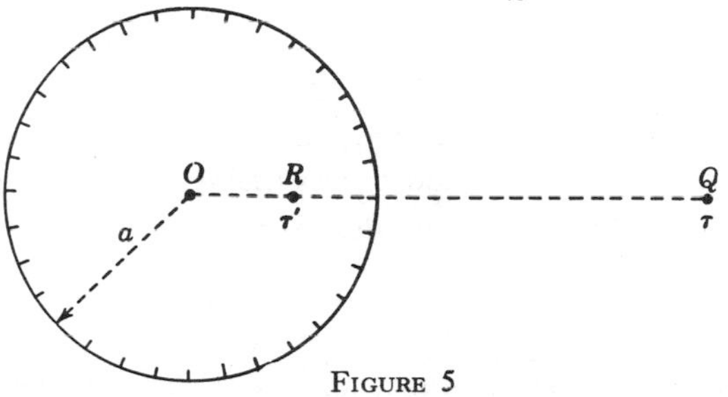

FIGURE 5

5. Point charge near a plane boundary between two semi-infinite dielectrics, Fig. 6; (Let $k = \epsilon_2/\epsilon_1$):

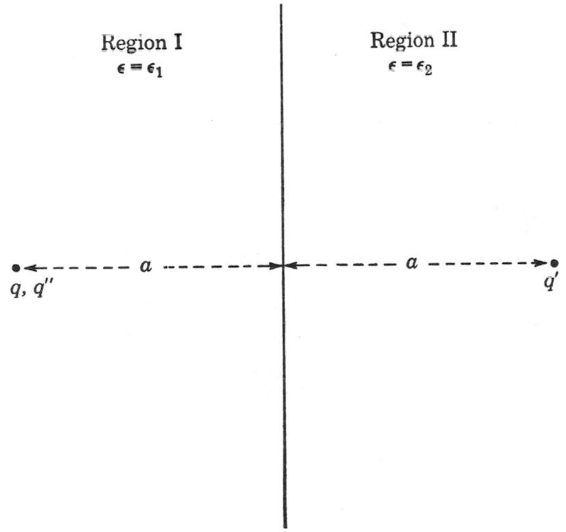

FIGURE 6

The field in region II is given by replacing region I with charge q'':

$$q'' = \frac{2k}{k+1} q$$

The field in region I is given by replacing region II with charge q':

$$q' = \left(\frac{k-1}{k+1}\right) q$$

2.7. Capacitors. A capacitor is a system consisting of two insulated conductors bearing equal and opposite charge. Let $\Delta\varphi$ be the difference of potential between the two conductors, and let q be the magnitude of the charge on each. Then the capacitance of the system is defined to be

$$C = \frac{q}{\Delta\varphi} \tag{1}$$

The capacitance of a single conductor is defined as the limit of the capacitance of a system of two such conductors as their separation becomes infinite. The work required to charge a capacitor is

$$W = \frac{1}{2} C(\Delta\varphi)^2 = \frac{1}{2} \frac{q^2}{C} \tag{2}$$

The capacitances of some simple capacitors may be calculated by the formulas shown below, where $\zeta = 1$ for the mks system, and $\zeta = 4\pi$ for the Gaussian system.

a. *Parallel plates:*

A = area of the plates

d = the separation of the plates

$$C = \frac{A\epsilon}{\zeta d}, \quad (A \gg d^2) \tag{3}$$

b. *Concentric spheres:*

r_1 = radius of inner sphere

r_2 = radius of outer sphere

$$C = \frac{4\pi\epsilon}{\zeta(1/r_1 - 1/r_2)} \tag{4}$$

c. *Concentric circular cylinders:*

r_1 = inner radius

r_2 = outer radius

L = length

$$C = \frac{4\pi\epsilon L}{\zeta 2 \ln r_2/r_1}, \quad (L \gg r_2 - r_1) \tag{5}$$

d. *Parallel circular cylinders:*

Same symbols as in example c. D = separation of centers; positive when the cylinders are external, negative when one cylinder is inside the other

$$\left. \begin{aligned} C = \frac{1}{2} \cdot \frac{4\pi\epsilon L}{\zeta} \left\{ \cosh^{-1} \left[\pm \frac{(D^2 - r_1^2 - r_2^2)}{2r_1 r_2} \right] \right\}^{-1} \\ (L \gg D \quad \text{and} \quad |r_1 - r_2|) \end{aligned} \right\} \tag{6}$$

e. *Two circular cylinders of equal radii:*

$$C = \frac{4\pi\epsilon}{\zeta} \cdot \frac{L}{4} \left(\cosh^{-1} \frac{D}{2r} \right)^{-1}, \quad (L \gg D) \tag{7}$$

f. *Cylinder and an infinite plane:*

$h =$ separation of center of cylinder and plane

$$C = \frac{4\pi\epsilon}{\zeta} \cdot \frac{L}{2} \left[\cosh^{-1} \frac{h}{r} \right]^{-1}, \quad (L \gg h) \tag{8}$$

g. *Circular disk of radius r:*

$$C = \frac{4\pi\epsilon}{\zeta} \cdot \frac{2r}{\pi} \tag{9}$$

2.8. Normal stress on a conductor. Consider an element of surface dA of a conductor. Let the charge density on this element be σ. Then there will be an outward force dF normal to dA:

mks Gaussian

$$dF = \left(\frac{1}{2} \cdot \frac{\sigma}{\epsilon} \right) (\sigma dA), \qquad dF = \left(\frac{\sigma 4\pi}{2\epsilon} \right) (\sigma dA)$$

The normal stress on such a surface is

mks Gaussian

$$\frac{dF}{dA} = \frac{\sigma^2}{2\epsilon}, \qquad \frac{dF}{dA} = \frac{4\pi\sigma^2}{2\epsilon} \tag{1}$$

2.9. Energy density of the electrostatic field. The energy of a static system of charges is just the sum over all the charge of the potential energy of each element of charge due to the field of all the other charge. This energy is related to the field by the equation

mks Gaussian

$$W = \int \frac{\mathbf{E} \cdot \mathbf{D}}{2} \, dv, \qquad W = \int \frac{\mathbf{E} \cdot \mathbf{D}}{8\pi} \, dv \tag{1}$$

The energy density of the electric field is therefore taken to be

mks Gaussian

$$u = \frac{\mathbf{E} \cdot \mathbf{D}}{2}, \qquad u = \frac{\mathbf{E} \cdot \mathbf{D}}{8\pi} \tag{2}$$

3. Magnetostatics

3.1. Fundamental laws. For magnetostatic situations Maxwell's equations for the magnetic field reduce to

$$\text{mks} \qquad\qquad \text{Gaussian}$$

$$\nabla \times \boldsymbol{H} = \boldsymbol{J}, \qquad \nabla \times \boldsymbol{H} = \frac{4\pi \boldsymbol{J}}{c} \tag{1}$$

$$\nabla \cdot \boldsymbol{B} = 0, \qquad \nabla \cdot \boldsymbol{B} = 0 \tag{2}$$

Since $\boldsymbol{B}$ is divergenceless it may be derived from a vector potential

$$\boldsymbol{B} = \nabla \times \boldsymbol{A} \tag{3}$$

Then for the case of a nonferromagnetic, isotropic, homogeneous medium, Eq. (1) becomes

$$\text{mks} \qquad\qquad \text{Gaussian}$$

$$\nabla^2 \boldsymbol{A} = -\mu \boldsymbol{J}, \qquad \nabla^2 \boldsymbol{A} = -\frac{4\pi\mu}{c}\boldsymbol{J} \tag{4}$$

where

$$\nabla^2 \boldsymbol{A} = \nabla\nabla \cdot \boldsymbol{A} - \nabla \times \nabla \times \boldsymbol{A}$$

and we choose $\nabla \cdot \boldsymbol{A} = 0$. Equation (4) can be integrated by the vector analogue of the Green's function method * to give

$$\left.\begin{aligned} \boldsymbol{A} = &-\frac{1}{4\pi}\int_V \frac{\nabla^2 \boldsymbol{A}\,dv}{r} \\ &-\frac{1}{4\pi}\int_S \left\{ \frac{\boldsymbol{n}\times(\nabla\times \boldsymbol{A})}{r} + (\boldsymbol{n}\times \boldsymbol{A})\times\nabla\left(\frac{1}{r}\right) + \boldsymbol{n}\cdot \boldsymbol{A}\nabla\left(\frac{1}{r}\right)\right\} dS \end{aligned}\right\} \tag{5}$$

where $\boldsymbol{A}$ is the vector potential at a point inside a volume V bounded by surface S, and $\boldsymbol{n}$ is the unit outward normal to the surface. Letting S approach infinity where we assume $\boldsymbol{A}$ must vanish reduces Eq. (5) to

$$\text{mks} \qquad\qquad \text{Gaussian}$$

$$\boldsymbol{A} = \frac{\mu}{4\pi}\int \boldsymbol{J}\frac{dv}{r}, \qquad \boldsymbol{A} = \frac{\mu}{c}\int \boldsymbol{J}\frac{dv}{r} \tag{6}$$

The potential of the current element $\boldsymbol{i}\,d\boldsymbol{s} = \boldsymbol{J}\,dv$ is then

$$\text{mks} \qquad\qquad \text{Gaussian}$$

$$d\boldsymbol{A} = \frac{\mu}{4\pi r}\,\boldsymbol{i}\,d\boldsymbol{s}, \qquad d\boldsymbol{A} = \frac{\mu}{cr}\,\boldsymbol{i}\,d\boldsymbol{s} \tag{7}$$

* Stratton, J. A., *Electromagnetic Theory*, McGraw-Hill Book Co., Inc., New York, 1941, p. 250.

so that

mks	Gaussian	
$dB = \frac{\mu i}{4\pi} \cdot \frac{ds \times r}{r^3}$,	$dB = \frac{\mu i}{c} \frac{ds \times r}{r^3}$	(8)

Equation (8) is called the Biot-Savart law.

3.2. Fields of some simple current distributions. From Eqs. (6) and (8) of § 3.1, we can deduce the fields caused by various simple current distributions. Let $\zeta = 1$ in mks units, $\zeta = 4\pi$ in Gaussian units; let $\xi = 1$ in mks units, $\xi = 1/c$ in Gaussian units.

a. *Infinite straight wire:*

$$H = \zeta\xi \frac{i}{2\pi R} \tag{1}$$

b. *On the axis of a circular current loop* (Fig. 7):

$$H = \zeta\xi \frac{iR^2}{2(x^2 + R^2)^{3/2}} \tag{2}$$

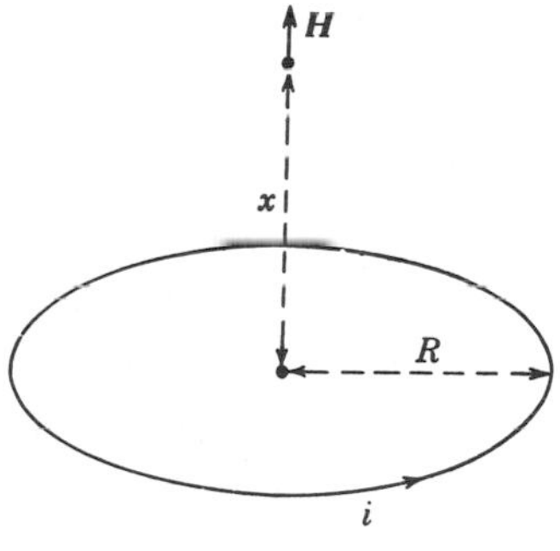

FIGURE 7

c. *On the axis of a solenoid* (Fig. 8):

$$H = \zeta\xi \frac{ni}{2} (\cos \theta_1 + \cos \theta_2) \tag{3}$$

where n = number of turns per unit length.

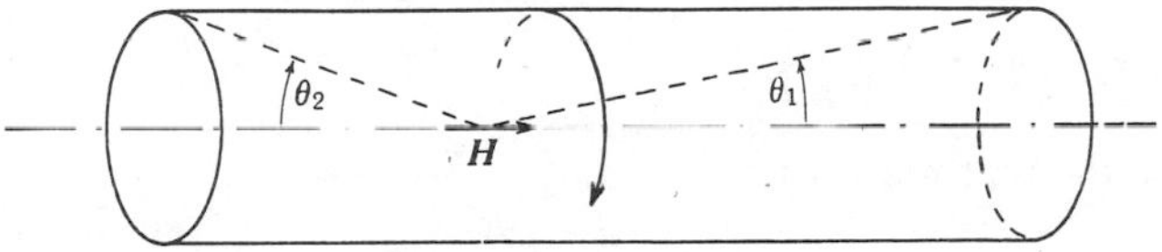

FIGURE 8

3.3. Scalar potential for magnetostatics; the magnetic dipole. For regions where $\boldsymbol{J}=0$, $\boldsymbol{H}$ is irrotational and may therefore be derived from a scalar potential function ψ.

$$\boldsymbol{H}=-\nabla\psi \tag{1}$$

The potential due to a loop of current is then

$$\text{mks} \qquad\qquad \text{Gaussian}$$
$$\psi=\frac{i\Omega}{4\pi}, \qquad \psi=\frac{i\Omega}{c} \tag{2}$$

where i is the current in the loop, and Ω is the solid angle determined by the loop and the point of observation. The sign of Ω is determined as follows. Choose an origin at the center of the loop (Fig. 9). Let $\boldsymbol{r}$ be the vector from the origin to the point of observation; let $\boldsymbol{a}$ be the vector from the origin to the current element $i\boldsymbol{ds}$. Then Ω has the same sign as $\boldsymbol{a}\times i\boldsymbol{ds}\cdot\boldsymbol{r}$.

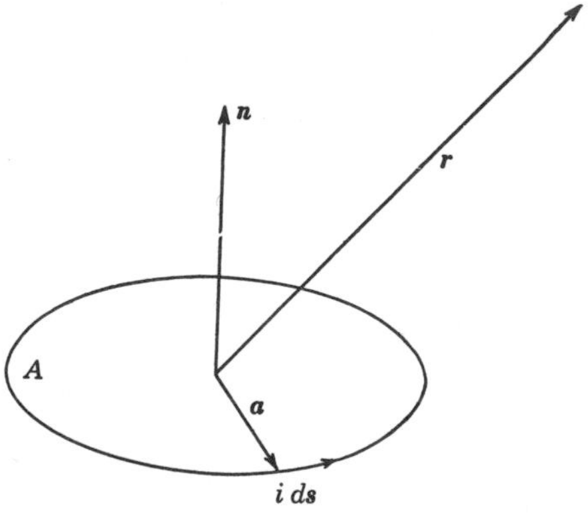

FIGURE 9

If the area A of the loop approaches zero and the current i in the loop, approaches infinity in such a way as to keep iA constant,

mks

$$\psi\rightarrow\frac{iA}{4\pi}\cdot\frac{\cos\theta}{r^2}=\frac{iA}{4\pi}\frac{\boldsymbol{n}\cdot\boldsymbol{r}}{r^3}=-\frac{iA}{4\pi}\boldsymbol{n}\cdot\nabla\left(\frac{1}{r}\right),$$

Gaussian

$$\psi\rightarrow\frac{iA}{c}\boldsymbol{n}\cdot\nabla\left(\frac{1}{r}\right) \tag{3}$$

where $\boldsymbol{n}$ is the unit normal to A.

$$n=\frac{\boldsymbol{a}\times i\boldsymbol{ds}}{|\boldsymbol{a}\times i\boldsymbol{ds}|}$$

By analogy with the expression for the potential due to the electric dipole, Eq. (4) of § 2.3, iA is called the magnetic dipole moment.

The preceding expressions are readily generalized to describe an arbitrary steady-state current distribution (which can always be regarded as consisting of a collection of current loops).

$$\left.\begin{aligned}&\text{mks} && \text{Gaussian}\\ &\psi = -\frac{iA}{4\pi}\mathbf{n}\cdot\nabla\left(\frac{1}{r}\right) && \psi = -\frac{\nabla(1/r)}{2c}\cdot\int \boldsymbol{\rho}\times \mathbf{J}dv\\ &\quad = -\frac{\nabla(1/r)}{8\pi}\cdot\oint \boldsymbol{\rho}\times i\mathbf{ds} &&\\ &\quad = -\frac{\nabla(1/r)}{8\pi}\cdot\int \boldsymbol{\rho}\times \mathbf{J}dv, &&\end{aligned}\right\} \quad (4)$$

where $\boldsymbol{\rho}$ is the radius vector from the origin to the current element $\mathbf{J}dv = i\mathbf{ds}$. To find the vector potential of a magnetic dipole:

$$B = \nabla\times A = \frac{\mu}{4\pi}\xi\zeta\nabla\left[\mathbf{m}\cdot\nabla\left(\frac{1}{r}\right)\right]$$

$$= \frac{+\mu}{4\pi}\xi\zeta\left\{(\mathbf{m}\cdot\nabla)\nabla\left(\frac{1}{r}\right) + \mathbf{m}\times\left[\nabla\times\nabla\left(\frac{1}{r}\right)\right]\right\}$$

$$= -\frac{\mu}{4\pi}(\zeta\xi)(\mathbf{m}\cdot\nabla)\nabla\left(\frac{1}{r}\right) = -\frac{\mu}{4\pi}\zeta\xi\nabla\times\left[\mathbf{m}\times\nabla\left(\frac{1}{r}\right)\right]$$

where $\mathbf{m} = iA\mathbf{n}$, and $\zeta\xi = 1$ in mks units; $\zeta\xi = 4\pi/c$ in Gaussian units. It is seen that the vector potential of a magnetic dipole is

$$\left.\begin{aligned}&\text{mks} && \text{Gaussian}\\ &A = -\frac{\mu}{4\pi}\mathbf{m}\times\nabla\left(\frac{1}{r}\right) && A = -\frac{\mu}{c}iA\mathbf{n}\times\nabla\left(\frac{1}{r}\right)\\ &\quad = -\frac{\mu}{4\pi}iA\mathbf{n}\times\nabla\left(\frac{1}{r}\right), &&\end{aligned}\right\} \quad (5)$$

3.4. Magnetic multiples. Let $\xi\zeta = 1$ for mks units; $\xi\zeta = 4\pi/c$ for Gaussian units. The vector potential dA at the point (x,y,z) due to the current element $\mathbf{J}dv = i\mathbf{ds}$ at the point (x_1,y_1,z_1) when

$$r = \sqrt{x^2+y^2+z^2} > r_1 = \sqrt{x_1^{\,2}+y_1^{\,2}+z_1^{\,2}}$$

is

$$dA = \frac{\mu\zeta\xi\mathbf{J}dv}{4\pi\sqrt{(x-x_1)^2+(y-y_1)^2+(z-z_1)^2}} = \sum_{n=0}^{\infty} dA_n \quad (1)$$

where

$$dA_n = \frac{\mu\zeta\xi}{4\pi} J(-1)^n \left[(\boldsymbol{r}_1 \cdot \nabla)_n (\boldsymbol{r}_1 \cdot \nabla)_{n-1} \cdots (\boldsymbol{r}_1 \cdot \nabla)_2 (\boldsymbol{r}_1 \cdot \nabla)_1 \left(\frac{1}{r} \right) \right] dv$$

and

$$\nabla = \boldsymbol{i}\frac{\partial}{\partial x} + \boldsymbol{j}\frac{\partial}{\partial y} + \boldsymbol{k}\frac{\partial}{\partial z}$$

If all the current is confined to the interior of a sphere of radius R, then for $r > R$

$$\left.\begin{aligned} \boldsymbol{A} &= \sum_{n=0}^{\infty} \boldsymbol{A}_n \\ &= \sum_{n=0}^{\infty} \frac{\mu\xi\zeta(-1)^n}{4\pi} \int \boldsymbol{J} \left[(\boldsymbol{r}_1 \cdot \nabla)_n (\boldsymbol{r}_1 \cdot \nabla)_{n-1} \cdots (\boldsymbol{r}_1 \cdot \nabla)_1 \left(\frac{1}{r} \right) \right] dv \end{aligned}\right\} \quad (2)$$

If the current distribution is stationary, the first term, $\boldsymbol{A}_0$, vanishes since in that case the current distribution can be analyzed into closed current loops and for the kth such loop

$$A_0{}^k = \frac{\mu_0\zeta\xi}{4\pi} \oint \frac{i}{r}\, \boldsymbol{ds} = \frac{\mu_0\zeta\xi}{4\pi r}\, i \oint \boldsymbol{dr}_1 = 0$$

The contribution of the kth current loop to A_1 is

$$\left.\begin{aligned} A_1{}^k &= -\frac{\mu\xi\zeta i_k}{4\pi} \oint \boldsymbol{r}_1 \cdot \nabla \left(\frac{1}{r} \right) \boldsymbol{dr}_1 \\ &= -\frac{\mu\xi\zeta i_k}{4\pi} \int \frac{(\boldsymbol{r}_1 \times \boldsymbol{dr}_1) \times \nabla(1/r)}{2} \\ &= -\frac{\mu\xi\zeta i_k}{4\pi} A_k \boldsymbol{n}_k \times \nabla \left(\frac{1}{r} \right) \end{aligned}\right\} \quad (3)$$

where A_k is the area of the kth loop and $\boldsymbol{n}_k$ is the unit normal to the loop. Thus by Eq. (5) of § 3.3,

$$A_1 = \Sigma_k A_1{}^k$$

is the vector potential of the magnetic dipole moment of the current distribution. Accordingly, A_n is identified with the vector potential of the magnetic 2^n pole moment of the distribution. Clearly, the magnetic monopole does not exist.

3.5. Magnetostatic boundary-value problems. In regions where $J = 0$ the scalar magnetic potential must satisfy the following conditions:

1. $\nabla^2\psi = 0$ at all points not on boundaries between two regions of different μ.
2. ψ is finite and continuous everywhere.
3. The normal derivatives of ψ across a surface are related by

$$\mu_2 \frac{\partial\psi_2}{\partial n} - \mu_1 \frac{\partial\psi_1}{\partial n} = 0$$

4. ψ vanishes at least as fast as $1/r^2$ as r becomes infinite, if all sources are confined to a finite region.

If $J \neq 0$ one must use the vectors B and H, which satisfy the following conditions:

1. $\nabla \times H = J$ (mks); $\nabla \times H = 4\pi/c\, J$ (Gaussian).
2. $B = \mu_0(H + M)$ (mks); $B = \mu_0(H + M)/4\pi$ (Gaussian), where M is the magnetization.
3. Across any boundary

$$\Delta(n \times H) = J_s \text{ (mks)}; \qquad \Delta(n \times H) = 4\pi/c\, J_s \text{ (Gaussian)}$$

$$\Delta(n \cdot B) = 0 \text{ (mks)}; \qquad \Delta(n \cdot B) = 0 \text{ (Gaussian)}$$

4. If sources are confined to a finite region, B vanishes at least as fast as $1/r^2$ as r becomes infinite.

These conditions determine the magnetic field uniquely.

In the absence of sources the solution ψ of a magnetostatic boundary value problem becomes identical with the solution φ of an electrostatic boundary value problem for the case of no charge when ϵ is replaced by μ.

As an example of a solution of a magnetostatic boundary value problem involving current, consider the field that results when an infinitely long straight wire of radius a, permeability μ_1, and with a current i is embedded in a medium of permeability μ_2, that contains an external field B_0 directed transverse to the axis of the wire.

$$\left.\begin{aligned} A_z &= \xi\left\{\frac{\mu_2}{2\pi} i \ln\frac{a}{r} - \frac{\mu_1}{4\pi} i + \left[r + \left(\frac{\mu_1 - \mu_2}{\mu_1 + \mu_2}\right)\frac{a^2}{r}\right] B_0 \sin\psi\right\}, \quad (r < a) \\ A_z &= \xi\left\{-\frac{\mu_1 i r^2}{4\pi a^2} + \frac{2\mu_1}{\mu_2 + \mu_1} B_0 r \sin\psi\right\} \qquad (r > a) \end{aligned}\right\} \quad (1)$$

$$A_x = A_y = 0$$

where $\xi = 1$ for mks units; $\xi = 1/c$ for Gaussian units.

The demagnetization factor L_1 is defined in a manner analogous to that for the depolarization factor, Eq. (5) of § 2.5,

$$L_1 M = H_0 - H_1$$

where M is the magnetization of the body in question, H_1 the field within the body, and H_0 the uniform external field.

3.6. Inductance. The quantity $\int \boldsymbol{B} \cdot d\boldsymbol{A} = n\Phi$ for a circuit is called the total flux linkage and is generally proportional to the current giving rise to $\boldsymbol{B}$. The self-inductance of a circuit is defined by the expression

$$L = \frac{n\Phi}{i} \tag{1}$$

where i is the current in the circuit. Similarly the mutual inductance between circuit 1 and circuit 2 is defined by

$$M_{12} = \frac{(n\Phi)_{12}}{i_2} \tag{2}$$

where $(n\Phi)_{12}$ is the total flux linkage of circuit 1 caused by current i_2 in circuit 2. It can be shown that

$$M_{12} = M_{21} \tag{3}$$

The inductances of some simple circuits are listed below, where $\xi\zeta = 1$ for mks units; $\xi\zeta = 4\pi/c$ for Gaussian units.

1. Circular loop, where radius of loop $= b$, radius of wire $= a$, $\mu' =$ magnetic permeability of the wire.

$$L = b\left[\mu\left(\ln\frac{8b}{a} - 2\right) + \frac{\mu'}{4}\right]\zeta\xi \tag{4}$$

2. Solenoid, where radius $= a$, length $= l$, number of turns per unit length $= n$.

$$L = \pi a^2 \mu n^2 [\sqrt{l^2 + a^2} - a]\zeta\xi \tag{5}$$

3. Parallel wires, where radii of the wires $= a, c$; separation of the centers $= b$; permeability of the wires $= \mu =$ permeability of the external medium.

$$L = \left(1 + 2\ln\frac{b^2}{ac}\right)\frac{\mu\xi\zeta}{4\pi} \tag{6}$$

4. Coaxial cable, where inner radius $= a$, outer radius $= b$.

$$L = 2\mu \ln\left(\frac{b}{a}\right)\zeta\xi = \text{self-inductance per unit length} \tag{7}$$

5. Two closely wound coils on a ring of permeability μ, where radius of the ring $= a$, radius of the cross section of the ring $= b$, and n and m are the respective numbers of turns in the two coils.

$$M = \mu nm[a - (a^2 - b^2)^{1/2}]\zeta\xi \tag{8}$$

6. Two circular loops (Fig. 10).

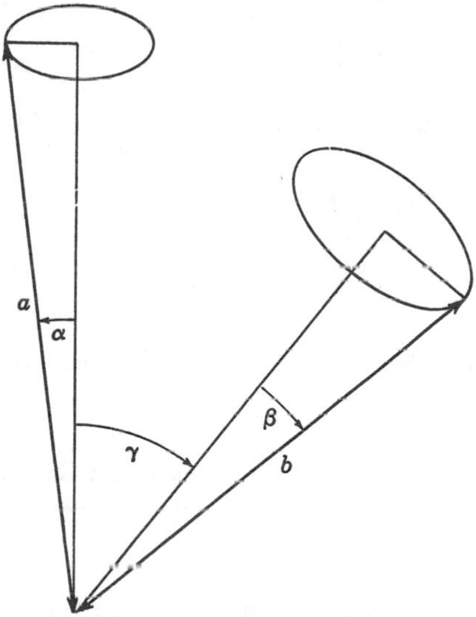

FIGURE 10

$$\left.\begin{aligned} M = \xi\zeta\pi\mu a \sin\alpha \sin\beta \\ \sum_{n=1}^{\infty} \frac{1}{n(n+1)} \left(\frac{a}{b}\right)^n P_n{}^0(\cos\alpha)\, P_n{}^0(\cos\beta)\, P_n{}^0(\cos\gamma) \end{aligned}\right\} \tag{9}$$

3.7. Magnetostatic energy density. The work required to establish current i in a circuit of self-inductance L is

$$W = \frac{1}{2} Li^2 \tag{1}$$

If we regard this energy as being stored in the magnetic field produced by the current i, the energy density of the magnetic field is

mks	Gaussian
$\dfrac{dW}{dv} = \dfrac{1}{2} \boldsymbol{H} \cdot \boldsymbol{B}$	$\dfrac{dW}{dv} = \dfrac{1}{8\pi} \boldsymbol{H} \cdot \boldsymbol{B}$ (2)

This type of energy is the electromagnetic analogue of mechanical kinetic energy.

4. Electric Circuits

4.1. The quasi-stationary approximation. The analysis of circuits is usually based on the assumption that the electric and magnetic fields at a given moment are essentially the same as would be produced by the charge and current distribution at that moment if these distributions were fixed in time. This is called the quasi-stationary approximation and is valid where the dimensions of the circuit are small compared to c/ω, where c is the velocity of light and ω is the angular frequency of current variation.

4.2. Voltage and impedance. The difference of potential between two points of a circuit is called the difference of voltage ΔV between these points. The impedance Z between two points of a circuit is the ratio of voltage drop to current between these points :

$$Z = \frac{\Delta V}{i} \tag{1}$$

It follows that the impedance of two circuit elements connected in series is

$$Z = Z_1 + Z_2 \tag{2}$$

and the impedance of two circuit elements connected in parallel is

$$Z = \frac{1}{1/Z_1 + 1/Z_2} \tag{3}$$

4.3. Resistors and capacitors in series and parallel connection.
Resistors :

series : $$R = R_1 + R_2 + R_3 + \dots = \text{total resistance} \tag{1}$$

parallel : $$\frac{1}{R} = \frac{1}{R_1} + \frac{1}{R_2} + \dots = (\text{total resistance})^{-1} \tag{2}$$

Capacitors :

parallel : $$C = C_1 + C_2 + C_3 + \dots = \text{total capacitance} \tag{3}$$

series : $$\frac{1}{C} = \frac{1}{C_1} + \frac{1}{C_2} + \dots = (\text{total capacitance})^{-1} \tag{4}$$

4.4. Kirchhoff's rules. Direct current circuit analysis is based on Kirchhoff's rules :

a. At any branch point of a circuit, the sum of the currents entering equals the sum of the currents leaving the junction.

b. The sum of the voltage drops around any closed loop of the network equals zero.

4.5. Alternating current circuits. Consider a circuit having a source of potential causing a voltage rise E between two points of the circuit. Let R be the resistance of the circuit, C be the capacitance of the circuit, and let L be its self inductance. Then by Faraday's law,

$$\left.\begin{aligned} \oint \mathbf{E} \cdot d\mathbf{r} &= -\int dV = -L\frac{di}{dt} = \left(\frac{q}{C} + iR - E\right) \\ \text{or} \qquad E &= iR + L\frac{di}{dt} + \frac{1}{C}\int idt \\ \frac{E}{L} &= \frac{d^2}{dt^2}q + \frac{R}{L}\cdot\frac{dq}{dt} + \frac{1}{CL}q \end{aligned}\right\} \quad (1)$$

The solution of this differential equation can be expressed as the sum of two parts

$$\left.\begin{aligned} q &= q_t + q_s \\ &= \text{transient solution} + \text{steady-state solution} \end{aligned}\right\} \quad (2)$$

There are three possibilities for q_t:

a. $R^2 > 4L/C$ (overdamping)

$$q_t = \frac{(i_0 - p_1 q_0)}{(p_2 - p_1)} e^{p_2 t} + \frac{(p_2 q_0 - i_0)}{(p_2 - p_1)} e^{p_1 t} \quad (3)$$

where

$$p_2 = -\frac{R}{2L} + \frac{1}{4L}\sqrt{R^2 - \frac{4L}{C}}$$

$$p_1 = -\frac{R}{2L} - \frac{1}{4L}\sqrt{R^2 - \frac{4L}{C}}$$

and q_0 and i_0 are the initial values of charge and current, respectively.

b. $R^2 = 4L/C$ (critical damping)

$$q_t = \left[q_0 + \left(i_0 + \frac{R}{2L}q_0\right)t\right]e^{-Rt/2L} \quad (4)$$

c. $R^2 < 4L/C$ (underdamping)

$$q_t = \left[\left(\frac{Rq_0}{2L\omega_1} + \frac{i_0}{\omega_1}\right)\sin\omega_1 t + q_0\cos\omega_1 t\right]e^{-Rt/2L} \quad (5)$$

where

$$\omega_1 = \frac{1}{\sqrt{LC}}\sqrt{1 - \frac{R^2C}{4L}}$$

The steady-state solution is

$$q_s = \frac{E_0 \sin(\omega t + \psi)}{L[(\omega_0{}^2 - \omega^2)^2 + (\omega_0\omega/Q)^2]^{1/2}} \quad (6)$$

where $E = E_0 \cos \omega t$; $Q = \omega_0 L/R$; $\omega_0 = 1/\sqrt{LC}$,

or

$$q_s = \frac{E_0 e^{j\omega t}}{L\left(\omega_0^2 - \omega^2 + j\,\dfrac{\omega\omega_0}{Q}\right)}, \quad (\text{if } E = E_0 e^{j\omega t}) \tag{7}$$

where $j = \sqrt{-1}$.

For the case where

$$E = E_0 e^{j\omega t}$$

the impedance for the steady state is

$$Z = \frac{E}{i} = R + j\left(\omega L - \frac{1}{\omega C}\right) \tag{8}$$

The instantaneous power delivered from the voltage source is

$$p = iE = i^2 Z = \frac{E^2}{Z} \tag{9}$$

The average power is

$$p_{\text{av}} = \frac{i^2 R}{2} = \frac{E_0 i_0}{2} \cos \theta_z \tag{10}$$

where E_0 and i_0 are the amplitudes of the respective periodic functions, and

$$\theta_z = \tan^{-1} \frac{\omega L - 1/\omega C}{R} \tag{11}$$

For a given circuit the current is greatest at the circular frequency ω_0,

$$\omega_0 = \sqrt{\frac{1}{LC}} \tag{12}$$

in which case $Z = R$.

5. Electromagnetic Radiation

5.1. Poynting's theorem. From Maxwell's equations and the vector identity

$$\nabla \cdot (\boldsymbol{E} \times \boldsymbol{H}) = \boldsymbol{H} \cdot \nabla \times \boldsymbol{E} - \boldsymbol{E} \cdot \nabla \times \boldsymbol{H}$$

it can be shown that

$$\left.\begin{aligned} &\text{mks} \\ &\nabla \cdot (\boldsymbol{E} \times \boldsymbol{H}) = -\frac{\partial}{\partial t} \cdot \frac{1}{2} (\boldsymbol{E} \cdot \boldsymbol{D} + \boldsymbol{B} \cdot \boldsymbol{H}) - \boldsymbol{E} \cdot \boldsymbol{J} \\ &\text{Gaussian} \\ &\nabla \cdot (\boldsymbol{E} \times \boldsymbol{H}) = -\frac{1}{c} \cdot \frac{\partial}{\partial t} \cdot \frac{1}{2} (\boldsymbol{E} \cdot \boldsymbol{D} + \boldsymbol{B} \cdot \boldsymbol{H}) - \boldsymbol{E} \cdot \boldsymbol{J} \end{aligned}\right\} \tag{1}$$

This is called Poynting's theorem and is interpreted by identifying

$$\text{mks} \qquad \text{Gaussian}$$

$$S = E \times H, \qquad S = \frac{c}{4\pi}(E \times H), \quad \text{(Poynting's vector)} \tag{2}$$

with the flux density of electromagnetic energy. In addition it can be shown that

$$\text{mks} \qquad \text{Gaussian}$$

$$S = \frac{1}{\sqrt{\epsilon\mu}}\left(\frac{E\cdot D}{2} + \frac{B\cdot H}{2}\right), \qquad S = \frac{c}{4\pi\sqrt{\epsilon\mu}}\left(\frac{E\cdot D}{2} + \frac{B\cdot H}{2}\right) \tag{3}$$

5.2. Electromagnetic stress and momentum. Suppose we have a charge distribution $\rho(x,y,z,t)$ in a homogeneous, isotropic medium. Then the total force of electromagnetic origin on the charge contained in volume V is

$$\text{mks} \qquad \text{Gaussian}$$

$$F = \int_V (\rho E + J \times B)dv, \qquad F = \int_V \left(\rho E + \frac{J \times B}{c}\right)dv \tag{1}$$

It is possible to reduce this equation to the following form.

$$F = -\int_V \epsilon\mu \frac{\partial S}{\partial t} dv + \zeta \iint_S (\bar{X}\cdot dAi + \bar{Y}\cdot dAj + \bar{Z}\cdot dAk) \tag{2}$$

where $\zeta = 1$ in mks units, $\zeta = 1/4\pi$ in Gaussian units, and

$$\left.\begin{aligned}
\zeta\bar{X}_x &= \zeta\tfrac{1}{2}[\epsilon(E_x^2 - E_y^2 - E_z^2) + \mu(H_x^2 - H_y^2 - H_z^2)] = T_{11}\\
\zeta\bar{X}_y &= \zeta(\epsilon E_xE_y + \mu H_xH_y) = T_{12}\\
\zeta\bar{X}_z &= \zeta(\epsilon E_xE_z + \mu H_xH_z) = T_{13}\\
\zeta\bar{Y}_x &= \zeta(\epsilon E_xE_y + \mu H_xH_y) = T_{21}\\
\zeta\bar{Y}_y &= \zeta\tfrac{1}{2}[\epsilon(E_y^2 - E_x^2 - E_z^2) + \mu(H_y^2 - H_x^2 - H_z^2)] = T_{22}\\
&\cdots \quad \cdots \quad \cdots \quad \cdots \quad \cdots \quad \cdots
\end{aligned}\right\} \tag{3}$$

For the case of stationary fields and charge distributions, the first term on the right of Eq. (3) vanishes, and it becomes apparent that T_{ij} is only the equilibrium electromagnetic stress tensor.

On the other hand, if the surface S is taken to be at infinity, the second term on the right of Eq. (3) vanishes, and

$$\text{mks} \qquad \text{Gaussian}$$

$$F = -\epsilon\mu \int_V \frac{\partial S}{\partial t} dv = \frac{dP}{dt}, \qquad F = -\frac{\epsilon\mu}{c^2}\int_V \frac{\partial S}{\partial t} dv = \frac{dP}{dt}$$

or

$$\frac{d}{dt}(P + G) = 0$$

mks Gaussian

where $$G = \epsilon\mu \int S dv, \qquad G = \frac{\epsilon\mu}{c^2} \int S dv \tag{4}$$

and P is the total mechanical momentum.

But since the boundaries of the system have been taken at infinity, the system must be closed. Hence its momentum must be conserved. One is thus led to identify G with the electromagnetic momentum. Accordingly, the electromagnetic momentum density is

mks Gaussian

$$g = \epsilon\mu S, \qquad g = \frac{\epsilon\mu}{c^2} S \tag{5}$$

5.3. The Hertz vector; electromagnetic waves. A completely general description of an electromagnetic field is provided by the specification of the four scalar functions that comprise the vector and scalar potentials A and φ. Very often one deals with fields that have symmetry properties rendering them susceptible to a simpler description involving less than four scalar functions. The basis for such a simplified description is provided by requiring the scalar potential or its counterpart to be some specific function of the vector potential. To see how this may best be done, perform a gauge transformation with an arbitrary function f yielding a new vector potential

$$A = A_{\text{old}} - \nabla f$$

and a new arbitrary scalar potential

$$\varphi = \varphi_{\text{old}} + \frac{\partial}{\partial t} f$$

for which Eqs. (4) and (5) of § 1.6 assume the form

$$\left.\begin{aligned} &\nabla \cdot \left(\epsilon \frac{\partial}{\partial t} A - P + \epsilon \nabla\varphi \right) = 0 && \text{(mks)} \\ &\nabla \cdot \left(\frac{\epsilon}{c} \cdot \frac{\partial}{\partial t} A - 4\pi P + \epsilon \nabla\varphi \right) = 0, && \text{(Gaussian)} \end{aligned}\right\} \tag{1}$$

$$\left.\begin{aligned} &\frac{\partial}{\partial t}\left(\nabla \times \nabla \times \int A \frac{dt}{\mu} + \epsilon \frac{\partial}{\partial t} A - P + \epsilon \nabla\varphi \right) = 0 && \text{(mks)} \\ &\frac{\partial}{\partial t}\left(\nabla \times \nabla \times \int A \frac{c\,dt}{\mu} + \frac{\epsilon}{c} \cdot \frac{\partial}{\partial t} A - 4\pi P + \epsilon \nabla\varphi \right) = 0, && \text{(Gaussian)} \end{aligned}\right\} \tag{2}$$

where $$P = \int J dt \tag{3}$$

so that $J = \partial P/\partial t$, and $\rho = -\nabla \cdot P$ by Eq. (1) of § 1.2.

When $\nabla \cdot \boldsymbol{P} = -\rho = 0$, Eqs. (1) and (2) of § 5.3 assume a simple form if f is chosen so that φ vanishes.

$$\left.\begin{array}{ll} \text{mks} & \text{Gaussian} \\ \nabla \cdot \boldsymbol{A} = 0 & \nabla \cdot \boldsymbol{A} = 0 \\ \nabla^2 \boldsymbol{A} - \mu\epsilon \dfrac{\partial^2}{\partial t^2} \boldsymbol{A} = -\mu \boldsymbol{J}, & \nabla^2 \boldsymbol{A} - \dfrac{\mu\epsilon}{c^2} \cdot \dfrac{\partial^2}{\partial t^2} \boldsymbol{A} = -\dfrac{4\pi\mu}{c} \boldsymbol{J} \end{array}\right\} \quad (4)$$

Thus the vector potential is useful for the simplified description of fields in the absence of charge.

Clearly Eqs. (1) and (2) of § 5.3 are fulfilled if it is required that

$$\left.\begin{array}{ll} \nabla \times \nabla \times \int \boldsymbol{A}\,dt + \epsilon\mu \dfrac{\partial}{\partial t} \boldsymbol{A} - \mu \boldsymbol{P} + \epsilon\mu \nabla\varphi = 0 & \text{(mks)} \\ \nabla \times \nabla \times \int \boldsymbol{A}\,dt + \dfrac{\epsilon\mu}{c^2} \dfrac{\partial}{\partial t} \boldsymbol{A} - \dfrac{4\pi\mu}{c} \boldsymbol{P} + \dfrac{\epsilon\mu}{c} \nabla\varphi = 0 & \text{(Gaussian)} \end{array}\right\} \quad (5)$$

In this way it is possible to reduce the four equations (4) and (5) of § 1.6 to three equations. Equation (5) has a simpler form when expressed in terms of the Hertz vector $\boldsymbol{\Pi}$.

$$\begin{array}{ll} \text{mks} & \text{Gaussian} \\ \boldsymbol{\Pi} = \dfrac{1}{\mu} \int \boldsymbol{A}\,dt, & \boldsymbol{\Pi} = \dfrac{c}{\mu} \int \boldsymbol{A}\,dt \end{array} \quad (6)$$

In terms of the Hertz vector, Eq. (5) becomes

$$\left.\begin{array}{ll} \nabla \times \nabla \times \boldsymbol{\Pi} + \epsilon\mu \dfrac{\partial^2}{\partial t^2} \boldsymbol{\Pi} - \boldsymbol{P} + \epsilon \nabla\varphi = 0 & \text{(mks)} \\ \nabla \times \nabla \times \boldsymbol{\Pi} + \dfrac{\epsilon\mu}{c^2} \dfrac{\partial^2}{\partial t^2} \boldsymbol{\Pi} - 4\pi \boldsymbol{P} + \epsilon \nabla\varphi = 0 & \text{(Gaussian)} \end{array}\right\} \quad (7)$$

If the gauge

$$\varphi = -\nabla \cdot \boldsymbol{\Pi}/\epsilon \quad (8)$$

is chosen, Eq. (7) reduces to a form particularly convenient when $\boldsymbol{\Pi}$ is expressed in rectangular coordinates.

$$\begin{array}{ll} \text{mks} & \text{Gaussian} \\ \nabla^2 \boldsymbol{\Pi} - \mu\epsilon \dfrac{\partial^2}{\partial t^2} \boldsymbol{\Pi} = -\boldsymbol{P}, & \nabla^2 - \dfrac{\mu\epsilon}{c^2} \cdot \dfrac{\partial^2}{\partial t^2} \boldsymbol{\Pi} = -4\pi \boldsymbol{P} \end{array} \quad (9)$$

where

$$\left.\begin{array}{ll} \boldsymbol{H} = \nabla \times \dfrac{\partial}{\partial t} \boldsymbol{\Pi}, & \boldsymbol{H} = \dfrac{1}{c} \nabla \times \dfrac{\partial}{\partial t} \boldsymbol{\Pi} \\ \boldsymbol{D} = \nabla\nabla \cdot \boldsymbol{\Pi} - \epsilon\mu \dfrac{\partial^2}{\partial t^2} \boldsymbol{\Pi}, & \boldsymbol{D} = \nabla\nabla \cdot \boldsymbol{\Pi} - \dfrac{\epsilon\mu}{c^2} \cdot \dfrac{\partial^2}{\partial t^2} \boldsymbol{\Pi} \end{array}\right\} \quad (10)$$

In a region where $\boldsymbol{P}$ vanishes, Maxwell's equations are symmetrical in $\boldsymbol{E}$ and $\boldsymbol{H}$ and in $\boldsymbol{B}$ and $\boldsymbol{D}$, so that in such regions there is a second possible field represented by the Hertz vector $\boldsymbol{\Pi}_1$.

mks	Gaussian
$\boldsymbol{E} = \nabla \times \frac{\partial}{\partial t} \boldsymbol{\Pi}_1,$	$\boldsymbol{E} = \frac{1}{4\pi} \nabla \times \frac{\partial}{\partial t} \boldsymbol{\Pi}_1$
$\boldsymbol{B} = -\nabla\nabla \cdot \boldsymbol{\Pi}_1 + \epsilon\mu \frac{\partial^2}{\partial t^2} \boldsymbol{\Pi}_1,$	$\boldsymbol{B} = \frac{c}{4\pi}\left(-\nabla\nabla \cdot \boldsymbol{\Pi}_1 + \frac{\epsilon\mu}{c^2} \frac{\partial^2}{\partial t^2} \boldsymbol{\Pi}_1\right)$

(11)

where

$$\nabla^2 \boldsymbol{\Pi}_1 - \mu\epsilon \frac{\partial^2}{\partial t^2} \boldsymbol{\Pi}_1 = 0, \qquad \nabla^2 \boldsymbol{\Pi}_1 - \frac{\mu\epsilon}{c^2} \frac{\partial^2}{\partial t^2} \boldsymbol{\Pi}_1 = 0 \tag{12}$$

Inside an isotropic, homogeneous conductor free from external sources, we again are able to have two distinct types of solutions, each of which can be expressed in terms of a Hertz vector.

mks

$$\left.\begin{aligned} \boldsymbol{E} &= \boldsymbol{E}_1 + \boldsymbol{E}_2 = -\frac{\nabla \times \nabla \times \boldsymbol{\Pi}_2}{\epsilon} + \mu \nabla \times \frac{\partial}{\partial t} \boldsymbol{\Pi}_1 \\ \boldsymbol{H} &= \boldsymbol{H}_1 + \boldsymbol{H}_2 = \nabla \times \nabla \times \boldsymbol{\Pi}_1 + \nabla \times \frac{\partial}{\partial t} \boldsymbol{\Pi}_2 + \frac{\sigma}{\epsilon} \nabla \times \boldsymbol{\Pi}_2 \end{aligned}\right\} \tag{13}$$

$$\left.\begin{aligned} \nabla^2 \boldsymbol{\Pi}_2 - \mu\epsilon \frac{\partial^2}{\partial t^2} \boldsymbol{\Pi}_2 - \mu\sigma \frac{\partial}{\partial t} \boldsymbol{\Pi}_2 &= 0 \\ \nabla^2 \boldsymbol{\Pi}_1 - \mu\epsilon \frac{\partial^2}{\partial t^2} \boldsymbol{\Pi}_1 - \mu\sigma \frac{\partial}{\partial t} \boldsymbol{\Pi}_1 &= 0 \end{aligned}\right\} \tag{14}$$

Gaussian

$$\left.\begin{aligned} \boldsymbol{E} &= \boldsymbol{E}_1 + \boldsymbol{E}_2 = \frac{\mu}{c^2} \nabla \times \frac{\partial \boldsymbol{\Pi}_1}{\partial t} - \frac{1}{\epsilon} \nabla \times \nabla \times \boldsymbol{\Pi}_2 \\ \boldsymbol{H} &= \boldsymbol{H}_1 + \boldsymbol{H}_2 = \frac{1}{c} \nabla \times \nabla \times \boldsymbol{\Pi}_1 + \frac{1}{c} \nabla \times \frac{\partial}{\partial t} \boldsymbol{\Pi}_2 + \frac{4\pi\sigma}{\epsilon c} \nabla \times \boldsymbol{\Pi}_2 \end{aligned}\right\} \tag{15}$$

$$\left.\begin{aligned} \nabla^2 \boldsymbol{\Pi}_2 - \frac{\mu\epsilon}{c^2} \cdot \frac{\partial^2}{\partial t^2} \boldsymbol{\Pi}_2 - \frac{4\pi\mu\sigma}{c^2} \cdot \frac{\partial}{\partial t} \boldsymbol{\Pi}_2 &= 0 \\ \nabla^2 \boldsymbol{\Pi}_1 - \frac{\mu\epsilon}{c^2} \cdot \frac{\partial^2}{\partial t^2} \boldsymbol{\Pi}_1 - \frac{4\pi\mu\sigma}{c^2} \cdot \frac{\partial}{\partial t} \boldsymbol{\Pi}_1 &= 0 \end{aligned}\right\} \tag{16}$$

Since the most general separable time dependence can be represented as a sum or integral over simple harmonic terms, let

$$\boldsymbol{\Pi} = \boldsymbol{\Pi}_0 e^{-j\omega t}, \quad (j = \sqrt{-1}) \tag{17}$$

Then Eqs. (14) and (16) become the vector wave equations

mks	Gaussian	
$\nabla^2\mathbf{\Pi}_0 + \mu\epsilon\omega^2\left(1 + \dfrac{j\sigma}{\omega\epsilon}\right)\mathbf{\Pi}_0 = 0,$	$\nabla^2\mathbf{\Pi}_0 + \dfrac{\mu\epsilon}{c^2}\omega^2\left(1 + \dfrac{j4\pi\sigma}{\omega\epsilon}\right)\mathbf{\Pi}_0 = 0$	(18)
$\nabla^2\mathbf{\Pi}_0 + k^2\left(1 + \dfrac{j\sigma}{\omega\epsilon}\right)\mathbf{\Pi}_0 = 0,$	$\nabla^2\mathbf{\Pi}_0 + k^2\left(1 + \dfrac{j4\pi\sigma}{\omega\epsilon}\right)\mathbf{\Pi}_0 = 0$	(19)
$\nabla^2\mathbf{\Pi}_0 + \kappa^2\mathbf{\Pi}_0 = 0,$	$\nabla^2\mathbf{\Pi}_0 + \kappa^2\mathbf{\Pi}_0 = 0$	(20)

where

$$\left.\begin{aligned} k &= \sqrt{\mu\epsilon}\,\omega & \qquad k &= \sqrt{\mu\epsilon}\,\frac{\omega}{c} \\ \kappa &= k\sqrt{1 + \frac{j\sigma}{\omega\epsilon}} & \qquad \kappa &= k\sqrt{1 + \frac{j4\pi\sigma}{\omega\epsilon}} \end{aligned}\right\} \qquad (21)$$

For $(\sigma/\omega\epsilon) \ll 1$ this equation represents a wave traveling with phase velocity

mks	Gaussian	
$v = \dfrac{1}{\sqrt{\epsilon\mu}} = \dfrac{1}{\sqrt{\epsilon_0\mu_0}\,\sqrt{\kappa_e\kappa_m}} = \dfrac{c}{n},$	$v = \dfrac{c}{\sqrt{\epsilon\mu}} = \dfrac{c}{\sqrt{\kappa_e\kappa_m}} = \dfrac{c}{n}$	(22)

where $c =$ velocity of propagation of light in free space and $n = \sqrt{\kappa_e\kappa_m}$ is called the index of refraction.

5.4. Plane waves. To find the expression for a plane wave traveling along the x axis in a source-free region, let

$$\mathbf{\Pi}_0 = o\mathbf{i} + o\mathbf{j} + \mathbf{\Pi}(x)\mathbf{k}$$

Then Eqs. (13), (14), and (18) of 5.3 have the solution

$$\left.\begin{aligned} &\text{mks} & &\text{Gaussian} \\ \mathbf{E}_1 &= -(\mu\omega\kappa)e^{\pm j\kappa x}\mathbf{j}, & \mathbf{E}_1 &= -\frac{\mu\omega\kappa}{c^2}\,e^{\pm j\kappa x}\mathbf{j} \\ \mathbf{E}_2 &= \frac{\kappa^2}{\epsilon}\,e^{\pm j\kappa x}\mathbf{k}, & \mathbf{E}_2 &= \frac{\kappa^2}{\epsilon}\,e^{\pm j\kappa x}\mathbf{k} \\ \mathbf{H}_1 &= -\kappa^2 e^{\pm j\kappa x}\mathbf{k}, & \mathbf{H}_1 &= -\frac{\kappa^2}{c}\,e^{\pm j\kappa x}\mathbf{k} \\ \mathbf{H}_2 &= -\left(\kappa\omega - j\frac{\kappa\sigma}{\epsilon}\right)e^{\pm j\kappa x}\mathbf{j}, & \mathbf{H}_2 &= -\left(\frac{\kappa\omega}{c} - j\frac{\kappa\sigma 4\pi}{\epsilon c}\right)e^{\pm j\kappa x}\mathbf{j} \end{aligned}\right\} \qquad (1)$$

The general solution is thus seen to satisfy the relations

$$\boldsymbol{E} = \boldsymbol{E}_1 + \boldsymbol{E}_2, \qquad \boldsymbol{H} = \boldsymbol{H}_1 + \boldsymbol{H}_2$$

$$\boldsymbol{E} \cdot \boldsymbol{H} = E_1 H_2 + E_2 H_1 = 0 \tag{2}$$

$$\boldsymbol{i} \times \boldsymbol{E} = Z\boldsymbol{H} \tag{3}$$

where

mks Gaussian

$$Z = \frac{\mu\omega}{\kappa} = \sqrt{\frac{\mu}{\epsilon + (j\sigma/\omega)}}, \qquad Z = \frac{4\pi}{c^2}\frac{\mu\omega}{\kappa} = \frac{4\pi}{c}\sqrt{\frac{\mu}{\epsilon + (j4\pi\sigma/\omega)}} \tag{4}$$

where Z is called the wave impedance.

5.5. Cylindrical waves. Let R, ψ, z be the parameters of a circular cylindrical coordinate system. To find the representation for cylindrical waves in a source-free region let the Hertz vector be

$$\boldsymbol{\Pi}_0 = o\boldsymbol{i} + o\boldsymbol{j} + \Pi(R\psi z)\boldsymbol{k}$$

Then Eq. (18) of § 5.13,

$$\nabla^2 \boldsymbol{\Pi}_0 + \kappa^2 \boldsymbol{\Pi}_0 = 0 \quad \text{becomes} \quad \nabla^2 \Pi + \kappa^2 \Pi = 0$$

which has the solutions

$$\Pi^{(n)} = e^{jn\psi}\, e^{\pm jhz} Z_n(\sqrt{\kappa^2 - h^2}\, R) \tag{1}$$

where h is a separation constant and Z_n is a solution of the nth order Bessel equation. From Eqs. (13) and (14) of § 5.3, we see that this gives two solutions :

Transverse magnetic

$$\left.\begin{array}{ll}
\text{mks} & \text{Gaussian} \\
E_R = jh\dfrac{\partial \Pi^{(n)}}{\partial R}, & E_R = \dfrac{jh}{4\pi}\cdot\dfrac{\partial \Pi^{(n)}}{\partial R} \\
E_\psi = -\dfrac{h}{R}n\Pi^{(n)}, & E_\psi = -\dfrac{h}{4\pi R}n\Pi^{(n)} \\
E_z = (\kappa^2 - h^2)\Pi^{(n)}, & E_z = \dfrac{(\kappa^2 - h^2)\Pi^{(n)}}{4\pi} \\
H_R = \dfrac{\kappa^2}{\mu\omega R}n\Pi^{(n)}, & H_R = \dfrac{\kappa^2 c n \Pi^{(n)}}{4\pi\mu\omega R} \\
H_\psi = \dfrac{j\kappa^2}{\mu\omega}\cdot\dfrac{\partial \Pi^{(n)}}{\partial R}, & H_\psi = \dfrac{j\kappa^2 c}{4\pi\mu\omega}\cdot\dfrac{\partial \Pi^{(n)}}{\partial R} \\
H_z = 0, & H_z = 0
\end{array}\right\} \tag{2}$$

Transverse electric

mks	Gaussian
$E_R = -\dfrac{\mu\omega n\Pi^{(n)}}{R}$,	$E_R = -\dfrac{\mu\omega n\Pi^{(n)}}{4\pi R}$
$E_\varphi = -j\mu\omega\dfrac{\partial\Pi^{(n)}}{\partial R}$,	$E_\varphi = -\dfrac{j\mu\omega}{c^2}\dfrac{\partial\Pi^{(n)}}{\partial R}$
$E_z = 0$,	$E_z = 0$
$H_R = jh\dfrac{\partial\Pi^{(n)}}{\partial R}$,	$H_R = \dfrac{jh}{c}\dfrac{\partial\Pi^{(n)}}{\partial R}$
$H_\varphi = -\dfrac{h}{R}n\Pi^{(n)}$,	$H_\varphi = -\dfrac{hn}{cR}\Pi^{(n)}$
$H_z = (\kappa^2 - h^2)\Pi^{(n)}$,	$H_z = \dfrac{(\kappa^2 - h^2)\Pi^{(n)}}{c}$

(3)

5.6. Spherical waves. Let r, θ, φ be the parameters of a spherical coordinate system. To find the representation for spherical electromagnetic waves in a source-free region, take the Hertz vector to be

$$\mathbf{\Pi}_0 = \Pi_0(r,\theta,\varphi)\boldsymbol{u}_r + o\boldsymbol{u}_\theta + o\boldsymbol{u}_\varphi$$

Then Eq. (8) of 5.3 becomes

$$\left.\begin{aligned}
&-\epsilon\frac{\partial\varphi_0}{\partial r} + \frac{1}{r^2\sin\theta}\cdot\frac{\partial}{\partial\theta}\sin\theta\frac{\partial\Pi_0}{\partial\theta} + \frac{1}{r^2\sin^2\theta}\frac{\partial^2\Pi_0}{\partial\varphi^2} + \kappa^2\Pi_0 = 0\\
&\frac{-\epsilon}{r}\cdot\frac{\partial\varphi_0}{\partial\theta} - \frac{1}{r}\frac{\partial}{\partial\theta}\left(\frac{\partial\Pi_0}{\partial r}\right) = 0\\
&\frac{-\epsilon}{r\sin\theta}\cdot\frac{\partial\varphi_0}{\partial\varphi} - \frac{1}{r\sin\theta}\frac{\partial}{\partial\varphi}\left(\frac{\partial\Pi_0}{\partial r}\right) = 0
\end{aligned}\right\}\qquad(1)$$

To satisfy the last two equations we need require only that the arbitrary scalar potential φ_0 be equal to $-(1/\epsilon)(\partial\Pi_0/\partial r)$. With this substitution, the first equation reduces to the form

$$\nabla^2\left(\frac{\Pi_0}{r}\right) + \kappa^2\left(\frac{\Pi_0}{r}\right) = 0 \qquad (2)$$

The general solution of this equation is

$$\left.\begin{aligned}
&\frac{\Pi_0}{r} = \frac{1}{\sqrt{\kappa r}}Z_{n+1/2}(\kappa r)P_n^{\,m}(\cos\theta)e^{\pm jm\varphi}\\
&(n = 0, 1, 2, 3, \ldots;\ m = 0, 1, 2, \ldots)
\end{aligned}\right\}\qquad(3)$$

where $Z_{n+(1/2)}$ is a solution of the $(n + \frac{1}{2})$ order Bessel equation, and P_n^m is the associated Legendre polynomial.

From Eq. (13) of § 5.3, we see how these solutions may be used to give two kinds of spherical waves.

Tranverse electric

mks	Gaussian
$E_r = 0,$	$E_r = E_r$ of mks system
$E_\theta = \frac{j\omega}{r \sin\theta} \cdot \frac{\partial \Pi_0}{\partial \varphi},$	$E_\theta = E_\theta$ of mks system
$E_\varphi = -\frac{j\omega}{r} \cdot \frac{\partial \Pi_0}{\partial \theta},$	$E_\varphi = E_\varphi$ of mks system
$H_r = \frac{1}{\mu}\left(\frac{\partial^2 \Pi_0}{\partial r^2} + \kappa^2 \Pi_0\right) = \frac{l(l+1)\Pi_0}{\mu r^2},$	$H_r = cH_r$ of mks system
$H_\theta = \frac{1}{\mu r} \frac{\partial^2 \Pi_0}{\partial r \partial \theta},$	$H_\theta = cH_\theta$ of mks system
$H_\varphi = \frac{1}{\mu r \sin\theta} \frac{\partial^2 (\Pi_0)}{\partial r \partial \varphi},$	$H_\varphi = cH_\varphi$ of mks system

(4)

Tranverse magnetic

mks	Gaussian
$E_r = \frac{1}{\epsilon}\left(\frac{\partial^2 \Pi_0}{\partial r^2} + \kappa^2 \Pi_0\right) = \frac{l(l+1)\Pi_0}{\epsilon r^2},$	$E_r = 4\pi E_r$ of mks system
$E_\theta = \frac{1}{r\epsilon} \cdot \frac{\partial^2 \Pi_0}{\partial r \partial \theta},$	$E_\theta = 4\pi E_\theta$ of mks system
$E_\varphi = \frac{1}{\epsilon r \sin\theta} \cdot \frac{\partial^2 \Pi_0}{\partial r \partial \varphi},$	$E_\varphi = 4\pi E_\varphi$ of mks system
$H_r = 0,$	$H_r = (4\pi/c) H_r$ of mks system
$H_\theta = -\frac{j\omega}{r \sin\theta} \cdot \frac{\partial \Pi_0}{\partial \varphi},$	$H_\theta = (4\pi/c) H_\theta$ of mks system
$H_\varphi = \frac{j\omega}{r} \cdot \frac{\partial \Pi_0}{\partial \theta},$	$H_\varphi = (4\pi/c) H_\varphi$ of mks system

(5)

where l is an integer.

5.7. Radiation of electromagnetic waves; the oscillating dipole. To find the expressions relating electromagnetic radiation in an isotropic medium to its sources we must solve the inhomogeneous vector wave equation for the Hertz vector.

$$\nabla^2\mathbf{\Pi} - \frac{1}{v^2}\cdot\frac{\partial^2\mathbf{\Pi}}{\partial t^2} = -\zeta\boldsymbol{P} \tag{1}$$

where $\zeta = 1$ for mks units; $\zeta = 4\pi$ for Gaussian units.

This is equivalent to the three inhomogeneous scalar wave equations for the rectangular components of $\boldsymbol{\pi}$ and $\boldsymbol{P}$.

$$\nabla^2\mathbf{\Pi}_i - \frac{1}{v^2}\cdot\frac{\partial^2\mathbf{\Pi}_i}{\partial t^2} = -\zeta P_i, \quad (i = 1,2,3) \tag{2}$$

A solution of this equation is

$$\left.\begin{aligned}\mathbf{\Pi}_i(x'y'z't) &= \frac{1}{4\pi}\int_V \frac{\zeta P_i^*}{r}\,dV \\ &+ \frac{1}{4\pi}\int_S\left[\frac{\nabla\Pi_i^*}{r} - \Pi_i^*\nabla\left(\frac{1}{r}\right) + \frac{1}{vr}\nabla\left(\frac{\partial\Pi_i^*}{\partial t}\right)\right]\cdot\boldsymbol{n}dS\end{aligned}\right\} \tag{3}$$

where (x',y',z') is the point of observation in volume V bounded by surface S, $r = \sqrt{(x-x')^2 + (y-y')^2 + (z-z')^2}$, $f^* = f(x,y,z,t-r/v)$, and $\boldsymbol{n}$ is the unit outward normal to surface element dS.

In an unbounded medium this becomes just

$$\mathbf{\Pi} = \frac{\zeta}{4\pi}\iiint\limits_{\substack{\text{all}\\\text{space}}} \frac{\boldsymbol{P}(x,y,z,t-r/v)}{r}\,dxdydz \tag{4}$$

If the time dependence of $\boldsymbol{P}$ is simple harmonic

$$\boldsymbol{P} = e^{-j\omega t}\boldsymbol{P}_0(x,y,z) \tag{5}$$

then

$$\mathbf{\Pi} = \frac{\zeta e^{-j\omega t}}{4\pi}\iiint\limits_{\substack{\text{all}\\\text{space}}} \boldsymbol{P}_0\frac{(xyz)e^{j\omega r/v}}{r}\,dxdydz \tag{6}$$

An important example is the radiation field of an oscillating electric dipole of moment $p = p_0e^{-j\omega t}$ along the z axis.

$$\begin{aligned}p &= e^{-j\omega t}\int z\rho dv = -e^{-j\omega t}\int z\nabla\cdot\boldsymbol{P}_0dv \\ &= e^{-j\omega t}\left[\int\boldsymbol{k}\cdot\boldsymbol{P}_0dv - \iint z\boldsymbol{P}_0\cdot\boldsymbol{n}dS\right]\end{aligned}$$

Taking the surface infinitely far from the origin

$$p = e^{-j\omega t}\int P_{0z}dv$$

And since $J_y = J_x = 0$, we must have $P_{0y} = P_{0x} = 0$ so

$$\boldsymbol{k}p = \int \boldsymbol{P} dv = \int \boldsymbol{P}_0 e^{-j\omega t} dv$$

Substituting this into Eq. (6) of 5.1, we get

$$\boldsymbol{\Pi} = \boldsymbol{k} \frac{\zeta}{4\pi} p_0 \frac{e^{-j\omega(t-r/v)}}{r}$$

Applying Eq. (10) of § 5.3 to the above Hertz vector yields the following field for the electric dipole of moment $p = p_0 e^{-j\omega t}$ along the z axis.

$$\left.\begin{aligned} E_r &= \frac{p_0 \kappa^3}{4\pi\epsilon} \zeta e^{-j(\omega t - \kappa r)} \cos\theta \left[\frac{-2j}{(\kappa r)^2} + \frac{2}{(\kappa r)^3}\right] \\ E_\theta &= \frac{p_0 \kappa^3}{4\pi\epsilon} \zeta e^{-j(\omega t - \kappa r)} \sin\theta \left[-\frac{1}{\kappa r} - \frac{j}{(\kappa r)^2} + \frac{1}{(\kappa r)^3}\right] \\ \boldsymbol{H}_\varphi &= -\frac{j\omega p_0 \kappa^2}{4\pi} \frac{\zeta}{\xi} e^{-j(\omega t - \kappa r)} \sin\theta \left[\frac{-j}{\kappa r} + \frac{1}{(\kappa r)^2}\right] \end{aligned}\right\} \quad (7)$$

where $\zeta = 1$ for mks units, $\zeta = 4\pi$ for Gaussian units; $\xi = 1$ for mks units, $\xi = 1/c$ for Gaussian units.

5.8. Huygen's principle. If ψ is a solution of the scalar equation

$$\nabla^2 \psi - \frac{1}{v^2} \ddot{\psi} = -Q \quad (1)$$

then ψ satisfies the equation

$$\left.\begin{aligned} \psi(x'y'z't) &= \frac{1}{4\pi} \int_V Q^* dV \\ &\quad + \frac{1}{4\pi} \int_S \frac{1}{r} \left[(\nabla\psi)^* - \frac{\psi^* \boldsymbol{r}}{r^2} + \frac{1}{v}\left(\frac{\partial}{\partial t}\psi\right)^* \frac{\boldsymbol{r}}{r}\right] \cdot \boldsymbol{n} dS \end{aligned}\right\} \quad (2)$$

where $\boldsymbol{r} = (x - x')\boldsymbol{i} + (y - y')\boldsymbol{j} + (z - z')\boldsymbol{k} = r \dfrac{\boldsymbol{r}}{|\boldsymbol{r}|} = r\boldsymbol{u}$

$f^* = f(x,y,z,t - r/v)$

$\boldsymbol{n}$ = unit outward normal to element of area dS

This is called Huygen's principle.

Suppose $Q = 0$ in V and ψ is the spherical wave.

$$\left.\begin{aligned} \psi &= \frac{e^{-j\omega(t - r_1/v)}}{r_1} \\ \text{where} \quad \boldsymbol{r}_1 &= (x - x_1)\boldsymbol{i} + (y - y_1)\boldsymbol{j} + (z - z_1)\boldsymbol{k} = r_1 \frac{\boldsymbol{r}_1}{|\boldsymbol{r}_1|} = r_1 \boldsymbol{u}_1 \end{aligned}\right\} \quad (3)$$

Then

$$\psi = \int_S \frac{e^{-j\omega[t-(r_1+r)/v]}}{4\pi r r_1}\left[\left(\frac{1}{r}+\frac{j\omega}{v}\right)(u\cdot n)-\left(\frac{1}{r_1}+\frac{j\omega}{v}\right)(u\cdot n)\right]dS \tag{4}$$

For light waves it is generally the case that $1/r$, $1/r_1 \ll \omega/v$, so that

$$\psi = \int_S \frac{e^{-j\omega[t-(r_1+r)/v]}}{4\pi r r_1}\frac{j\omega}{v}(u\cdot n - u_1\cdot n)dS \tag{5}$$

5.9. Electromagnetic waves at boundaries in dielectric media. Given a plane electromagnetic wave

$$\boldsymbol{E} = \boldsymbol{E}_0 e^{-j(\omega t-\boldsymbol{k}\cdot\boldsymbol{r})} = (\boldsymbol{E}_y + \boldsymbol{E}_{xz})e^{-j(\omega t-\boldsymbol{k}\cdot\boldsymbol{r})}$$

whose propagation vector $\boldsymbol{k}$ lies in the x,z plane impinging on the boundary between two dielectrics which coincides with the x,y plane, let $n = n_1$ for $z < 0$, $n = n_2$ for $z > 0$, and let $\boldsymbol{k}\cdot\boldsymbol{u} > 0$ where $\boldsymbol{u}$ is a unit vector parallel to the z axis (Fig. 11).

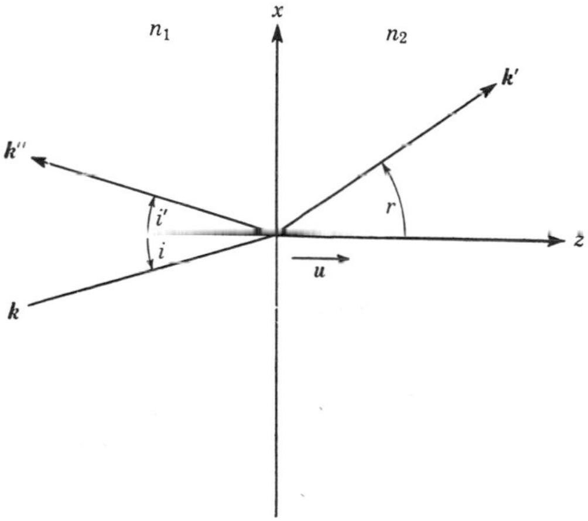

FIGURE 11

Let

$$i = \cos^{-1}\frac{\boldsymbol{k}\cdot\boldsymbol{u}}{|\boldsymbol{k}|}$$

be the angle of $\boldsymbol{k}$ with respect to $\boldsymbol{u}$. Then there can only result a refracted wave ($\boldsymbol{k}'$) a reflected wave ($\boldsymbol{k}''$) or both. The angles these waves make with the normal to the surface are related by

$$i' = i \tag{1}$$

$$n_1 \sin i = n_2 \sin r, \quad \text{(Snell's law)} \tag{2}$$

The amplitudes of the refracted and reflected waves are related to that of the incident beam by Fresnel's equations :

$$\frac{E''_y}{E_y} = \frac{\mu_1 \tan i - \mu_2 \tan r}{\mu_1 \tan i + \mu_2 \tan r} \quad \left(= \frac{\sin(i-r)}{\sin(i+r)} \text{ if } \mu_1 = \mu_2 \right) \tag{3}$$

$$\frac{E''_{xz}}{E_{xz}} = \frac{\mu_1 \sin i \cos i - \mu_2 \sin r \cos r}{\mu_1 \sin i \cos i + \mu_2 \sin r \cos r} \quad \left(= \frac{\tan(i-r)}{\tan(i+r)} \text{ if } \mu_1 = \mu_2 \right) \tag{4}$$

$$\frac{E'_y}{E_y} = \frac{2\mu_2 \sin r \cos i}{\mu_1 \sin i \cos r + \mu_2 \sin r \cos i} \quad \left(= \frac{2 \sin r \cos i}{\sin(i+r)} \text{ if } \mu_1 = \mu_2 \right) \tag{5}$$

$$\left.\begin{aligned} \frac{E'_{xz}}{E_{xz}} &= \frac{2\mu_2 \cos i \sin r}{\mu_2 \sin r \cos r + \mu_1 \sin i \cos i} \\ &\qquad \left(= \frac{2 \sin r \cos i}{\sin(i+r)\cos(i-r)} \text{ if } \mu_1 = \mu_2 \right) \end{aligned}\right\} \tag{6}$$

where E_y is the component of $\boldsymbol{E}_0$ normal to the plane of incidence (x,z plane) and E_{xz} is the component of $\boldsymbol{E}_0$ tangent to the plane of incidence.

When the angle of incidence equals

$$i_B = \tan^{-1} \frac{n_2}{n_1}, \quad \text{(Brewster's angle)} \tag{7}$$

$$\frac{E''_{xz}}{E_{xz}} = 0$$

so that the reflected light is polarized. When the angle of incidence exceeds

$$i_r = \sin^{-1} \frac{n_2}{n_1} \tag{8}$$

total reflection occurs, and the refracted wave vanishes.

For further optical formulas, see Chapters 16 and 17.

5.10. Propagation of electromagnetic radiation in wave guides. Let the z axis be the axis of the wave guide. Take the Hertz vector to be

$$\Pi_0 = o\boldsymbol{i} + o\boldsymbol{j} + \Pi_0 \boldsymbol{k} \tag{1}$$

Then Π_0 must be a solution of the scalar wave equation

$$\nabla^2 \Pi_0 + \kappa \Pi_0 = 0 \tag{2}$$

Since there are no obstacles to propagation along the z axis, one can separate out the z dependence to get

$$\Pi_0 = e^{\pm j k_1 z} u(x,y) \tag{3}$$

where u is a solution of the equation

$$\left[\frac{\partial^2}{\partial x^2} + \frac{\partial^2}{\partial y^2} + \left(\frac{2\pi}{\lambda_c}\right)^2\right] u = 0 \tag{4}$$

$$\left(\frac{2\pi}{\lambda_c}\right)^2 = \kappa^2 - k_1^2 = (2\pi)^2\left(\frac{1}{\lambda_0^2} - \frac{1}{\lambda_g^2}\right)$$

The expressions for the field vectors are derived from the Hertz vector by means of Eqs. (10) and (11) of § 5.3.

Transverse electric (TE)

$$\left.\begin{aligned} \boldsymbol{E} &= -j\omega\mu e^{jk_1 z}\left(\boldsymbol{i}\frac{\partial u}{\partial y} - \boldsymbol{j}\frac{\partial u}{\partial x} + \boldsymbol{k}o\right)\xi^2 \\ \boldsymbol{H} &= e^{jk_1 z}\left[\boldsymbol{i}jk_1\frac{\partial u}{\partial x} + \boldsymbol{j}jk_1\frac{\partial u}{\partial y} + \boldsymbol{k}(\kappa^2 - k_1^2)u\right]\xi \end{aligned}\right\} \tag{5}$$

Transverse magnetic (TM)

$$\left.\begin{aligned} \boldsymbol{E} &= -\frac{e^{jk_1 z}}{\epsilon}\left[\boldsymbol{i}jk_1\frac{\partial u}{\partial x} + \boldsymbol{j}jk_1\frac{\partial u}{\partial y} + \boldsymbol{k}(\kappa^2 - k_1^2)u\right] \\ \boldsymbol{H} &= j\omega e^{jk_1 z}\left(\boldsymbol{i}\frac{\partial u}{\partial y} - \boldsymbol{j}\frac{\partial u}{\partial x} + \boldsymbol{k}o\right)\xi \end{aligned}\right\} \tag{6}$$

where $\xi = 1$ for mks units, $\xi = 1/c$ for Gaussian units.

It is seen that

$$\boldsymbol{E}\cdot\boldsymbol{H} = 0, \qquad \boldsymbol{H}_t = \frac{\boldsymbol{k}\times\boldsymbol{E}_t}{Z_0} \tag{7}$$

where the subscript t identifies the transverse components and

$$Z_0 = \frac{\lambda_0}{\lambda_g}\sqrt{\frac{\mu}{\epsilon}}, \quad \text{(for the TM mode)} \tag{8}$$

$$Z_0 = \frac{\lambda_g}{\lambda_0}\sqrt{\frac{\mu}{\epsilon}}, \quad \text{(for the TE mode)} \tag{9}$$

The function $u(x,y)$ depends on the shape of the cross section of the wave guide and on the material of its walls. If we suppose the walls to be perfect conductors,

$$\boldsymbol{E}_{\text{tangent}} = \boldsymbol{B}_{\text{normal}} = 0$$

at the boundaries so that for TM waves

$$\left.\begin{aligned} u &= 0 \\ (\nabla u)_{\text{tan}} &= 0 \end{aligned}\right\} \text{ at the boundaries} \tag{10}$$

and for TE waves

$$(\nabla u)_{\text{normal}} = 0 \quad \text{at the boundaries} \tag{11}$$

5.11. The retarded potentials; the Lienard-Wiechert potentials; the self-force of electric charge. If the gauge

mks	Gaussian	
$\nabla \cdot A = -\epsilon\mu \dfrac{\partial\varphi}{\partial t}$,	$\nabla \cdot A = -\dfrac{\epsilon\mu}{c} \cdot \dfrac{\partial\varphi}{\partial t}$	(1)

is chosen, Eqs. (4) and (5) of § 1.6 take the form of inhomogeneous wave equations:

mks	Gaussian	
$\nabla^2 A - \dfrac{1}{v^2} \cdot \dfrac{\partial^2 A}{\partial t^2} = -\mu J$,	$\nabla^2 A - \dfrac{1}{v^2} \dfrac{\partial^2 A}{\partial t^2} = -\dfrac{4\pi}{c}\mu J$	(2)
$\nabla^2 \varphi - \dfrac{1}{v^2} \cdot \dfrac{\partial^2 \varphi}{\partial t^2} = -\dfrac{\rho}{\epsilon}$,	$\nabla^2 \varphi - \dfrac{1}{v^2} \dfrac{\partial^2 \varphi}{\partial t^2} = -\dfrac{4\pi}{\epsilon}\rho$	

where $v = 1/\sqrt{\mu\epsilon}$ for the mks system, $v = c/\sqrt{\mu\epsilon}$ for the Gaussian system. We obtain the general solution of these equations by adding a special solution of the inhomogeneous wave equation to the general solution of the corresponding homogeneous wave equation. The special solution of the above equations which relates the electromagnetic field to its sources is

$$\varphi(x,y,z,t) = \frac{\zeta}{4\pi\epsilon} \iiint_{\substack{\text{all}\\ \text{space}}} \frac{\rho(x_1,y_1,z_1,t - r/v)}{r} dx_1 dy_1 dz_1 \tag{4}$$

$$A(x,y,z,t) = \frac{\zeta\xi\mu}{4\pi} \iiint_{\substack{\text{all}\\ \text{space}}} \frac{J(x_1,y_1,z_1,t - r/v)}{r} dx_1 dy_1 dz_1 \tag{5}$$

where r is the magnitude of the vector

$$r = (x_1 - x)i + (y_1 - y)j + (z_1 - z)k$$

The expressions above are called the retarded potentials.

When the source of the field is a point charge of magnitude q, velocity $\dot{r} = (d/dt)r$, and acceleration $\ddot{r} = (d^2/dt^2)r$; then the expressions for the retarded potentials can be integrated to give the Lienard and Wiechert potentials.

$$\varphi = \frac{\zeta}{4\pi\epsilon} \frac{q}{(r + \dot{r} \cdot r/c)} \bigg|_{t-r/c} \tag{6}$$

$$A = \frac{\zeta\xi\mu}{4\pi} \frac{q\dot{r}}{(r + \dot{r} \cdot r/c)} \bigg|_{t-r/c} \tag{7}$$

The fields resulting from these potentials are

$$E = \frac{\zeta q}{4\pi\epsilon}\left[\frac{(1-\dot{r}^2/c^2)(r+r\dot{r}/c)}{(r+r\cdot\dot{r}/c)^3}+\frac{r}{c^2}\times\frac{(r+r\dot{r}/c\times\ddot{r})}{(r+r\cdot\dot{r}/c^3)}\right]_{t-r/c} \tag{8}$$

$$H = \xi\epsilon\mu\,\frac{(E\times r)_{t-r/c}}{r} \tag{9}$$

The expressions above can be used to compute the interaction of a distribution of charge with its own field. For this purpose we postulate a rigid distribution of charge such that all elements of charge dq have the same velocity $\dot{r}$ and acceleration $\ddot{r}$. It is necessary to assume that the average linear dimension of the charge distribution r_0 is so small and the motion of the charge varies so slowly that the change of acceleration in the period of time it takes a light wave to pass the charge distribution is small compared to the acceleration itself.

$$\left|\frac{r_0}{c}\dddot{r}\right| \ll |\ddot{r}|$$

On the basis of this assumption the charge distribution is found to exert the force F on itself:

$$F = F_0 + F_1 + F_2 + \ldots$$

where
$$F_0 = -\frac{2}{3}\cdot\frac{\ddot{r}}{c^2}\iint\frac{dq\,dq'}{2r}\cdot\frac{\zeta}{4\pi\epsilon} \tag{10}$$

$$F_1 = \frac{2}{3}\cdot\frac{q^2}{c^3}\dddot{r}\,\frac{\zeta}{4\pi\epsilon},\qquad q = \int dq$$

$$F_2 \sim \frac{\ddddot{r}}{c^4}\iint r\,dq\,dq' \sim \frac{q^2\ddddot{r}\,r_0}{c^4} \tag{11}$$

where F_0 is the term representing the inertia of the field of the charge, F_1 is independent of the form of the charge distribution and accounts for the damping resulting from the radiation of energy.

Perhaps the most important point is that this chapter has been prepared with the view of giving the reader an insight into the coherent structure of the whole subject of electromagnetic theory, rather than presenting him with an unrelated set of fundamental formulas in the field.

TABLE 1

Conversion Table for Symbols

To convert an expression from its form in the mks system to its proper form in the Gaussian system, and vice versa, make the following substitutions

mks		Gaussian	mks		Gaussian
E	$\rightarrow$	E	J	$\rightarrow$	J
B	$\rightarrow$	B/c	φ	$\rightarrow$	φ
D	$\rightarrow$	$D/4\pi$	A	$\rightarrow$	A/c
H	$\rightarrow$	$Hc/4\pi$	$\mathbf{S}$	$\rightarrow$	$\mathbf{S}$
ϵ	$\rightarrow$	$\epsilon/4\pi$	$\mathbf{\Pi}$	$\rightarrow$	$\mathbf{\Pi}/4\pi$
μ	$\rightarrow$	$\mu 4\pi/c^2$	C	$\rightarrow$	$C/4\pi$
ρ	$\rightarrow$	ρ	L	$\rightarrow$	L/c

TABLE 2

Conversion Table for Units

	mks units	Gaussian units
time, t	1 sec	= 1 sec
length, l	1 m	= 10^2 cm
mass, m	1 kg	= 10^3 gm
force, F	1 newton	= 10^5 dynes
energy, w	1 joule	= 10^7 ergs
power, p	1 watt	= 10^7 ergs/sec
charge, q	1 coulomb	= 3×10^9 statcoulombs
electric field, E	1 v/m	= $1/3 \times 10^{-4}$ statvolt/cm
electric displacement, D	1 coulomb/m	= $4\pi \times 3 \times 10^5$ statcoulombs/cm^2
potential, φ	1 v	= $1/3 \times 10^{-2}$ statvolt
capacitance, C	1 farad	= 9×10^{11} cm
current, i	1 amp	= 3×10^9 statamperes
resistance, R	1 ohm	= $1/9 \times 10^{-11}$ statohm
magnetic field, H	1 amp/m	= $4\pi \times 10^{-3}$ oersted
magnetic induction, B	1 weber/m^2	= 10^4 gausses
magnetic flux, BdS	1 weber	= 10^8 maxwells
inductance, L	1 henry	= 10^9 abhenrys

TABLE 3

THE FUNDAMENTAL CONSTANTS

	mks	Gaussian
permitivity in vacuum, ϵ_0	$= 8.854 \times 10^{-12}$ farad/m	$= 1$
permeability in vacuum, μ_0	$= 4\pi \times 10^{-7}$ henry/m	$= 1$
velocity of light, c	$= 2.998 \times 10^{8}$ m/sec	$= 2.998 \times 10^{10}$ cm/sec
charge of the electron, e. .	$= 1.601 \times 10^{-19}$ coulomb	$= 4.803 \times 10^{-10}$ statcoulomb
mass of the electron, m_0. .	$= 9.108 \times 10^{-31}$ kg	$= 9.108 \times 10^{-28}$ g

TABLE 4

COORDINATES AND SYMBOLS

Unless indicated otherwise, the following systems of coordinates and symbols were used in this chapter.

Rectangular coordinates (Fig. 12), $\mathbf{r}_1 = \mathbf{i}x_1 + \mathbf{j}y_1 + z_1\mathbf{k}$:

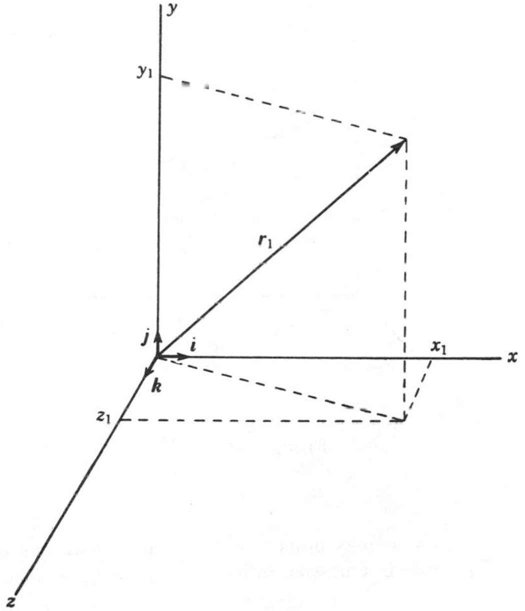

FIGURE 12

Circular cylindrical coordinates (Fig. 13):

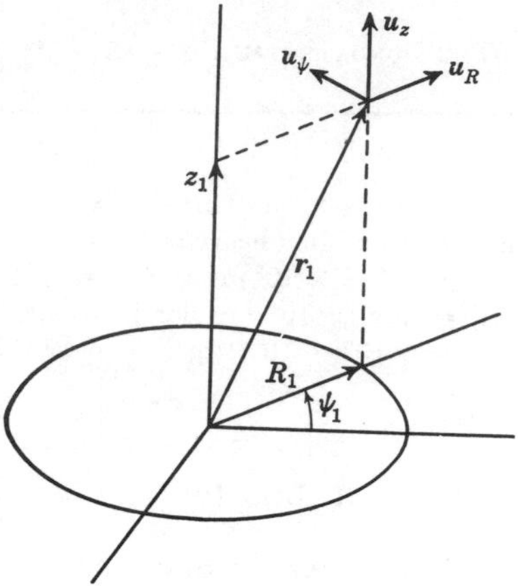

FIGURE 13

Spherical coordinates (Fig. 14):

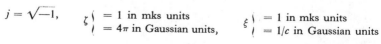

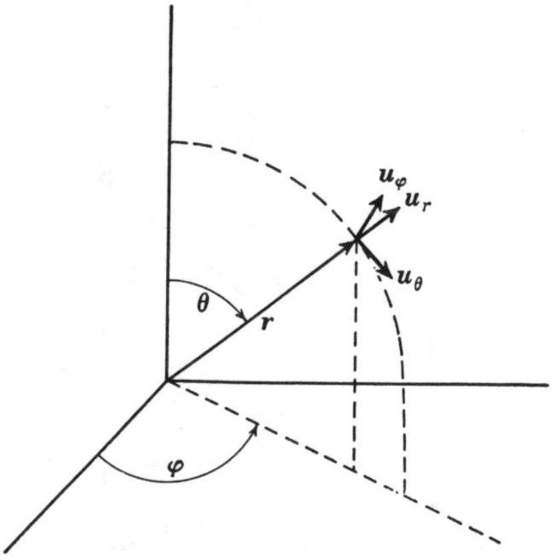

FIGURE 14

$j = \sqrt{-1}$, $\quad \zeta \begin{cases} = 1 \text{ in mks units} \\ = 4\pi \text{ in Gaussian units,} \end{cases} \quad \xi \begin{cases} = 1 \text{ in mks units} \\ = 1/c \text{ in Gaussian units} \end{cases}$

Bibliography

1. Abraham, M. and Becker, R., *The Classical Theory of Electricity and Magnetism*, Blackie & Son, Ltd., London, 1932. A standard intermediate text.
2. Becker, R., *Theorie der Elektrizität*, Band II, *Elektronentheorie*, B. G. Teubner, Berlin, 1933. The classical theory of the electron, containing much material not to be found elsewhere.
3. Born, M., *Optik*, Julius Springer, Berlin, 1933. An advanced comprehensive book based on electromagnetic theory.
4. Frank, N. H., *Introduction to Electricity and Optics*, McGraw-Hill Book Company, Inc., New York, 1950.
5. Jeans, J., *The Mathematical Theory of Electricity and Magnetism*, Cambridge University Press, London, 1948. A classical treatment of electrostatics and magnetostatics, presented without the use of vector analysis. (Dover reprint)
6. Slater, J. C., *Microwave Electronics*, D. Van Nostrand Company, Inc., New York, 1950. Contains material on transmission line theory and circuit analysis, as well as electromagnetic waves.
7. Smythe, W. R., *Static and Dynamic Electricity*, McGraw-Hill Book Company, Inc., New York, 1950. An excellent reference text for methods of solution of advanced problems.
8. Stratton, J. A., *Electromagnetic Theory*, McGraw-Hill Book Company, Inc., New York, 1941. A valuable treatise on the subject.
9. Blatt, J. M. and Weisskopf, *Theoretical Nuclear Physics*, John Wiley and Sons, New York, 1952. Appendix B provides an excellent discussion of multipole radiation.

Chapter 14

ELECTRONICS

By Emory L. Chaffee

Professor of Applied Physics
Harvard University

1. Electron Ballistics

Since the field of electronics extends over the areas of physics and engineering, many of the basic formulas in electronics dealing with specific fields will be found listed under these fields, such as atomic physics, nuclear theory, particle acceleration, conduction in gases, electron microscope, solid state, microwave spectroscopy, etc. This wide spread of the subject of electronics makes it difficult to select basic formulas not to be found duplicated elsewhere. The formulas chosen are those which are fundamental and of general application in the broad field of electronics.

1.1. Current. A stream of charged particles composed of n particles per unit length of the path, moving with a velocity $\boldsymbol{v}$ and each particle having a charge q constitutes a current $\boldsymbol{I}$.

$$\boldsymbol{I} = nq\boldsymbol{v} \tag{1}$$

The current density $\boldsymbol{J}$ is

$$\boldsymbol{J} = \rho\boldsymbol{v} \tag{2}$$

where ρ is the charge density.

1.2. Forces on electrons. An electron having a charge $-e_s$ located in an electric field $\boldsymbol{E}_s$ in esu is acted upon by a force $\boldsymbol{F}$.

$$\boldsymbol{F} = -e_s\boldsymbol{E}_s \quad \text{dynes} \qquad e_s = 4.803 \times 10^{-10} \quad \text{esu} \tag{1}$$

or

$$\boldsymbol{F} = -e_m\boldsymbol{E}_m \quad \text{dynes} \qquad e_m = 1.602 \times 10^{-20} \quad \text{emu} \tag{2}$$

where $\boldsymbol{E}_m$ is in emu.

An electron having a velocity v cm/sec in a magnetic field of flux density B in emu is acted upon by a force F.

$$\boldsymbol{F} = -e_m\boldsymbol{v} \times \boldsymbol{B} \quad \text{dynes} \qquad e_m = 1.602 \times 10^{-20} \quad \text{emu} \tag{3}$$

1.3. Energy of electron. The loss in potential energy or gain in kinetic energy U of an electron having a charge $-e_m$ while moving from point x_1 to point x_2 in an electric field E_m in emu is

$$U = -e_m \int_{x_1}^{x_2} \boldsymbol{E}_m \cdot \boldsymbol{dx} = -e_m(V_1 - V_2) \quad \text{ergs} \tag{1}$$

where V_1 and V_2 are the potentials in emu at x_1 and x_2, respectively.

The kinetic energy of an electron moving with velocity v cm/sec is

$$\left.\begin{aligned} U &= m_0c^2\left[\frac{1}{\sqrt{1-(v/c)^2}} - 1\right] \quad \text{ergs} \\ &= \frac{1}{2}m_0v^2\left[1 + \frac{3}{4}\left(\frac{v}{c}\right)^2 + \frac{5}{8}\left(\frac{v}{c}\right)^4 + \dots\right] \quad \text{ergs} \end{aligned}\right\} \tag{2}$$

where m_0 = rest mass of the electron $= 9.106 \times 10^{-28}$ gram, and c = velocity of light $= 2.9978 \times 10^{10}$ cm/sec.

The transverse mass m_t of an electron moving with velocity v is

$$m_t = \frac{m_0}{\sqrt{1-(v/c)^2}} \tag{3}$$

The longitudinal mass m_l of an electron is

$$m_l = \frac{m_0}{[1-(v/c)^2]^{3/2}} \tag{4}$$

1.4. Electron orbit. An electron having a velocity of v cm/sec in a magnetic field of flux density B moves in a stable circular orbit of radius r cm, when $F = -m_tv^2/r$ (see § 1.2), giving for the cyclotron condition

$$\frac{m_tv}{r} = e_mB \tag{1}$$

where $m_0 = 9.108 \times 10^{-28}$ gram (see § 1.3).

$$t = 10^4\,\frac{2m_t}{e_m}\cdot\frac{V}{B} \quad \text{cms} \tag{2}$$

where V is the electron velocity v expressed in equivalent volts.

The periodic time T of one revolution of an electron in the cyclotron orbit is

$$T = \frac{2\pi m_t}{e_mB} \tag{3}$$

2. Space Charge

2.1. Infinite parallel planes. When space charge exists between parallel planes d centimeters apart, one being an emitter of electrons having no initial velocity, the space charge limited current is

$$\left.\begin{aligned} J &= \frac{1}{9\pi}\cdot\frac{2e_s}{m}\cdot\frac{V^{3/2}}{d^2} \quad \text{(esu)} \\ &= 2.336\times 10^{-6}\,\frac{V_b^{\,3/2}}{d^2} \quad \text{amp/cm}^2, \quad \text{(Child's law)} \end{aligned}\right\} \quad (1)$$

where V_b is the potential difference between the planes in volts. Relativity change in mass is neglected.

Variation of field, velocity, and space charge with distance x from emitter is

$$E_x = -\,(\text{constant})\cdot x^{1/3} \quad (2)$$

$$v_x = (\text{constant})\cdot x^{2/3} \quad (3)$$

$$\rho_x = (\text{constant})\cdot x^{-2/3} \quad (4)$$

where ρ is the space-charge density.

When initial velocity of emission is not neglected, the space-charge-limited current is

$$J = 2.336\times 10^{-6}\frac{(V_b - V_m)^{3/2}}{(d - x_m)^2}\left(1 + 2.66\,\frac{kT}{e_s(V_b - V_m)} + \cdots\right) \text{amp/cm}^2 \quad (5)$$

where V_m is the potential with respect to the emitter of the potential minimum x_m cm from the emitter; k being Boltzmann's constant (1.3804×10^{-16} erg/deg), and T the absolute temperature of the emitter.

2.2. Cylindrical electrodes. The space-charge-limited current from a cylindrical central emitter of radius a cm to a concentric cylindrical anode of radius r cm, neglecting initial velocity of emission, is

$$\left.\begin{aligned} I &= \frac{2}{9}\sqrt{\frac{2e_s}{m}}\,\frac{lV^{3/2}}{\beta^2 r} \quad \text{(esu)} \\ &= 14.68\times 10^{-6}\,\frac{lV_p^{\,3/2}}{\beta^2 r} \quad \text{amp} \end{aligned}\right\} \quad (1)$$

where l is the length of the plate,

and

$$\beta = \log \frac{r}{a} - \frac{2}{5}\left(\log \frac{r}{a}\right)^2 + \frac{11}{120}\left(\log \frac{r}{a}\right)^3 - \frac{47}{3.300}\left(\log \frac{r}{a}\right)^4 + \dots *$$

Variation of field, velocity, and space-charge density with r is

$$E_r = (\text{constant}) \cdot (r\beta^2)^{2/3} \tag{2}$$

$$v_r = (\text{constant}) \cdot (r\beta^2)^{1/3} \tag{3}$$

$$\rho_r = (\text{constant}) \cdot (r^2\beta)^{-2/3} \tag{4}$$

3. Emission of Electrons

3.1. Thermionic emission. The saturation current density is

$$J = AT^2(1 - r)\epsilon^{-\Phi e/kT}\epsilon^{-E^{1/2}e^{3/2}/kT} \quad \text{amp/cm}^2 \tag{1}$$

where $A = 4\pi mek^2/h^3 = 120$ amp/cm² deg²
T = absolute temperature in degrees C
r = reflection coefficient of electrons at the potential barrier of the surface
Φ = electron affinity or potential barrier in esu
e = electronic charge in esu
k = Boltzmann's constant (1.3804×10^{-16} erg/deg)
E = electric field at the surface in esu
h = Planck's constant (6.625×10^{-27} erg sec)

The factor $\epsilon^{-E^{1/2}e^{3/2}/kT}$ gives the effect upon emission caused by the electric field at the surface, i.e., the "Schottky effect."

3.2. Photoelectric emission

$$\frac{1}{2} mv^2 = h\nu - \Phi e + U_k \quad \text{ergs} \tag{1}$$

where m = mass of the electron
v = maximum velocity of ejected electron in cm/sec
h = Planck's constant
ν = frequency of incident radiation
Φ = electron affinity or potential barrier in esu
e = charge of electron in esu
U_k = kinetic energy of electron inside metal

* For table of β^2 see LANGMUIR, *Phys. Rev.*, **21**, 435 (1923) and LANGMUIR and BLODGETT, *Phys. Rev.*, **22**, 347 (1923).

4. Fluctuation Effects

4.1. Thermal noise. The mean-square voltage in frequency interval $d\nu$ across an impedance having a real component of R ohms is

$$d\bar{V}_\nu^2 = 4 \times 10^{-7} RkT\, d\nu \quad \text{volts}^2 \text{ sec} \tag{1}$$

$$d\bar{V}_\omega^2 = \frac{2}{\pi} \times 10^{-7} RkT\, d\omega \quad \text{volts}^2 \text{ sec} \tag{2}$$

where T is the temperature in degrees Kelvin and k is Boltzmann's constant, $(1.3804 \times 10^{-16}$ erg/deg), and $\omega = 2\pi\nu$.

4.2. Shot noise. The mean-square voltage across a pure resistance R ohms in the plate circuit of a diode operating under space current saturation is

$$\bar{V}^2 = \frac{eIR}{2C_{pf}} \quad \text{volts}^2 \tag{1}$$

where e is the electronic charge in coulombs, I is the average plate current in amperes, and C_{pf} is the capacitance between plate and cathode in farads.

If the impedance in the plate circuit is Z in parallel with C_{pf}, the mean-square voltage in the frequency range $2\pi\, d\nu = d\omega$ is

$$d\bar{V}_\omega^2 = \frac{eIZ_\omega^2}{\pi}\left[\left(\int_0^\infty i(\lambda)\sin\omega\lambda\, d\lambda\right)^2 + \left(\int_0^\infty i(\lambda)\cos\omega\lambda\, d\lambda\right)^2\right] d\omega \tag{2}$$

where $i(\lambda)$ is the current in Z_ω caused by unit change in C_{pf}.

Bibliography

1. Dow, W. G., *Fundamentals of Engineering Electronics*, John Wiley & Sons, Inc. New York, 1937.
2. Fink, D. F., *Engineering Electronics*, McGraw-Hill Book Company, Inc., New York, 1938.
3. Hughes, A. L. and DuBridge, L. A., *Photoelectric Phenomena*, McGraw-Hill Book Company, Inc., New York, 1932.
4. Maxfield, F. A. and Benedict, R. R., *Theory of Gaseous Conduction and Electronics*, McGraw-Hill Book Company, Inc., New York, 1941.
5. Millman, J. and Seely, S., *Electronics*, McGraw-Hill Book Company, Inc., New York, 1951.
6. M. I. T. Electrical Engineering Staff, *Applied Electronics*, John Wiley & Sons, Inc., New York, 1943.
7. Spangenberg, K. R., *Vacuum Tubes*, McGraw-Hill Book Company, Inc., New York, 1948. *Advances in Electronics*, edited by L. Marton, Academic Press, New York; Vol. I, 1948; Vol. 2, 1950; Vol. 3, 1951.

Chapter 15

SOUND AND ACOUSTICS

By Philip M. Morse

Professor of Physics
Massachusetts Institute of Technology

1. Sound and Acoustics

The science of acoustics bears a close relationship to may other fields of physics, as covered in this book of fundamental formulas. For example, hydrodynamics provides the basic equations that govern the flow of sound in an acoustic system. There are problems of dynamics and boundary values must be satisfied. Thermodynamics, statistical mechanics, and kinetic theory have a relationship in terms of the microscopic picture of acoustics. Reference to certain of these other chapters, therefore, may assist the student. The following formulas were chosen as being the most fundamental to the physical side of acoustics. The purely engineering phases, of course, properly lie outside the scope of this work, if for no other reason than the fact that many of them are empirical.

1.1. Wave equation, definitions. For a gas or fluid with negligible viscosity, where $\boldsymbol{v}(\boldsymbol{r},t)$ is the velocity of displacement of the fluid, at $\boldsymbol{r},t$, from equilibrium, P_0 is the equilibrium pressure and $p(\boldsymbol{r},t)$ the deviation from this, the equation determining the velocity potential ψ, for fluctuations of p small compared with P_0, and the relationship between ψ and other quantities is

$$c^2\nabla^2\psi = (\partial^2\psi/\partial t^2), \quad c^2 = (K/\rho) \tag{1}$$

$$\boldsymbol{v} = -\operatorname{grad}\psi, \quad p = \rho(\partial\psi/\partial t), \quad \Delta\rho = (\rho/c^2)(\partial\psi/\partial t) \tag{2}$$

where $K = \rho c^2$ is the compressional modulus of the fluid and $\Delta\rho$ is its departure from equilibrium density ρ. For a gas, vibrating rapidly enough, the departures from equilibrium are adiabatic, every change in pressure p brings about a change in temperature ΔT from the equilibrium temperature T. For a gas having ratio of specific heat $(c_p/c_v) = \gamma$, the compressional modulus and temperature change are

$$K = \rho c^2 = \gamma P_0, \quad \Delta T = (T/c^2)(\gamma - 1)(\partial\psi/\partial t) = (T/\rho c^2)(\gamma - 1)p \tag{3}$$

1.2. Energy, intensity. The energy density of sound in the fluid is

$$\left.\begin{aligned} E &= \tfrac{1}{2}\rho v^2 + \tfrac{1}{2}(c^2/\rho)p^2 \\ &= \tfrac{1}{2}\rho(|\operatorname{grad}\psi|^2 + c^2|\partial\psi/\partial t|^2), \quad \text{(energy per unit volume)} \end{aligned}\right\} \tag{1}$$

The power flow through the fluid, the intensity of sound

$$\boldsymbol{S} = p\boldsymbol{v} = -\rho(\partial\psi/\partial t)\operatorname{grad}\psi, \quad \text{(power per unit area)} \tag{2}$$

When intensity can be measured directly (when *both* velocity and pressure are measured) the magnitude of $\boldsymbol{S}$ is often given in terms of the *decibel scale.*

$$\text{Intensity level} = 10\log(\text{intensity in microwatts per sq cm}) + 100 \tag{3}$$

An intensity of 10^{-10} microwatt per square centimeter is zero intensity level, of 1 watt per square centimeter is 160 intensity level, etc. Sound of zero intensity level at 1000 cycles per second frequency is just at the threshold of hearing for the average person. The ear is less sensitive to frequencies either higher or lower than 1000 c. For example, the threshold of hearing is at about 40 decibels for 100 c. and at about 10 decibels for 10,000 c. Above about 20,000 c. and below about 20 c. sound is not heard.

1.3. Plane wave of sound. The " standard " wave motion is the simple-harmonic, plane wave, of frequency $\nu = (\omega/2\pi) = (kc/2\pi)$, wavelength $\lambda = (2\pi/k) = (c/\nu)$ and direction parallel to the propagation vector $\boldsymbol{k}$.

$$\left.\begin{aligned} &\psi = Ae^{i(\boldsymbol{k}\cdot\boldsymbol{r}-\omega t)}, \quad p = Pe^{i(\boldsymbol{k}\cdot\boldsymbol{r}-\omega t)}, \quad \boldsymbol{v} = U_{\max}(\boldsymbol{k}/k)e^{i(\boldsymbol{k}\cdot\boldsymbol{r}-\omega t)} \\ &P_{\max} = -i\rho\omega A, \quad U_{\max} = (P_{\max}/\rho c) = ikA, \\ &\qquad (\Delta T)_{\max} = (T/\rho c^2)(\gamma - 1)P_{\max} \\ &E = (P_{\max}^2/2\rho c^2) = \tfrac{1}{2}\rho U_{\max}^2, \quad S = (P_{\max}^2/2\rho c) = \tfrac{1}{2}\rho c U_{\max}^2 = Ec \end{aligned}\right\} \tag{1}$$

In most cases intensity is not measured directly; pressure amplitude $P_{\max}$ is measured. Sound is then measured in terms of *sound pressure level,* which is adjusted to be equal to the intensity level for a plane wave in air at standard conditions.

$$\text{Pressure level} = 20\log(P_{\text{rms}}/2 \times 10^{-4}) = 20\log P_{\text{rms}} + 74 \tag{2}$$

where $P_{\text{rms}} = P_{\max}/\sqrt{2}$ is the root-mean-square pressure amplitude measured in microbars (dynes per square centimeter). Note that for any other condition except a plane wave in air at standard conditions, pressure level is *not equal* to intensity level.

1.4. Acoustical constants for various media. The pertinent sound propagation constants in cgs units for various media are as follows. Air at 760 mm mercury pressure, 20° C temperature; $\rho = 0.00121$, $c = 34{,}400$, $\rho c = 42$, $\rho c^2 = K = 1{,}420{,}000$, $\gamma = 1.40$, $P_0 = 1.013 \times 10^3$. Sea water, at 15° C temperature, 31.6 g salt per 1000 g water (standard conditions); $\rho = 1.02338$, $c = 150{,}000$, $\rho c = 1.53 \times 10^5$, $\rho c^2 = K = 2.3 \times 10^{10}$.

1.5. Vibrations of sound producers; simple oscillator. A simple oscillator has mass M grams, resistance R dyne second per centimeter, and spring compliance C centimeter per dyne. The equation of motion is

$$M(d^2x/dt^2) + R(dx/dt) + (x/C) = \text{force} \tag{1}$$

The free vibration is

$$x = Ae^{-kt} \sin(\omega_r t + \varphi) \tag{2}$$

where the damping constant $k = (R/2M)$ and the vibration frequency is $(\omega_r/2\pi)$, where $\omega_r{}^2 = (1/MC) - k^2$. The Q of the oscillator, the number of cycles for the amplitude of motion to reduce to $e^{-\pi}$ of its initial value is $Q = \sqrt{M/CR^2}$. For forced motion, if the driving force is $Fe^{-i\omega t}$, the displacement and velocity of the oscillator are $x = (F/-i\omega Z)e^{-i\omega t}$ and $(dx/dt) = (F/Z)e^{-i\omega t}$, respectively, where Z is the *mechanical impedance* of the oscillator.

$$Z = -i\omega M + R - (1/i\omega C) = j\omega M + R + (1/j\omega C), \quad (i = -j) \tag{3}$$

where M is analogous to electric inductance, and C to electric capacitance in a series RLC circuit. When M or ω is large enough for ωM to predominate, the oscillator is *mass-controlled* and $Z \simeq j\omega M$; when C or ω is small enough, the oscillator is *stiffness-controlled* and $Z \simeq (1/j\omega C)$.

1.6. Flexible string under tension. For mass ϵ gram per centimeter length, tension T dynes, the transverse displacement $y(x,t)$ of a point x on the string, from equilibrium, is given by the wave equation (neglecting friction)

$$c^2(\partial^2 y/\partial x^2) = (\partial^2 y/\partial t^2), \quad c^2 = (T/\epsilon) \tag{1}$$

where c is wave velocity. For a string clamped at $x = 0$ and $x - l$, the possible free vibrations are

$$y(x,t) = \sum_{n=1}^{\infty} A_n \sin(n\pi x/l) \cos[(n\pi ct/l) - \psi_n] \tag{2}$$

The allowed frequencies are $\nu_n = (\omega_n/2\pi) = (nc/2l)$, the nth being called *n*th *harmonic* of the *fundamental frequency* $\nu_1 = (c/2l)$. The amplitudes A_n and phases φ_n are determined by the initial shape $y_0(x)$ and velocity $v_0(x)$ of the string.

$$A_n \cos \psi_n = (2/l) \int_0^l y_0(x) \sin (\pi n x/l) dx \tag{3}$$

$$A_n \sin \psi_n = (2/n\pi c) \int_0^l v_0(x) \sin (\pi n x/l) dx \tag{4}$$

For a transverse driving force $F = F_0 e^{i\omega t}$ dynes per centimeter applied uniformly along the string of length l, the steady-state driven motion of the string is

$$y(x,t) = \frac{F_0 e^{-i\omega t}}{\epsilon \omega^2} \left[\frac{\cos (\omega/c)(x - \frac{1}{2} l)}{\cos (\omega l/2c)} - 1 \right] \tag{5}$$

The mean displacement is

$$\bar{y}(t) = (F_0 e^{-i\omega t}/\epsilon\omega^2) [(2c/\omega l) \tan (\omega l/2c) - 1] \tag{6}$$

having resonances at the fundamental ω_1 and at all odd harmonics $\omega_3, \omega_5, \ldots$.

1.7. Circular membrane under tension. For mass σ per unit area and tension T per unit length, the transverse displacement $y(r,\varphi,t)$ from equilibrium of a circular membrane clamped at $r = a$, for free vibration, is

$$\left. \begin{aligned} y(r,\varphi,t) = \sum_{m,n} [A_{mn} \cos (m\varphi) + B_{mn} \sin (m\varphi)] J_m(\pi \beta_{mn} r/a) \cdot \\ \cos [(\pi \beta_{mn} ct/a) - \psi_{mn}] \end{aligned} \right\} \tag{1}$$

where the term $J_m(z)$ is the Bessel function of order m, and β_{mn} is the nth root of the equation $J_m(\pi\beta) = 0$. The resonance frequencies are $\nu_{mn} = (\omega_{mn}/2\pi) = (\beta_{mn} c/2a)$. The first few values of β are: $\beta_{01} = 0.7655$, $\beta_{02} = 1.7571$, $\beta_{11} = 1.2197$, $\beta_{12} = 2.2330$, $\beta_{mn} \to n + \frac{1}{2}m - \frac{1}{4}$ where $n \gg 1$.

For a transverse driving force $F = F_0 e^{-i\omega t}$ dynes per square centimeter, uniform over the membrane, the steady-state displacement, after transients have disappeared, is

$$y(r,t) = \frac{F_0 e^{-i\omega t}}{\sigma \omega^2} \left[\frac{J_0(\omega r/c)}{J_0(\omega a/c)} - 1 \right] \tag{2}$$

and the mean displacement is

$$\bar{y}(t) = (F_0 e^{-i\omega t}/\sigma\omega^2) [J_2(\omega a/c)/J_0(\omega a/c)] \tag{3}$$

Resonance occurs at $\omega = \omega_{0n} = (\pi \beta_{0n} c/a)$.

1.8. Reflection of plane sound waves, acoustic impedance. If a plane sound wave, of pressure amplitude P_{inc}, is incident on a plane surface at angle of incidence θ, a reflected wave is produced, with amplitude dependent on the reaction of the surface to the sound pressure. This reaction may be expressed in terms of a ratio z between the pressure at the surface, p at $x = 0$, to the normal velocity of the air at the surface, v_x at $x = 0$ (in many cases z is independent of θ). The pressure wave is

$$\left.\begin{aligned} p &= P_{inc}e^{ik(x\cos\theta + y\sin\theta - ct)} \\ &\quad + P_{inc}\left[\frac{(z/\rho c)\cos\theta - 1}{(z/\rho c)\cos\theta + 1}\right]e^{ik(-x\cos\theta + y\sin\theta - ct)} \end{aligned}\right\} \quad (1)$$

If S_i is the incident intensity $(P_{inc}^2/2\rho c)$, the reflected intensity is

$$S_r = S_i\left|\frac{(z/\rho c)\cos\theta - 1}{(z/\rho c)\cos\theta + 1}\right|^2 \quad (2)$$

which is equal to S_i when z is imaginary (surface impedance a pure reactance). When z has a real part, S_r is less than S_i and the surface absorbs some of the incident power.

1.9. Sound transmission through ducts. For a pipe, with axis parallel to the x axis, of cross-sectional area $A(x)$ and length of cross-sectional perimeter $L(x)$, the inner surface of which has acoustic impedance $z_w(x)$, when the logarithmic derivatives (A'/A), (L'/L), and (z'/z) are small compared to $(1/L)$, the following wave equation is approximately valid.

$$\frac{1}{c^2}\left(\frac{\partial^2 p}{\partial t^2}\right) + \left(\frac{L}{cA}\right)\left(\frac{\rho c}{z_w}\right)\left(\frac{\partial p}{\partial t}\right) = \left(\frac{1}{A}\right)\frac{\partial}{\partial x}\left(A\frac{\partial p}{\partial x}\right), \quad (c^2 = K/\rho) \quad (1)$$

For a simple-harmonic wave $p = P(x)e^{-i\omega t}$ where $(\omega = kc)$, this reduces to

$$\frac{1}{A}\cdot\frac{d}{dx}\left(A\frac{dP}{dx}\right) + k^2n^2P = 0, \quad n^2 = 1 + \left(\frac{icL}{\omega A}\right)\left(\frac{\rho c}{z_w}\right) \quad (2)$$

When A is independent of x but (A/L) is small compared to $(c/\omega) = 2\pi\lambda$, the motion for a wave proceeding in the positive x direction is

$$p = P_{max}e^{ik(nx - ct)} \simeq P_{max}e^{-(L/2A)(\rho c/z_w) + ik(x - ct)} \quad (3)$$

The reciprocal of z_w, $(1/z_w) = g - is$, is the acoustic admittance of the duct surface; the wave is attenuated if the conductance g is greater than zero; the effective phase velocity of the wave in the duct is $c/[1 + (L/2A)(\rho cs)]$, which is greater than c if s is negative (mass reactance), less than c if s is positive (stiffness reactance of wall).

The attenuation of the wave in decibels per centimeter is $(4.34L\rho cg/A)$, with L, A, etc., given in centimeters (since ρcg is dimensionless, if L and A

are given in feet, $4.34\rho cgL/A$ will be the attenuation per foot of duct). If the tube is closed at $x = l$ by material (or an abrupt change of the duct) of effective impedance $(p/v_x)_{x=l} = z_l$, there will be a reflected as well as an incident wave in the region $x < l$. The pressure and velocity at x,ϵ can be written

$$\left.\begin{aligned} p &= P_0 \sinh\,[ikn(x-l)+\psi]e^{-i\omega t} \\ v &= (nP_0/\rho c)\cosh\,[ikn(x-l)+\psi]e^{-i\omega t} \end{aligned}\right\} \tag{1}$$

where $\psi = \tanh^{-1}(nz_l/\rho c)$. The ratio of p to v at $x = 0$ is then

$$z_0 = (\rho c/n)\tanh\,[\tanh^{-1}(nz_l/\rho c) - ikn_l] \tag{2}$$

which would be proportional to the load on a diaphragm, driving the air in the tube, placed at $x = 0$.

1.10. Transmission through long horn. When z_w is large enough so that n is nearly unity, but when A varies with x, we have the case of the horn. For the conical horn $A = A_0[(x/d)+1]^2$ the solution for an outgoing wave is

$$p = P_0[(xA)+1]^{-1}e^{ik(x-ct)}, \quad z_0 = \left(\frac{p}{v}\right)_{x=0} = \frac{\rho c}{1+i(c/\omega d)} \tag{1}$$

For an exponential horn $S = S_0 e^{2x/d}$ and the solution is

$$p = P_0 e^{-x/d+ik(\tau x-ct)}, \quad z_0 = \left(\frac{p}{v}\right)_{x=0} = \rho c\tau - i\rho c(c/\omega d)\;(\omega > c/d) \tag{2}$$

where $\tau = \sqrt{1-(c/\omega d)^2}$ and $kc = \omega$. If the duct of the previous section is terminated with a conical or exponential horn at $x = l$, the z_0 of this section is to be inserted for z_l in the previous equations.

1.11. Acoustical circuits. A duct with various constrictions, openings, and divisions is analogous to an a-c electric circuit as long as the wavelength of the sound is long compared with the dimensions of the tube. The total flow of air, $A(x)v$, past any cross section, is analogous to the current, and the pressure is analogous to the voltage. The analogous impedance $Z_a = (z/A)$ may be computed in terms of the impedance of the analogous electric circuit. A constriction in the tube, consisting of a hole of area A in a baffle plate, is analogous to a series inductance of value

$$L_a = (\rho l_e/A), \quad l_e = l + 0.8\sqrt{A} \tag{1}$$

where l is the length of the constriction ($l = 0$ when the plate is of negligible thickness). A chamber of larger cross section (a tank), of volume V, is analogous to a shunt capacitance of value

$$C_a = (V/\rho c^2) \tag{2}$$

A Helmholtz resonator is a tank with a single hole, connecting to the outside, is analogous to a series LC circuit. The resonance frequency is

$$\nu_r = (1/2\pi\sqrt{L_a C_a}) = (c/2\pi)\sqrt{A/l_e V} \tag{3}$$

A muffler, consisting of a sequence of tanks, connected one to the next by constrictions, is analogous to a low-pass filter network of series inductances alternating with shunt capacitances. Frequencies above the cutoff frequency

$$\nu_c = (1/\pi\sqrt{L_a C_a}) = (c/\pi)\sqrt{A/l_e V} \tag{4}$$

are attenuated, those below ν_c are transmitted.

1.12. Radiation of sound from a vibrating cylinder. When a long cylinder of radius a oscillates with a velocity $U_0 e^{-i\omega t}$ transverse to its axis, the sound wave radiated from its surface into an infinite medium is

$$\left.\begin{aligned} P &= \left(\frac{\pi\omega^2\rho a^2 U_0}{2c}\right)(\cos\phi)\,[H_1^{(1)}(kr)] \\ &\to \left(\frac{\pi\omega^2\rho a^2 U_0}{2ic}\right)\sqrt{\frac{2}{i\pi kr}}\cos\phi\, e^{ik(r-ct)}, \quad (kr \gg 1) \end{aligned}\right\} \tag{1}$$

when a is small compared to $(2\pi/\lambda) = (1/k)$. Cylindrical coordinates, r, ϕ are used, ϕ being the angle between radius r and the plane of vibration of the cylinder. The intensity of the sound several wavelengths from the cylinder, and the total power radiated per unit length of cylinder are

$$S_r \simeq (\pi\omega^3\rho a^4 U_0^2/4c^2 r)\cos^2\phi, \quad \Pi \simeq (\pi^2\omega^3\rho a^4 U_0^2/4c^2) \tag{2}$$

The reaction force F of the medium back on the moving cylinder, per unit length, is proportional to its transverse velocity U; the ratio (F/U) is the radiation impedance per unit length of the vibrating cylinder.

$$Z_{\text{rad}} \simeq -i\omega(\pi a^2\rho) + (\pi^2\omega^2\rho a^4/2c^2), \quad (\omega a \ll c) \tag{3}$$

1.13. Radiation from a simple source. Any source of sound of dimensions much smaller than a wavelength of the radiated sound sends out a wave which is almost symmetric spherically if there are no reflecting objects nearby. The *source strength* Q_0 is the amplitude of the total in-and-out flow of fluid from the source, in cubic centimeters per second. The radiated wave, a distance r from the radiator is

$$p \simeq -i\omega(\rho/4\pi r)Q_0 e^{ik(r-ct)}, \quad (kc = \omega) \tag{1}$$

The corresponding intensity and total power radiated are

$$S_r \simeq (\rho\omega^2 Q_0^2/32\pi^2 c r^2), \quad \Pi \simeq (\rho\omega^2 Q_0^2/8\pi c) \tag{2}$$

1.14. Radiation from a dipole source. A sphere of radius a oscillating back and forth along the spherical coordinate axis gives rise to dipole radiation when $2\pi r$ is small compared to the wavelength. If the linear velocity of the sphere is $U_0 e^{-i\omega t}$, the radiated pressure, intensity, and power, several wavelengths from the source, are

$$\left.\begin{aligned} p \simeq -\left(\frac{\omega^2 \rho a^3 U_0}{2cr}\right) \cos\vartheta\, e^{ik(r-ct)}, \quad S_r \simeq \left(\frac{\omega^4 \rho a^6 U_0^2}{8c^3}\right)\left(\frac{\cos^2\vartheta}{r^2}\right), \\ \Pi = \left(\frac{\pi\omega^4 \rho a^6 U_0^2}{6c^3}\right) \end{aligned}\right\} \quad (1)$$

The radiation impedance back on the sphere, the ratio of the net reaction force to the velocity of the sphere, is

$$Z_{\text{rad}} \simeq -i\omega(\tfrac{2}{3}\pi\rho_0 a^3) + (\pi\omega^4\rho a^6/3c^3), \quad (\omega a \gg c) \quad (2)$$

1.15. Radiation from a piston in a wall. If a flat-topped piston, of circular cross section with radius a, set in a flat, rigid wall, vibrates parallel to its axis with velocity $U_0 e^{-i\omega t}$, the radiated pressure and power at a distancer $(r \gg a)$ from the center of the piston, r at angle θ to the piston axis, and the total radiated power are

$$\left.\begin{aligned} & p \simeq -i\omega\rho U_0 a^2 \frac{e^{ik(r-ct)}}{r} [J_1(ka\sin\theta)/ka\sin\theta] \\ & \xrightarrow[ka\to 0]{} -i\omega\rho(\pi a^2 U_0/2\pi r)e^{ik(r-ct)} \\ & S \simeq \tfrac{1}{2}(\rho\omega^2 U_0^2 a^4/r^2)[J_1(ka\sin\theta)/ka\sin\theta]^2 \xrightarrow[ka\to 0]{} (\rho\omega^2/8\pi^2 c r^2)(\pi a^2 U_0)^2 \\ & \Pi = \tfrac{1}{2}(\pi a^2 \rho c U_0^2)[1-(1/ka)J_1(2ka)], \quad (ka = \omega a/c = 2\pi a/\lambda) \end{aligned}\right\} \quad (1)$$

The radiation impedance, the ratio of reaction force on the piston front to the piston transverse velocity, is

$$\left.\begin{aligned} & Z_{\text{rad}} = \pi a^2 \rho c[R_0 - iX_0] \\ & R_0 = 1 - \left(\frac{1}{ka}\right) J_1(2ka) \to \begin{cases} \tfrac{1}{2}(ka)^2, & (ka \ll 1) \\ 1 & (ka \gg 1) \end{cases} \\ & X_0 = \frac{4}{\pi}\int_0^{1/2\pi} \sin(2ka\cos\alpha)\sin^2\alpha\, d\alpha \to \begin{cases} (8ka/3\pi), & (ka \ll 1) \\ (2/\pi ka), & (ka \gg 1) \end{cases} \end{aligned}\right\} \quad (2)$$

1.16. Scattering of sound from a cylinder. A plane wave of sound, of frequency $(\omega/2\pi)$, incident at right angles to the axis of a rigid circular cylinder of radius a is in part underflected (continuing as a plane wave) and in part scattered. For wavelengths long compared with the cylinder radius a, the intensity S_s of the wave scattered at angle φ to the direction of the incident wave and the total power scattered per unit length of cylinder are.

$$S_s \simeq (\pi\omega^3 a^4/8c^3 r)S_0(1-2\cos\phi)^2, \quad \Pi_s \simeq (3\pi^2\omega^3 a^4 S_0/4c^3), \quad (ka \ll 1) \qquad (1)$$

where S_0 is the intensity of the incident plane wave. For wavelengths short compared with a, half the scattered wave interferes with the undeflected plane wave, forming the shadow of the cylinder (with diffraction at the edge of the shadow) and the other half is reflected with distribution of intensity

$$S_r \simeq (aS_0/2r)\sin(\phi/2), \quad \Pi_r \simeq 2aS_0, \quad (ka \gg 1) \qquad (2)$$

The net force on the cylinder because of the wave, per unit length of cylinder, is in the direction of the incident wave of magnitude

$$F \to \begin{cases} i\omega(4\pi^2 a^2/c)P_0 e^{-i\omega t}, & (ka \ll 1) \\ \sqrt{8\pi ac/\omega}\, P_0 e^{-(\omega/c)(a-ct)i\pi/4}, & (ka \gg 1) \end{cases} \qquad (3)$$

where P_0 is the pressure amplitude of the incident plane wave.

1.17. Scattering of sound from a sphere. The corresponding scattered intensity and power from a plane wave of frequency $(\omega/2\pi) = (kc/2\pi)$ incident on a rigid sphere of radius a, at distance r and angle ϑ to the incident wave are

$$S_s \simeq (\omega^4 a^6 S_0/9c^4 r^2)(1-3\cos\vartheta)^2, \quad \Pi_s \simeq (16\pi\omega^4 a^6/9c^4)S_0 \qquad (1)$$

for wavelengths long compared with a, where $ka \ll 1$. For very short wavelengths half of the scattered wave produces the shadow and the other half has intensity and power scattered,

$$S_r \simeq (a^2/4r^2)S_0, \quad \Pi_r \simeq \pi a^2 S_0, \quad (ka \gg 1) \qquad (2)$$

The net pressure on the sphere, caused by the incident wave is

$$p_a \simeq (1 + \tfrac{3}{2}\, ika\cos\vartheta)P_0 e^{i\omega t}, \quad (ka \ll 1) \qquad (3)$$

1.18. Room acoustics. When sound is produced in a room the average intensity eventually comes to a constant value when the power produced is just balanced by the power lost, most of which is lost by absorption at the room walls (for the acoustic frequencies; high-frequency sound is absorbed in the air). If the room walls are sufficiently irregular in shape the steady-state sound is fairly uniformly distributed in direction and position in the room. The absorbing properties of the wall, in this case, may be expressed in terms of the *absorption coefficient* α, the fraction of the incident intensity which is absorbed and not reflected. This coefficient is related to the acoustic impedance of the wall, as defined previously. If an area A_s of the wall (or ceiling or floor) has absorption coefficient α_s, the sum $(\alpha_1 A_1 + \alpha_2 A_2 + \ldots)$ over the whole surface of the room is called a, the *total absorption of the room.* If A_s is given in square feet, the units of a are called *sabines.* The mean intensity level in the room, for steady state, is

$$\text{Intensity level} \simeq 10 \log (\Pi/a) + 130 \quad \text{decibels} \tag{1}$$

if Π is the power output in watts and a is in sabines. If, after reaching steady state, the sound is turned off, the intensity decays exponentially, on the average; the time required for the intensity to diminish by 60 decibels,

$$T = (0.049 V/a) \quad \text{seconds} \tag{2}$$

is called the *reverberation time* (V is the volume of the room in cubic feet). The reverberation time is a useful criterion (not the only one, however) for the acoustical properties of an auditorium. Satisfactory values lie between 0.7 second and 1.5 seconds, the higher values being more appropriate for rooms used for music, the lower for rooms used primarily for speaking. A reverberation time less than 0.5 second indicates an undue amount of absorption and a corresponding lack of room sonority.

Bibliography

1. Beranek, L. L., *Acoustic Measurements*, John Wiley & Sons, Inc., New York, 1949. A good modern text on experimental acoustics.
2. Bergmann, L., *Ultrasonics*, John Wiley & Sons, Inc., New York, 1944. A discussion of the newly developing field of very high frequency sound. Contains a good bibliography.
3. Fletcher, H., *Speech and Hearing*, D. Van Nostrand Company, Inc., New York, 1929. An excellent discussion of the physiological aspects of acoustics.
4. Knudsen, V. O. and Harris, C. M., *Acoustical Design in Architecture*, John Wiley & Sons, Inc., New York, 1950. A recent review of an important and rapidly developing application of acoustics.
5. Morse, P. M., *Vibration and Sound*, 2d ed., McGraw-Hill Book Company, Inc., New York, 1948. A text on the recent developments of the theory of wave motion and its application to acoustics.
6. Randall, R. H., *Introduction to Acoustics*, Addison-Wesley Press, Inc., Cambridge, Mass., 1951. An up-to-date and easily readable introductory text.
7. Rayleigh, J. W. S., *The Theory of Sound* (reprint), Dover Publications, New York, 1945. The classic of the science of acoustics.

INDEX

This comprehensive index covers the two volumes of the book. Volume One contains pages 1 through 364 and Volume Two contains pages 365 through 741.

A CATALOG OF SELECTED

DOVER BOOKS

IN ALL FIELDS OF INTEREST

THE BOOK OF BEASTS: Being a Translation from a Latin Bestiary of the Twelfth Century, T. H. White. Wonderful catalog real and fanciful beasts: manticore, griffin, phoenix, amphivius, jaculus, many more. White's witty erudite commentary on scientific, historical aspects. Fascinating glimpse of medieval mind. Illustrated. 296pp. 5⅜ × 8¼. (Available in U.S. only) 24609-4 Pa. $5.95

FRANK LLOYD WRIGHT: ARCHITECTURE AND NATURE With 160 Illustrations, Donald Hoffmann. Profusely illustrated study of influence of nature—especially prairie—on Wright's designs for Fallingwater, Robie House, Guggenheim Museum, other masterpieces. 96pp. 9¼ × 10¾. 25098-9 Pa. $7.95

FRANK LLOYD WRIGHT'S FALLINGWATER, Donald Hoffmann. Wright's famous waterfall house: planning and construction of organic idea. History of site, owners, Wright's personal involvement. Photographs of various stages of building. Preface by Edgar Kaufmann, Jr. 100 illustrations. 112pp. 9¼ × 10. 23671-4 Pa. $7.95

YEARS WITH FRANK LLOYD WRIGHT: Apprentice to Genius, Edgar Tafel. Insightful memoir by a former apprentice presents a revealing portrait of Wright the man, the inspired teacher, the greatest American architect. 372 black-and-white illustrations. Preface. Index. vi + 228pp. 8¼ × 11. 24801-1 Pa. $9.95

THE STORY OF KING ARTHUR AND HIS KNIGHTS, Howard Pyle. Enchanting version of King Arthur fable has delighted generations with imaginative narratives of exciting adventures and unforgettable illustrations by the author. 41 illustrations. xviii + 313pp. 6⅛ × 9¼. 21445-1 Pa. $6.50

THE GODS OF THE EGYPTIANS, E. A. Wallis Budge. Thorough coverage of numerous gods of ancient Egypt by foremost Egyptologist. Information on evolution of cults, rites and gods; the cult of Osiris; the Book of the Dead and its rites; the sacred animals and birds; Heaven and Hell; and more. 956pp. 6⅛ × 9¼. 22055-9, 22056-7 Pa., Two-vol. set $20.00

A THEOLOGICO-POLITICAL TREATISE, Benedict Spinoza. Also contains unfinished *Political Treatise*. Great classic on religious liberty, theory of government on common consent. R. Elwes translation. Total of 421pp. 5⅜ × 8½. 20249-6 Pa. $6.95

INCIDENTS OF TRAVEL IN CENTRAL AMERICA, CHIAPAS, AND YUCATAN, John L. Stephens. Almost single-handed discovery of Maya culture; exploration of ruined cities, monuments, temples; customs of Indians. 115 drawings. 892pp. 5⅜ × 8½. 22404-X, 22405-8 Pa., Two-vol. set $15.90

LOS CAPRICHOS, Francisco Goya. 80 plates of wild, grotesque monsters and caricatures. Prado manuscript included. 183pp. 6⅝ × 9⅝. 22384-1 Pa. $4.95

AUTOBIOGRAPHY: The Story of My Experiments with Truth, Mohandas K. Gandhi. Not hagiography, but Gandhi in his own words. Boyhood, legal studies, purification, the growth of the Satyagraha (nonviolent protest) movement. Critical, inspiring work of the man who freed India. 480pp. 5⅜ × 8½. (Available in U.S. only) 24593-4 Pa. $6.95

ILLUSTRATED DICTIONARY OF HISTORIC ARCHITECTURE, edited by Cyril M. Harris. Extraordinary compendium of clear, concise definitions for over 5,000 important architectural terms complemented by over 2,000 line drawings. Covers full spectrum of architecture from ancient ruins to 20th-century Modernism. Preface. 592pp. 7½ × 9⅜. 24444-X Pa. $14.95

THE NIGHT BEFORE CHRISTMAS, Clement Moore. Full text, and woodcuts from original 1848 book. Also critical, historical material. 19 illustrations. 40pp. 4⅝ × 6. 22797-9 Pa. $2.25

THE LESSON OF JAPANESE ARCHITECTURE: 165 Photographs, Jiro Harada. Memorable gallery of 165 photographs taken in the 1930's of exquisite Japanese homes of the well-to-do and historic buildings. 13 line diagrams. 192pp. 8⅝ × 11¼. 24778-3 Pa. $8.95

THE AUTOBIOGRAPHY OF CHARLES DARWIN AND SELECTED LETTERS, edited by Francis Darwin. The fascinating life of eccentric genius composed of an intimate memoir by Darwin (intended for his children); commentary by his son, Francis; hundreds of fragments from notebooks, journals, papers; and letters to and from Lyell, Hooker, Huxley, Wallace and Henslow. xi + 365pp. 5⅜ × 8. 20479-0 Pa. $6.95

WONDERS OF THE SKY: Observing Rainbows, Comets, Eclipses, the Stars and Other Phenomena, Fred Schaaf. Charming, easy-to-read poetic guide to all manner of celestial events visible to the naked eye. Mock suns, glories, Belt of Venus, more. Illustrated. 299pp. 5¼ × 8¼. 24402-4 Pa. $7.95

BURNHAM'S CELESTIAL HANDBOOK, Robert Burnham, Jr. Thorough guide to the stars beyond our solar system. Exhaustive treatment. Alphabetical by constellation: Andromeda to Cetus in Vol. 1; Chamaeleon to Orion in Vol. 2; and Pavo to Vulpecula in Vol. 3. Hundreds of illustrations. Index in Vol. 3. 2,000pp. 6⅛ × 9¼. 23567-X, 23568-8, 23673-0 Pa., Three-vol. set $38.85

STAR NAMES: Their Lore and Meaning, Richard Hinckley Allen. Fascinating history of names various cultures have given to constellations and literary and folkloristic uses that have been made of stars. Indexes to subjects. Arabic and Greek names. Biblical references. Bibliography. 563pp. 5⅜ × 8½. 21079-0 Pa. $7.95

THIRTY YEARS THAT SHOOK PHYSICS: The Story of Quantum Theory, George Gamow. Lucid, accessible introduction to influential theory of energy and matter. Careful explanations of Dirac's anti-particles, Bohr's model of the atom, much more. 12 plates. Numerous drawings. 240pp. 5⅜ × 8½. 24895-X Pa. $4.95

CHINESE DOMESTIC FURNITURE IN PHOTOGRAPHS AND MEASURED DRAWINGS, Gustav Ecke. A rare volume, now affordably priced for antique collectors, furniture buffs and art historians. Detailed review of styles ranging from early Shang to late Ming. Unabridged republication. 161 black-and-white drawings, photos. Total of 224pp. 8⅜ × 11¼. (Available in U.S. only) 25171-3 Pa. $12.95

VINCENT VAN GOGH: A Biography, Julius Meier-Graefe. Dynamic, penetrating study of artist's life, relationship with brother, Theo, painting techniques, travels, more. Readable, engrossing. 160pp. 5⅜ × 8½. (Available in U.S. only) 25253-1 Pa. $3.95

HOW TO WRITE, Gertrude Stein. Gertrude Stein claimed anyone could understand her unconventional writing—here are clues to help. Fascinating improvisations, language experiments, explanations illuminate Stein's craft and the art of writing. Total of 414pp. 4⅜ × 6⅜. 23144-5 Pa. $5.95

ADVENTURES AT SEA IN THE GREAT AGE OF SAIL: Five Firsthand Narratives, edited by Elliot Snow. Rare true accounts of exploration, whaling, shipwreck, fierce natives, trade, shipboard life, more. 33 illustrations. Introduction. 353pp. 5⅜ × 8½. 25177-2 Pa. $7.95

THE HERBAL OR GENERAL HISTORY OF PLANTS, John Gerard. Classic descriptions of about 2,850 plants—with over 2,700 illustrations—includes Latin and English names, physical descriptions, varieties, time and place of growth, more. 2,706 illustrations. xlv + 1,678pp. 8½ × 12¼. 23147-X Cloth. $75.00

DOROTHY AND THE WIZARD IN OZ, L. Frank Baum. Dorothy and the Wizard visit the center of the Earth, where people are vegetables, glass houses grow and Oz characters reappear. Classic sequel to *Wizard of Oz*. 256pp. 5⅜ × 8. 24714-7 Pa. $4.95

SONGS OF EXPERIENCE: Facsimile Reproduction with 26 Plates in Full Color, William Blake. This facsimile of Blake's original "Illuminated Book" reproduces 26 full-color plates from a rare 1826 edition. Includes "The Tyger," "London," "Holy Thursday," and other immortal poems. 26 color plates. Printed text of poems. 48pp. 5¼ × 7. 24636-1 Pa. $3.50

SONGS OF INNOCENCE, William Blake. The first and most popular of Blake's famous "Illuminated Books," in a facsimile edition reproducing all 31 brightly colored plates. Additional printed text of each poem. 64pp. 5¼ × 7. 22764-2 Pa. $3.50

PRECIOUS STONES, Max Bauer. Classic, thorough study of diamonds, rubies, emeralds, garnets, etc.: physical character, occurrence, properties, use, similar topics. 20 plates, 8 in color. 94 figures. 659pp. 6⅛ × 9¼. 21910-0, 21911-9 Pa., Two-vol. set $15.90

ENCYCLOPEDIA OF VICTORIAN NEEDLEWORK, S. F. A. Caulfeild and Blanche Saward. Full, precise descriptions of stitches, techniques for dozens of needlecrafts—most exhaustive reference of its kind. Over 800 figures. Total of 679pp. 8⅛ × 11. Two volumes. Vol. 1 22800-2 Pa. $11.95
Vol. 2 22801-0 Pa. $11.95

THE MARVELOUS LAND OF OZ, L. Frank Baum. Second Oz book, the Scarecrow and Tin Woodman are back with hero named Tip, Oz magic. 136 illustrations. 287pp. 5⅜ × 8½. 20692-0 Pa. $5.95

WILD FOWL DECOYS, Joel Barber. Basic book on the subject, by foremost authority and collector. Reveals history of decoy making and rigging, place in American culture, different kinds of decoys, how to make them, and how to use them. 140 plates. 156pp. 7⅞ × 10¾. 20011-6 Pa. $8.95

HISTORY OF LACE, Mrs. Bury Palliser. Definitive, profusely illustrated chronicle of lace from earliest times to late 19th century. Laces of Italy, Greece, England, France, Belgium, etc. Landmark of needlework scholarship. 266 illustrations. 672pp. 6⅛ × 9¼. 24742-2 Pa. $14.95

ILLUSTRATED GUIDE TO SHAKER FURNITURE, Robert Meader. All furniture and appurtenances, with much on unknown local styles. 235 photos. 146pp. 9 × 12. 22819-3 Pa. $7.95

WHALE SHIPS AND WHALING: A Pictorial Survey, George Francis Dow. Over 200 vintage engravings, drawings, photographs of barks, brigs, cutters, other vessels. Also harpoons, lances, whaling guns, many other artifacts. Comprehensive text by foremost authority. 207 black-and-white illustrations. 288pp. 6 × 9. 24808-9 Pa. $8.95

THE BERTRAMS, Anthony Trollope. Powerful portrayal of blind self-will and thwarted ambition includes one of Trollope's most heartrending love stories. 497pp. 5⅜ × 8½. 25119-5 Pa. $8.95

ADVENTURES WITH A HAND LENS, Richard Headstrom. Clearly written guide to observing and studying flowers and grasses, fish scales, moth and insect wings, egg cases, buds, feathers, seeds, leaf scars, moss, molds, ferns, common crystals, etc.—all with an ordinary, inexpensive magnifying glass. 209 exact line drawings aid in your discoveries. 220pp. 5⅜ × 8½. 23330-8 Pa. $3.95

RODIN ON ART AND ARTISTS, Auguste Rodin. Great sculptor's candid, wide-ranging comments on meaning of art; great artists; relation of sculpture to poetry, painting, music; philosophy of life, more. 76 superb black-and-white illustrations of Rodin's sculpture, drawings and prints. 119pp. 8⅜ × 11¼. 24487-3 Pa. $6.95

FIFTY CLASSIC FRENCH FILMS, 1912–1982: A Pictorial Record, Anthony Slide. Memorable stills from Grand Illusion, Beauty and the Beast, Hiroshima, Mon Amour, many more. Credits, plot synopses, reviews, etc. 160pp. 8¼ × 11. 25256-6 Pa. $11.95

THE PRINCIPLES OF PSYCHOLOGY, William James. Famous long course complete, unabridged. Stream of thought, time perception, memory, experimental methods; great work decades ahead of its time. 94 figures. 1,391pp. 5⅜ × 8½. 20381-6, 20382-4 Pa., Two-vol. set $19.90

BODIES IN A BOOKSHOP, R. T. Campbell. Challenging mystery of blackmail and murder with ingenious plot and superbly drawn characters. In the best tradition of British suspense fiction. 192pp. 5⅜ × 8½. 24720-1 Pa. $3.95

CALLAS: PORTRAIT OF A PRIMA DONNA, George Jellinek. Renowned commentator on the musical scene chronicles incredible career and life of the most controversial, fascinating, influential operatic personality of our time. 64 black-and-white photographs. 416pp. 5⅜ × 8¼. 25047-4 Pa. $7.95

GEOMETRY, RELATIVITY AND THE FOURTH DIMENSION, Rudolph Rucker. Exposition of fourth dimension, concepts of relativity as Flatland characters continue adventures. Popular, easily followed yet accurate, profound. 141 illustrations. 133pp. 5⅜ × 8½. 23400-2 Pa. $3.95

HOUSEHOLD STORIES BY THE BROTHERS GRIMM, with pictures by Walter Crane. 53 classic stories—Rumpelstiltskin, Rapunzel, Hansel and Gretel, the Fisherman and his Wife, Snow White, Tom Thumb, Sleeping Beauty, Cinderella, and so much more—lavishly illustrated with original 19th century drawings. 114 illustrations. x + 269pp. 5⅜ × 8½. 21080-4 Pa. $4.50

SUNDIALS, Albert Waugh. Far and away the best, most thorough coverage of ideas, mathematics concerned, types, construction, adjusting anywhere. Over 100 illustrations. 230pp. 5⅜ × 8½. 22947-5 Pa. $4.50

PICTURE HISTORY OF THE NORMANDIE: With 190 Illustrations, Frank O. Braynard. Full story of legendary French ocean liner: Art Deco interiors, design innovations, furnishings, celebrities, maiden voyage, tragic fire, much more. Extensive text. 144pp. 8⅜ × 11¾. 25257-4 Pa. $9.95

THE FIRST AMERICAN COOKBOOK: A Facsimile of "American Cookery," 1796, Amelia Simmons. Facsimile of the first American-written cookbook published in the United States contains authentic recipes for colonial favorites—pumpkin pudding, winter squash pudding, spruce beer, Indian slapjacks, and more. Introductory Essay and Glossary of colonial cooking terms. 80pp. 5⅜ × 8½. 24710-4 Pa. $3.50

101 PUZZLES IN THOUGHT AND LOGIC, C. R. Wylie, Jr. Solve murders and robberies, find out which fishermen are liars, how a blind man could possibly identify a color—purely by your own reasoning! 107pp. 5⅜ × 8½. 20367-0 Pa. $2.50

THE BOOK OF WORLD-FAMOUS MUSIC—CLASSICAL, POPULAR AND FOLK, James J. Fuld. Revised and enlarged republication of landmark work in musico-bibliography. Full information about nearly 1,000 songs and compositions including first lines of music and lyrics. New supplement. Index. 800pp. 5⅜ × 8¼. 24857-7 Pa. $14.95

ANTHROPOLOGY AND MODERN LIFE, Franz Boas. Great anthropologist's classic treatise on race and culture. Introduction by Ruth Bunzel. Only inexpensive paperback edition. 255pp. 5⅜ × 8½. 25245-0 Pa. $5.95

THE TALE OF PETER RABBIT, Beatrix Potter. The inimitable Peter's terrifying adventure in Mr. McGregor's garden, with all 27 wonderful, full-color Potter illustrations. 55pp. 4¼ × 5½. (Available in U.S. only) 22827-4 Pa. $1.75

THREE PROPHETIC SCIENCE FICTION NOVELS, H. G. Wells. *When the Sleeper Wakes, A Story of the Days to Come* and *The Time Machine* (full version). 335pp. 5⅜ × 8½. (Available in U.S. only) 20605-X Pa. $5.95

APICIUS COOKERY AND DINING IN IMPERIAL ROME, edited and translated by Joseph Dommers Vehling. Oldest known cookbook in existence offers readers a clear picture of what foods Romans ate, how they prepared them, etc. 49 illustrations. 301pp. 6⅛ × 9¼. 23563-7 Pa. $6.50

SHAKESPEARE LEXICON AND QUOTATION DICTIONARY, Alexander Schmidt. Full definitions, locations, shades of meaning of every word in plays and poems. More than 50,000 exact quotations. 1,485pp. 6½ × 9¼. 22726-X, 22727-8 Pa., Two-vol. set $27.90

THE WORLD'S GREAT SPEECHES, edited by Lewis Copeland and Lawrence W. Lamm. Vast collection of 278 speeches from Greeks to 1970. Powerful and effective models; unique look at history. 842pp. 5⅜ × 8½. 20468-5 Pa. $11.95

THE BLUE FAIRY BOOK, Andrew Lang. The first, most famous collection, with many familiar tales: Little Red Riding Hood, Aladdin and the Wonderful Lamp, Puss in Boots, Sleeping Beauty, Hansel and Gretel, Rumpelstiltskin; 37 in all. 138 illustrations. 390pp. 5⅜ × 8½. 21437-0 Pa. $5.95

THE STORY OF THE CHAMPIONS OF THE ROUND TABLE, Howard Pyle. Sir Launcelot, Sir Tristram and Sir Percival in spirited adventures of love and triumph retold in Pyle's inimitable style. 50 drawings, 31 full-page. xviii + 329pp. 6½ × 9¼. 21883-X Pa. $6.95

AUDUBON AND HIS JOURNALS, Maria Audubon. Unmatched two-volume portrait of the great artist, naturalist and author contains his journals, an excellent biography by his granddaughter, expert annotations by the noted ornithologist, Dr. Elliott Coues, and 37 superb illustrations. Total of 1,200pp. 5⅜ × 8.
Vol. I 25143-8 Pa. $8.95
Vol. II 25144-6 Pa. $8.95

GREAT DINOSAUR HUNTERS AND THEIR DISCOVERIES, Edwin H. Colbert. Fascinating, lavishly illustrated chronicle of dinosaur research, 1820's to 1960. Achievements of Cope, Marsh, Brown, Buckland, Mantell, Huxley, many others. 384pp. 5¼ × 8¼. 24701-5 Pa. $6.95

THE TASTEMAKERS, Russell Lynes. Informal, illustrated social history of American taste 1850's–1950's. First popularized categories Highbrow, Lowbrow, Middlebrow. 129 illustrations. New (1979) afterword. 384pp. 6 × 9.
23993-4 Pa. $6.95

DOUBLE CROSS PURPOSES, Ronald A. Knox. A treasure hunt in the Scottish Highlands, an old map, unidentified corpse, surprise discoveries keep reader guessing in this cleverly intricate tale of financial skullduggery. 2 black-and-white maps. 320pp. 5⅜ × 8½. (Available in U.S. only) 25032-6 Pa. $5.95

AUTHENTIC VICTORIAN DECORATION AND ORNAMENTATION IN FULL COLOR: 46 Plates from "Studies in Design," Christopher Dresser. Superb full-color lithographs reproduced from rare original portfolio of a major Victorian designer. 48pp. 9¼ × 12¼. 25083-0 Pa. $7.95

PRIMITIVE ART, Franz Boas. Remains the best text ever prepared on subject, thoroughly discussing Indian, African, Asian, Australian, and, especially, Northern American primitive art. Over 950 illustrations show ceramics, masks, totem poles, weapons, textiles, paintings, much more. 376pp. 5⅜ × 8. 20025-6 Pa. $6.95

SIDELIGHTS ON RELATIVITY, Albert Einstein. Unabridged republication of two lectures delivered by the great physicist in 1920–21. *Ether and Relativity* and *Geometry and Experience*. Elegant ideas in non-mathematical form, accessible to intelligent layman. vi + 56pp. 5⅜ × 8½. 24511-X Pa. $2.95

THE WIT AND HUMOR OF OSCAR WILDE, edited by Alvin Redman. More than 1,000 ripostes, paradoxes, wisecracks: Work is the curse of the drinking classes, I can resist everything except temptation, etc. 258pp. 5⅜ × 8½. 20602-5 Pa. $4.50

ADVENTURES WITH A MICROSCOPE, Richard Headstrom. 59 adventures with clothing fibers, protozoa, ferns and lichens, roots and leaves, much more. 142 illustrations. 232pp. 5⅜ × 8½. 23471-1 Pa. $3.95